COMPUTATIONAL AND
ALGORITHMIC LINEAR
ALGEBRA AND *n*-DIMENSIONAL
GEOMETRY

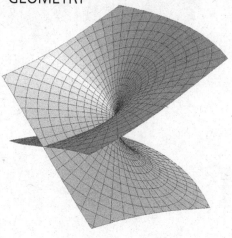

COMPUTATIONAL AND ALGORITHMIC LINEAR ALGEBRA AND *n*-DIMENSIONAL GEOMETRY

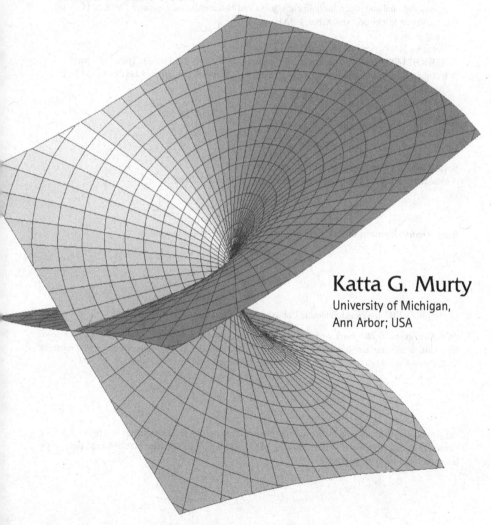

Katta G. Murty
University of Michigan,
Ann Arbor; USA

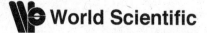
World Scientific

NEW JERSEY · LONDON · SINGAPORE · BEIJING · SHANGHAI · HONG KONG · TAIPEI · CHENNAI

Published by

World Scientific Publishing Co. Pte. Ltd.

5 Toh Tuck Link, Singapore 596224

USA office: 27 Warren Street, Suite 401-402, Hackensack, NJ 07601

UK office: 57 Shelton Street, Covent Garden, London WC2H 9HE

Library of Congress Cataloging-in-Publication Data
Murty, Katta G., 1936– author.
    Computational and algorithmic linear algebra and N-dimensional geometry / by Katta G Murty
(University of Michigan, Ann Arbor, USA).
    pages cm
    Includes bibliographical references and index.
    ISBN 978-9814366625 (hardcover : alk. paper) -- ISBN 978-9814366632 (pbk : alk. paper)
    1. Algebras, Linear--Data processing--Textbooks. 2. Geometry, Analytic--Textbooks. I. Title.
    QA185.D37M87 2014
    512'.5--dc23

                                                                    2014024554

British Library Cataloguing-in-Publication Data
A catalogue record for this book is available from the British Library.

Image credit: Riemann surface sqrt by Jan Homann from Wikipedia commons

Printed in Singapore

Contents

Preface

The Importance of the Subject

The subject of **linear algebra** originated more than 2000 years ago in an effort to develop tools for modeling real world problems in the form of systems of linear equations, and to solve and analyze these problems using those models. Linear algebra and its twentieth century extensions, **linear and integer programming**, are the most useful and most heavily used branches of mathematics. A thorough knowledge of the most fundamental parts of linear algebra is an essential requirement for anyone in any technical job these days in order to carry out job duties at an adequate level. In this internet age with computers playing a vital and increasing role in every job, making the most effective use of computers requires a solid background in at least the basic parts of linear algebra.

Recent Changes in the Way Linear Algebra is Taught

Until about the 1970's there used to be a full semester course on computational linear algebra at the sophomore level in all undergraduate engineering curricula. That course used to serve the purpose of providing a good background in the basics of linear algebra very well. However, due to increasing pressure to introduce more and more material in the undergraduate mathematics sequence of engineering curricula, beginning in the 1970's many universities started combining this first level linear algebra course with the course on differential equations and shaping it into a combined course on differential equations and linear algebra. On top of this, linear algebra is being squeezed out of this combined course gradually, with the result that now-a-days most engineering undergraduates get only about four weeks of instruction on linear algebra. The net effect of all this is that many undergraduates of science and engineering are not developing the linear algebra skills that they need.

The Goals of This Book

There are many outstanding books on linear algebra, but most of them have their focus on undergraduates in mathematics departments. They concentrate on developing theorem-proving skills in the readers, and not so much on helping them develop mathematical modeling, computational, and algorithmic skills. Students who miss a few classes, particularly those already not doing well in the class, often find it very difficult to catch up with the help of such text books. The net result after about half the term is an unhealthy situation in which only the top half of the class is comfortably following the teachers lectures, with the rest falling more and more behind.

Also, with mounting pressure to pack more and more material into undergraduate mathematics courses, it is highly unlikely that more class time can be found to allocate to linear algebra. So, an important component of a solution strategy to this problem is to see whether students can be helped to learn these concepts to a large extent from their own efforts.

An Incident Providing Evidence of Self-Learning:

As evidence of self-learning of higher mathematics, I will mention the strangest of mathematical *talks* ever given, by Cole, in October 1903 at an AMS meeting, on Mersenne's claim that $2^{67} - 1$ is a prime number. This claim fascinated mathematicians for over 250 years. Not a single word was spoken at the seminar by anyone, but everyone there understood the result obtained by Cole. Cole wrote on the blackboard: $2^{67} - 1 = 147573952589676412927 = a$. Then he wrote 761838257287, and underneath it 193707721. Without uttering a word, he multiplied the two numbers together to get a, and sat down to the wild applause from the audience. ⋈

In the same way, I strongly believe that students can learn the concepts of numerical and algorithmic matrix algebra mostly on their own using a carefully designed book with illustrative examples, exercises, figures and well placed summaries. My primary goal is to make this such a book.

Our focus is to help the students develop the mathematical modeling, computational, and algorithmic skills that they need to bring linear algebra tools and facts to solving real world problems. Proofs of simple results are given without being specially labeled as proofs. Proofs of more complex facts are left out, for the reader to look up in any of the traditional textbooks on the subject, some of which are listed in the bibliography.

Importance of Learning the Details of the Methods

Students often ask me why it is important for them to learn the details of the methods for solving systems of linear equations these days when computers and high quality software for solving them are so widely available everywhere, and so easy to use. I keep telling them that many complex real world applications demand deeper knowledge than that of pressing a button to solve a given system of linear equations. First, someone has to construct the system of equations from information available on the real world problem: this part of the job is called mathematical modeling. If the model constructed yields a strange and unrealistic solution, or if it turns out to have no solution, one has to analyze the model and decide what changes should be made in it to make sure that it represents the real world problem better. This type of analysis can only be carried out efficiently by someone who has a good understanding of the methods of linear algebra. We have all seen the total helplessness of some grocery store checkout persons who cannot add two simple numbers when their computerized cash register fails. A person without the proper knowledge of linear algebra methods will be in the same position when he/she has to analyze and fix an invalid model giving strange results to the real world problem. Fortunately for grocery store checkout persons, the occurrence of cash register terminal failures is rare. However, for technical people trying to model their problems, the need to analyze and fix an invalid model is a daily event. Typically, models for real world problems are revised several times before they produce an acceptable solution to the problem.

For these reasons, I believe that it is very important for all undergraduates to learn the basics of linear algebra well.

The Poetic Beauty of the Subject

In classical Indian literature in Sanskrit, Telugu, etc., whenever they wanted to praise something or someone, they would compare them to the mighty ocean. In an effort to create an even higher pedestal of excellence, they developed the concept of **"kshiira saagar"** (literally "milk ocean", the mightiest ocean of pure milk goodness), and if they wanted to say that something is very good, they would say that it is like kshiira saagar. Actually, in terms of its applicability, sheer elegance and depth, linear algebra is very much like kshiira saagar. In this book you will get acquainted with the most basic tools and facts of linear algebra that are called for most frequently in applications. Those of you who have developed a taste for the beauty of the subject can delve deeper into the subject by looking up some of the other books mentioned in the bibliography.

Distinct Features of this Book

Chapter 1 explains the main difference between the classical subject of linear algebra (study of linear equations without any linear inequalities), and the modern 20th century subjects of linear programming (study of linear constraints including linear inequalities) and integer programming and combinatorial optimization (study of linear constraints and integer restrictions on some variables). This important distinction is rarely explained in other linear algebra books. When trying to handle a system of linear constraints including linear inequalities and integer requirements on some variables by classical linear algebra techniques only, we essentially have to ignore the linear inequalities and integer requirements. Chapter 1 deals with constructing mathematical models of systems of linear equations for real world problems, and the Gauss–Jordan (GJ) and Gaussian (G) elimination methods for solving these systems and analyzing the solution sets for such systems, including practical infeasibility analysis techniques for modifying an infeasible system into a feasible one. To keep the new terminology introduced to a minimum, the concept of a "matrix" is not even used in Chapter 1. All the algorithmic steps are stated using the fundamental tool of row operations on the detached coefficient tableau for the system with the variables entered in a top row in every tableau. This makes it easier for the readers to see that the essence of these methods is to take linear combinations of constraints in the original system to get an equivalent but simpler system from which a solution can be read out. Chapter 1 shows many instances of what sets this book apart from other mathematics books on linear algebra. In the descriptions of GJ, G methods in most of these books, the variables are usually left out. Also, they state the termination condition to be that of reaching the RREF (reduced row echelon form) or the REF (row echelon form). A tableau is defined to be in RREF [REF] if it contain a full set of unit vectors [vectors of an upper triangular matrix] in proper order at the left end.

OR people have realized that it is not important that all unit vectors (or vectors of the upper triangular matrix) be at the left end of the tableau (they can be anywhere and can be scattered all over); also it is not important that they be in proper order from left to right. They use the very simple **data structure** (this phrase means a strategy for storing information generated during the algorithm, and using it to improve the efficiency of that algorithm) of associating the variable corresponding to the rth unit vector (or the rth vector of the upper triangular matrix) in the final tableau as the rth basic variable or basic variable in the rth row; and they store these basic variables in a column on the tableau as the algorithm progresses. This data structure makes it easier to read the solution directly from the final tableau of the GJ method by making all nonbasic variables $= 0$; and the rth basic variable $=$ the rth RHS constant, for all r. We present the GJ, G methods using this data structure. Students find it easier to understand the main ideas behind the elimination methods using it. It also opens the possibility of pivot column selection strategies instead of always selecting the leftmost eligible column in these methods.

I believe it is George Dantzig who introduced the data structure of storing basic variables in his pioneering paper, the 1947 paper on the simplex method. He called the final tableau the *canonical tableau* to distinguish it from the mathematical concepts RREF, REF.

Another important difference that appears in Chapter 1, between this book and all the other books on linear algebra, is worth mentioning. A system of linear equations is infeasible (i.e., has no solution) iff the fundamental inconsistent equation "$0 = 1$" can be expressed as a linear combination of equations in the system. Thus all the linear algebra books state that if the equation "$0 = a$" where a is a nonzero number, shows up in one of the tableaus during the application of the GJ or G methods on the system, then the system is infeasible. As an example, consider the system

$$x_1 + x_2 + x_3 = 2$$

$$-x_1 - x_2 - x_3 = -1$$

$$2x_1 + 4x_2 + 6x_3 = 12$$

The linear combination of equations in this system with coefficients (1, 1, 0) is the equation $0x_1 + 0x_2 + 0x_3 = 1$ or $0 = 1$, hence this system is infeasible.

It is possible for the equation "$0 = 1$" to appear in the GJ or G methods due to computational errors. In hand computation these may be human errors, in digital computation on a computer these may be due to accumulation of roundoff errors. So, whenever the equation "$0 = a$" for some $a \neq 0$ appears in the GJ or G methods, it is also helpful for the methods to produce the row vector of coefficients in a linear combination of constraints in the original system that yields this "$0 = a$" equation. In the above example that vector of coefficients is (1, 1, 0). This vector has been called by the obscure name **supervisor principle for infeasibility** (the idea behind this name is that if your supervisor is suspicious about your claim of infeasibility of the system, he/she can personally verify the claim by obtaining the linear combination of constraints in the system with coefficients in this vector and verify that it is "$0 = a$" where $a \neq 0$). But we will call this vector of coefficients by the simpler name **evidence (or certificate) of infeasibility**. Given this evidence one can easily check whether it leads to the inconsistent equation "$0 = a$" where $a \neq 0$.

In the same way when an equation "$0 = 0$" appears during the GJ or G methods, we know that the corresponding original constraint is a redundant constraint that can be deleted from the system without changing its set of solutions. This constraint is redundant becuse it can be obtained as a linear combination of other constraints on which pivot steps have been carried out so far. The vector of coefficients in this linear combination is called the **evidence (or certificate) of redundancy** for this redundant constraint.

Other books do not discuss how these evidences can be obtained. These evidences are obtained easily as byproducts of the GJ, G methods as explained in Section 1.12, if the memory matrix is added to the system at the beginning of these methods. Also, in the computationally efficient versions of the GJ and G methods discussed in Section 4.11, we show that these evidences are automatically obtained without any additional expense.

Chapter 2 then introduces the concept of a matrix, the evolution of matrix arithmetic, and the study of determinants of square matrices.

Chapter 3 introduces the fundamentals of n-dimensional geometry starting with the representation of vectors as points using a coordinate frame of reference and the connections between algebra and geometry that this representation makes possible. We include a thorough treatment of important geometric objects such as subspaces, affine spaces, convex sets, hyperplanes, half-spaces, straight lines, half-lines, directions, rays, cones, and orthogonal projections that appear most frequently in courses that the student is likely to take following this study, and in real world applications.

Chapter 4 on numerical linear algebra discusses the fundamental concepts of linear dependence and independence, rank, inverse, factorizations; and efficient algorithms to check or compute these. Major differences in the mathematical properties of linear constraints involving equations only, and those involving some inequalities; come to the surface in the study of linear optimization problems involving them. This topic, not discussed in other books, is the subject of Section 4.10. In this section we also discuss the duality theorem for linear optimization subject to linear equality constraints (a specialization of the duality theorem of linear programming) and its relationship to the theorem of alternatives for linear equation systems. In Section 4.11 we discuss the efficient memory matrix versions of the GJ, G methods (based on the ideas introduced by George Dantzig in his revised simplex method for linear programming), these versions have the additional advantage of producing evidences of redundancy or infeasibility automatically without any additional work.

Chapter 5 deals with representing quadratic functions using matrix notation, and efficient algorithms to check whether a given quadratic function is convex, concave, or neither. The treatment of positive (negative) (semi)definiteness and indefiniteness of square matrices, and diagonalization of quadratic forms is included.

Chapter 6 includes the definitions and review of the main properties of eigen values and eigen vectors of square matrices, and their role in matrix diagonalizations.

Finally in the brief Chapter 7, we review the features of commercial software resources available these days to solve many of the problems discussed in earlier chapters.

Acknowledgements

MATLAB Computations in Chapter 7:

Jeanne Whalen helped me with the MATLAB computations in Chapter 7 on a Marian Sarah Parker Scholarship during Summer 2001. My thanks to her and to the University of Michigan Marian Sarah Parker Scholarship Administration (Women in Engineering Office at the University of Michigan College of Engineering).

Suggestions, corrections, and many other kinds of help have been received from several others too numerous to mention by name, and I express my heartfelt thanks to all of them.

About the Results and Excercises

Exercises are numbered serially beginning with number 1 in each section. Exercise $i.j.k$ refers to the kth exercise in Section $i.j$. In the same way Result $i.j.k$ refers to the kth result in Section $i.j$.

About the Numerical Exercises

At this level most students cannot really appreciate the details of an algorithm until they actually solve several numerical problems by themselves by hand. That's why I included many carefully constructed numerical exercises classified by the final outcome (infeasible system, system with unique solution, system with many solutions, redundant constraints, etc.) so that they know what to expect when they solve each exercise.

Given proper instructions, a digital computer can execute an algorithm on any data without complaining. But hand computation becomes very tedious if complicated fractions show up. That's why the numerical exercises are carefully constructed so that a pivot element of ± 1 is available for most of the steps of the algorithms. I hope that this makes it more enjoyable to understand the algorithms by working out the numerical exercises without encountering the tedium brought on by ugly fractions.

Katta G. Murty

Ann Arbor, MI, December 2010

Glossary of Symbols and Abbreviations

R^n — The n-dimensional real Euclidean vector space. The space of all vectors of the form $x = (x_1, \ldots, x_n)^T$ (written either as a row vector as here, or as a column vector) where each x_j is a real number.

$=, \geq, \leq$ — Symbols for equality, greater than or equal to, less than or equal to, which must hold for each component.

x^T, A^T — Transpose of vector x, matrix A.

(a_{ij}) — Matrix with a_{ij} as the general element in it.

$\backslash$ — Set difference symbol. If D, E are two sets, $D \backslash E$ is the set of all elements of D which are not in E.

A^{-1} — Inverse of the nonsingular square matrix A.

$\langle u, v \rangle$. — Dot (inner) product of vectors u, v.

$\mathrm{diag}(a_1, \ldots, a_n)$ — The $n \times n$ square matrix whose diagonal elements are $a_1, \ldots, a_n$ in that order, and all off-diagonal entries are 0.

$A_{i.}, A_{.j}$ — The ith row vector, jth column vector of matrix A.

$\det(A)$ — Determinant of square matrix A.

$||x||$ — Euclidean norm of vector $x = (x_1, \ldots, x_n)$, it is $\sqrt{x_1^2 + \ldots + x_n^2}$. Euclidean distance between two vectors x, y is $||x - y||$.

$\mathrm{rank}(A)$ — Rank of a matrix A, same as the rank of its set of row vectors, or its set of column vectors.

WRT	With respect to.
RHS	Right hand side.
PC, PR	Pivot column, pivot row.
RC , IE	Redundant constraint, inconsistent equation identified.
RI, CI	Row interchange, column interchange.
BV	Basic variable selected for that row.
GJ method	The Gauss–Jordan elimination method for solving systems of linear equations.
GJ pivot	The Gauss–Jordan pivot step on a tableau or a matrix.
G pivot	The Gaussian pivot step on a tableau or a matrix.
G method	The Gaussian elimination method for solving systems of linear equations.
PSD, NSD	Positive semidefinite, negative semidefinite.
PD, ND	Positive definite, negative definite.
Scalar	A real number. "Scalar multiplication" means multiplication by a real number.
⋈	This symbol indicates the end of the present portion of text (i.e., example, exercise, comment etc.). Next line either resumes what was being discussed before this portion started, or begins the next portion.

A Poem on Linear Algebra

To put you in a receptive mood as you embark on the study of this wonderful subject, I thought it fitting to begin this book with a lovely Telugu poem on the virtues of linear algebra composed by Dhulipalla V. Rao with help from my brother Katta Sai, daughter Madhusri Katta and myself. The original poem in roman transliteration first, followed by the English translation.

- -

OhO lIniyar AljIbrA nIkivE mAjOhArlu!

sAgaramagAdhamani caduvukunnaM
nIlOtiMkAminnani telusukunnAM

aMdAniki marOpEru araviMdaM
gaNitulaku nIiMpu marimari aMdaM

jIvanAniki maMchinIru eMtO muKyaM
mAnavALi pragatiki nuvvaMtakannA muKyaM

maMciruciki baMginipalli mAmiDi maharAju
gaNitameriginavAriki nuvvE rArAju

paMTalaku tEneTIgalostAyi tODu
nuvulEkuMTE mAnavajIvitamavunu bIDu

medaDuku padunu nI adhyayanaM
pragatiki sOpAnaM nI paTanaM.

- -

Oh, linear algebra! You are
Deep as the mighty ocean
As exciting as the pink lotus is beautiful
As satisfying to my brain as the sweet banginapalli* mango is to my tongue
As essential to science as water is to life
As useful in applications as bees are for the cultivation of crops[+]
I cannot describe in words how much richer human life has become because
 we discovered you!.

---

* A variety of mango grown in Andhra Pradesh, India, with crisp sweet
  flesh that has no fibrous threads in it, that is considered by some to be
  the most delicious among fruits.
[+] Bees carry out the essential task of pollinating the flowers in fruit, veg-
  etable, and some other crops. Without bees, we will not have the abun-
  dance of fruits and vegetables we enjoy today.

Chapter 1

Systems of Simultaneous Linear Equations

1.1 What Are Algorithms?

Algorithms are procedures for solving computational problems. Sometimes algorithms are also called **methods**.

An algorithm for a computational problem is a systematic step-by-step procedure for solving it that is so clearly stated without any ambiguities that it can be programmed for execution by a machine (a computer).

Historical note on algorithms: The word "algorithm" originated from the title of the Latin translation of a book

Algoritmi de Numero Indorum

meaning *Al-Khwarizmi Concerning the Hindu Art of Reckoning.* It was written in 825 AD by the Arabic scholar "Muhammad Ibn Musa Al-Khwarizmi" who lived in Iraq, Iran (mostly Baghdad); and was based on earlier Indian and Arabic treatizes. This book survives only in its Latin translation, as all the copies of the original Arabic version have been lost or destroyed.

Algorithms seem to have originated in the work of ancient Indian mathematicians on rules for solving linear and quadratic equations.

As an illustrative example, we provide the following problem and discuss an algorithm for solving it:

Problem 1: Input: Given positive integer n. Required output: Find all prime numbers $\leq n$.

A **prime number** is a positive integer >1 that is divisible only by itself and by 1.

We provide an algorithm for this problem dating back to the 3rd century BC, known as the **Eratosthenes Sieve**.

Algorithm: Eratosthenes Sieve for Finding All Prime Numbers $\leq n$

BEGIN

Initial Step: Write all integers 1 to n in increasing order from left to right. Strike off 1 and put a pointer at 2, the first prime.

General Step: Suppose pointer is now on r. Counting from r, strike off every rth number to the right.

If all numbers to the right of r are now struck off, collect all the numbers in the list that are not struck off. These are all the primes $\leq n$, terminate.

Otherwise move the pointer right to the next number not struck off, and repeat this general step with the pointer in its new position.

END

Example 1: Illustration of Eratosthenes Sieve for $n = 10$

We will now apply this algorithm on the instance of the problem with $n = 10$. At the end of Step 1, the sequence of numbers written looks like this:

$$\not{1}, \overset{2}{\underset{\uparrow}{}}, 3, 4, 5, 6, 7, 8, 9, 10.$$

After the General step is applied once, the sequence of numbers looks like this:

$$\not{1}, 2, \overset{3}{\underset{\uparrow}{}}, \not{4}, 5, \not{6}, 7, \not{8}, 9, \not{10}.$$

Applying the General step again, we get this:

$$\not{1}, 2, 3, \not{4}, \overset{5}{\underset{\uparrow}{}}, \not{6}, 7, \not{8}, \not{9}, \not{10}.$$

Applying the General step again, we get this:

$$\not{1}, 2, 3, \not{4}, 5, \not{6}, \overset{7}{\underset{\uparrow}{}}, \not{8}, \not{9}, \not{10}.$$

Now all the numbers to the right of the pointer are struck off, so we can terminate. The set of numbers not struck off is $\{2, 3, 5, 7\}$, this is the set of all primes ≤ 10. ⋈[1]

There may be several different algorithms for solving a problem. The mathematical characterization of the most efficient algorithm for a problem is quite complicated. Informally, we will say that an algorithm for solving a problem is more efficient than another if it is faster or requires less computational effort.

It usually requires a deep study of a problem to uncover details of its structure and find good characterizations of its solution, in order to develop efficient algorithms for it. We will illustrate this statement with an example of a problem in networks and an efficient algorithm for it that came out of a deep study of the problem.

A **network** consists of points called **nodes**, and **edges** each of which is a line joining a pair of nodes. Such networks are used to model and represent the system of streets in a city, in transportation, routing, traffic flow, and distribution studies. In this representation street intersections or traffic centers are represented by nodes in the network and two-way street segments joining adjacent intersections are represented by edges.

A network is said to be **connected** if it is possible to travel from any node to any other node in the network using only the edges in the network. Figure 1.1 is a network in which nodes are numbered and drawn

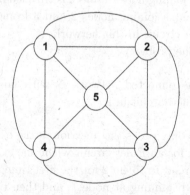

Figure 1.1

---

[1]This symbol indicates the end of this portion of text, here Example 1.

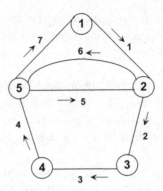

Figure 1.2

as little circles, and each edge joins a pair of nodes. This network is clearly connected.

An **Euler circuit** (named after the 18th century Swiss Mathematician Leonhard Euler) in a connected network is a route which begins at a node, say node 1, goes through *each edge of the network exactly once, and returns to the starting node, node 1, at the end.* Euler circuits play an important role in problems such as those of finding least distance routes for postmen's beats, good routes for school buses and garbage collection vehicles, etc.

As an example, in the network with 5 nodes and 7 edges in Figure 1.2, traveling along the edges in the order and direction indicated on the edges in Figure 1.2 (the edge joining nodes 1 and 2 is traveled first in the direction from 1 to 2, then the edge joining nodes 2 and 3 from node 2 to node 3, etc.) leads to one Euler circuit in this network.

Now we describe the problem.

Problem 2: Given a connected network G with nodes $1, \ldots, n$, check whether there is an Euler circuit in it.

Notice that Problem 2 only poses the question of the existence of an Euler circuit, it does not ask for an Euler circuit when one exists.

Someone who has read Problem 2 for the first time may try to develop an algorithm for it by beginning at node 1, and then tracing the edges in some order without repeating any edge in an effort to construct an Euler circuit. But the most efficient algorithm for it is an indirect algorithm developed by L. Euler in the year 1736. The algorithm described below, only answers the existence question, and does not output an Euler circuit.

Euler's algorithm for Problem 2

BEGIN

Define the degree of a node in G to be the number of edges containing it. Find the degree of each node in G.

If the degree of every node in G is an even number, there is an Euler circuit in G. If at least one node in G has odd degree, there is no Euler circuit in G. Terminate.

END

Example 2: Illustration of Euler's Algorithm to check the existence of an Euler circuit

Let d_i denote the degree of node i. For the network in Figure 1.2, we have $(d_1, d_2, d_3, d_4, d_5) = (2, 4, 2, 2, 4)$. Since all these node degrees are even, this network has an Euler circuit. An Euler circuit in it was given above.

For the network in Figure 1.1, we have $(d_1, d_2, d_3, d_4, d_5) = (5, 5, 5, 5, 4)$. Since there are some odd numbers among these node degrees, this network has no Euler circuit by the above algorithm.

The basis for this algorithm is a very beautiful theorem proved by L. Euler in the year 1736 that a connected network has an Euler circuit iff the degree of every node in it is even.

The branch of science dealing with the development of efficient algorithms for a variety of computational problems is the most challenging area of research today.

Exercises

1.1.1: Apply the Eratostenes Sieve to find all prime numbers $\leq n = 22$.

1.1.2: Consider the network in Figure 1.3. Check whether this network has an Euler circuit using the algorithm discussed above. Verify the correctness of the answer by actually trying to construct an Euler circuit beginning with node 1 and explain your conclusion carefully. $\bowtie$

In this book we will discuss algorithms for solving systems of linear equations, and other related linear algebra problems.

Systems of simultaneous linear equations pervade all technical areas. Knowledge about models involving simultaneous linear equations, and the fundamental concepts behind algorithms for solving and analyzing them,

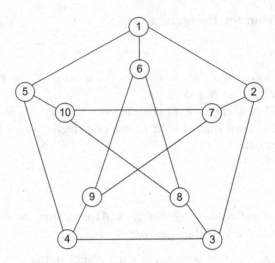

Figure 1.3   Network for Exercise 1.1.2

is extremely important for everyone with aspirations of getting any technical job. Also, systems of simultaneous linear equations lie at the heart of computational methods for solving most mathematical models, as many of these methods include steps that require the solution of such systems. Hence the material discussed in this chapter is very fundamental, and extremely important for all students of engineering and the applied sciences.

We begin with some example applications that lead to models involving simultaneous linear equations.

1.2   Some Example Applications Involving Linear Equations

Example 1: Scrap Metal Blending Problem

Consider the following problem. A steel company has four different types of scrap metal (called SM-1 to SM-4) with the following compositions:

Type	% in type, by weight, of element			
	Al	Si	C	Fe
SM-1	5	3	4	88
SM-2	7	6	5	82
SM-3	2	1	3	94
SM-4	1	2	1	96

They need to blend these four scrap metals into a mixture for which the composition by weight is:

$$Al - 4.43\%, \quad Si - 3.22\%, \quad C - 3.89\%, \quad Fe - 88.46\%.$$

How should they prepare this mixture? To answer this question, we need to determine the proportions of the 4 scrap meatals SM-1, SM-2, SM-3, SM-4 in the blend to be prepared.

Historical note on algebra, linear algebra: The most fundamental idea in linear algebra which was discovered more than 5000 years ago by the Chinese, Indians, Iranians, and Babylonians, is to represent the quantities that we like to determine by symbols; usually letters of the alphabet like x, y, z; and then express the relationships between the quantities represented by these symbols in the form of equations, and finally use these equations as tools to find out the true values represented by the symbols. The symbols representing the unknown quantities to be determined are nowadays called **unknowns** or **variables** or **decision variables**.

The process of representing the relationships between the variables through equations or other functional relationships is called **modeling** or **mathematical modeling**.

Linear algebra gradually evolved into algebra, one of the chief branches of mathematics. Even though the subject originated more than 5000 years ago, the name *algebra* itself came much later, it is derived from the title of an Arabic book *Hisab al-jabr w'almuqabalah* written by the same mathematician Al-Khowarizmi discussed in Section 1.1, around the same year 825 AD. The term "al-jabr" in Arabic means "restoring" in the sense of solving an equation. In Latin translation the title of this book became *Ludus Algebrae*, the second word in this title surviving as the modern word *algebra* for the subject.

The word *linear* in "linear angebra" refers to the "linear combinations" (see Sections 1.5, 3.5) in the spaces studied, and the *linearity* of "linear functions" (see Section 1.7) and "linear equations" studied in the subject.

Model for scrap metal blending problem: Returning to our scrap metal blending problem, in order to model it, we first define the decision variables, denoted by x_1, x_2, x_3, x_4, where for $j = 1$ to 4

$$x_j = \text{proportion of SM-}j \text{ by weight in the mixture}$$

Then the percentage by weight, of the element Al in the mixture will be, $5x_1 + 7x_2 + 2x_3 + x_4$, which is required to be 4.43. Arguing the same way for the elements Si, C, and Fe, we find that the decision variables x_1 to x_4 must satisfy each equation in the following system of **linear equations** in order to lead to the desired mixture:

$$5x_1 + 7x_2 + 2x_3 + x_4 = 4.43$$
$$3x_1 + 6x_2 + x_3 + 2x_4 = 3.22$$
$$4x_1 + 5x_2 + 3x_3 + x_4 = 3.89$$
$$88x_1 + 82x_2 + 94x_3 + 96x_4 = 88.46$$
$$x_1 + x_2 + x_3 + x_4 = 1.$$

The last equation in the system stems from the fact that the sum of the proportions of various ingradients in a blend must always be equal to 1. This system of equations is the mathematical model for our scrap metal blending problem. By solving it, we will find the values of the decision variables x_1 to x_4 that define the desired blend.

Discussion

RHS Constants, Coefficients

It is customary to write the equations as above with the terms involving the variables on the **left hand side (LHS)** of the "=" sign, and the constant terms on the **right hand side (RHS)**. The constant term appearing on the RHS of each equation is known as the **RHS constant** in that equation. The RHS constant in the first equation is 4.43, etc.

The number "5" appearing on the LHS of the first equation in the system is known as the **coefficient** of the variable x_1 in that equation, etc.

In this system of equations, every variable appears with a nonzero coefficient in every equation. If a variable does not appear in an equation, we will say that its coefficient in that equation is zero, and vice versa.

A variable is considered to be in the system iff it has a nonzero coefficient in at least one of the constraints. Therefore, every variable in the system must have a nonzero coefficient in at least one equation.

Moving Terms from Left to Right and Vice Versa

Terms can be moved from left to right (or vice versa) across the "=" symbol by multiplying them by a "−" sign. The justification for this comes from the principle that *when you add equal amounts to equal quantities, the resulting quantities will be equal.* In other words, when you add the same quantity to both sides of an equation, the equation continues to be valid. For example, the first equation in the above system is:

$$5x_1 + 7x_2 + 2x_3 + x_4 = 4.43.$$

Suppose we want to move the $7x_2$ term in this equation from the left side to the right side of the "=" symbol. Add $-7x_2$ to both sides of this equation. By the above principle this leads to:

$$5x_1 + 7x_2 + 2x_3 + x_4 - 7x_2 = 4.43 - 7x_2.$$

On the left hand side the $+7x_2$ and the $-7x_2$ terms cancel with each other, leading to:

$$5x_1 + 2x_3 + x_4 = 4.43 - 7x_2.$$

Notice that the sign of this term has changed as it moved from the left to the right of the equality symbol in this equation. In the same way, we can write the first constraint in an equivalent manner by moving the constant term from the right side of "=" to the left side as:

$$5x_1 + 7x_2 + 2x_3 + x_4 - 4.43 = 0.$$

Another equivalent way of writing the first constraint is:

$$5x_1 = 4.43 - 7x_2 - 2x_3 - x_4.$$

Nonnegativity Restrictions on the Variables

The system in this example consists of 5 equations in 4 variables. From the definition of the variables given above, it is clear that a solution to this system of equations makes sense for the blending application under consideration, only if all the variables in the system have nonnegative values in it. Each nonnegative solution to this system of equations leads to a desirable mixture. The nonnegativity restrictions on the variables are **linear inequality constraints**. They cannot be expressed in the form of linear equations, and since the focus of this book is linear equations, for the moment we will ignore them.

System of Equations, Solutions, Vectors

In the system of equations obtained above, each equation is associated with a specific requirement in the problem. For example, equations 1 to 4 are associated with the requirements of the elements Al, Si, C, Fe respectively in the mixture; and equation 5 must be satisfied by any blend, hence, we will call it the blending requirement.

In the same way, each decision variable in the system is associated with a unique scrap metal that can be included in the blend.

A system like this is called a **system of linear equations**, or more precisely, a **system of simultaneous linear equations** because a solution is required to satisfy all the equations in the system.

Using the Gauss–Jordan (GJ) elimination method discussed later in Section 1.16, it can be determined that this system has the unique solution:

$$\bar{x} = (\bar{x}_1, \bar{x}_2, \bar{x}_3, \bar{x}_4) = (0.25, 0.34, 0.39, 0.02)$$

which happens to be also nonnegative. Thus a mixture of desired composition can only be obtained by mixing the scrap metals in the following way:

Scrap metal	% by weight in blend
SM1	25
SM2	34
SM3	39
SM4	2

The solution $\bar{x} = (\bar{x}_1, \bar{x}_2, \bar{x}_3, \bar{x}_4) = (0.25, 0.34, 0.39, 0.02)$ is called a **solution vector** to the system of equations. A solution, or solution vector for a system of linear equations, gives specific numerical values to each variable in the system that together satisfy all the equations in the system.

The word **vector** in this book refers to a set of numbers arranged in a specified order, that's why it is called an **ordered set of numbers**. It is an ordered set because the first number in the vector is the proportion by weight of the specific raw material SM1, the second number is the proportion by weight of SM2, etc. Hence, if we change the order of numbers in a vector, the vector changes; i.e., for example $(0.25, 0.34, 0.39, 0.02)$ and $(0.34, 0.25, 0.39, 0.02)$ are not the same vector.

A vector giving specific numerical values to each variable in a system of linear equations is said to be a **solution** or **feasible solution** for the system if it satisfies every one of the equations in the system. A vector violating at least one of the equations in the system is said to be **infeasible** to the system.

Since it has four entries, $\bar{x}$ above is said to be a four dimensional vector. An n **dimensional vector** will have n entries.

The definition of vector that we are using here is standard in all mathematical literature. The word "vector" is also used to represent a directed line segment joining a pair of points in the two dimensional plane in geometry and physics. Using the same word for two different concepts confuses young readers. To avoid confusion, we will use the word "vector" to denote an ordered set of numbers only in this book. For the other concept "the direction from one point towards another point" in space of any dimension we will use the word "direction" itself or the phrase "direction vector" (see Section 3.5).

As we will learn later, vectors can be of two types, **row vectors** and **column vectors**. If the entries in a vector are written horizontally one after the other, the vector is called a row vector. The solution vector $\bar{x}$ is a row vector. If the entries in a vector are written vertically one below the other, the vector is called a column vector.

The symbol R^n denotes the set of all possible n dimensional vectors (either row vectors or column vectors). If u is an n dimensional vector with entries $u_1, \ldots, u_n$ in that order, each u_i is a real number; and we indicate this by $u \in R^n$ (i.e., u is contained in R^n). An n dimensional vector is also called an n-**tuple** or an n-**tuple vector**.

In the n-dimensional vector $x = (x_1, \ldots, x_{j-1}, x_j, x_{j+1}, \ldots, x_n)$, x_j is known as the j-**th entry** or j-**th component** or j-**th coordinate** of the vector x, for $j = 1$ to n.

R^n is called the n-**dimensional vector space**, or the **space of all** n-**dimensional vectors**.

Historical note on vectors: The original concept of a vector (ordered set of numbers, also called a sequence now-a-days) goes back to prehistoric times in ancient India, in the development of the **position notation** for representing numbers by an ordered set or sequence of digits. There are 10 digits, these are $0, 1, \ldots, 9$. When we write the number 2357 as a sequence of digits ordered from left to right, we are using a notation for the number which is actually $7 + 5(10) + 3(10^2) + 2(10^3)$.

In the symbol "2357" the rightmost digit 7 is said to be in the "digits position". The next digit 5 to the left is in the "10s position". The next one 3 to the left is in the "100s position", or "10^2s position"; and so on. Clearly changing the order of digits in a number changes its value, that's why it is important to treat the digits in a number as an ordered set or sequence, which we are now calling a vector.

In the number $904 = 4 + 9(10^2)$, there is no entry in the 10s position. For representing such numbers the people who developed the position notation in ancient India realized that they had to introduce a symbol for "nothing" ("sunya" in Sanskrit), and introduced the symbol "0" (sunna) for **zero** or "nothing". It is believed that zero got introduced officially as a digit through this process, even though it was not recognized as a number until then.

Example 2: Athlete's Supplemental Diet Formulation

The nutritionist advising a promising athlete preparing for an international competition has determined that his client needs 110 units of Vitamin E, 250 units of Vitamin K, and 700 calories of stored muscular energy (SME) per day in addition to the normal nutrition the client picks up from a regular diet. These additional nutrients can be picked up by including five different supplemental foods with the following compositions (remember that the units of measurement of each supplemental food and each nutrient may be different, for example one may be measured in liters, another in Kg, etc.).

Nutrient	Nutrient units/unit of food				
	Food 1	Food 2	Food 3	Food 4	Food 5
Vit E	50	0	80	90	4
Vit K	0	0	200	100	30
SME	100	300	50	250	350

A supplemental diet specifies the quantity of each of foods 1 to 5 to include in the athlete's daily intake. In order to express mathematically the conditions that a supplemental diet has to satisfy to meet all the requirements, we define the decision variables: for $j = 1$ to 5

$$x_j = \text{units of food } j \text{ in the supplemental diet}$$

The total amount of Vit E contained in the supplemental diet represented by the vector $x = (x_1, x_2, x_3, x_4, x_5)$ is $50x_1 + 80x_3 + 90x_4 + 4x_5$ units. Unfortunately, not all of this quantity is absorbed by the body, only a fraction of it is absorbed and the rest discarded. The fraction absorbed varies with the food and the nutrient, and is to be determined by very careful (and tedious) metabolic analysis. This data is given below.

Nutrient	Prop. nutrient absorbed, in food				
	Food 1	Food 2	Food 3	Food 4	Food 5
Vit E	0.6	—	0.8	0.7	0.9
Vit K	—	—	0.4	0.6	1.0
SME	0.5	0.6	0.9	0.4	0.8

So the units of Vit E absorbed by the athlete's body in the supplemental diet represented by x is $50 \times 0.6x_1 + 80 \times 0.8x_3 + 90 \times 0.7x_4 + 4 \times 0.9x_5$, which is required to be 110. Arguing the same way with Vit K, and Calories of SME, we find that the decision variables x_1 to x_5 must satisfy the following system of linear equations in order to yield a supplemental diet satisfying all the nutritionist's requirements:

$$30x_1 + 64x_3 + 63x_4 + 36x_5 = 110$$
$$80x_3 + 60x_4 + 30x_5 = 250$$
$$50x_1 + 180x_2 + 45x_3 + 100x_4 + 280x_5 = 700.$$

This is a system of three equations in five variables. Of course, here also, in addition to these equations, the variables have to satisfy the nonnegativity restrictions, which we ignore.

Example 3: A Coin Assembly Problem

A class in an elementary school was assigned the following project by their teacher. They are given an unlimited supply of US coins consisting of: pennies (1 cent), nickels (5 cents), dimes (10 cents), quarters (25 cents), half dollars (50 cents), and dollars. They are required to assemble a set of these coins satisfying the following constraints:

Total value of the set has to be equal to \$31.58
Total number of coins in the set has to be equal to 171
Total weight of the set has to be equal to 735.762 grams.

Here is the weight data:

Coin	Weight in grams
Penny	2.5
Nickel	5.0
Dime	2.268
Quarter	5.67
Half Dollar	11.34
Dollar	8.1

Formulate the class's problem as a system of linear equations.

Let the index $j = 1$ to 6 refer to penny, nickel, dime, quarter, half dollar, and dollar, respectively. For $j = 1$ to 6 define the decision variable:

$x_j = $ number of coins of coin j in the assembly.

Then the constraints to be satisfied by the solution vector $x = (x_1, x_2, x_3, x_4, x_5, x_6)$ corresponding to an assembly satisfying the teacher's requirements are:

$$0.01x_1 + 0.05x_2 + 0.10x_3 + 0.25x_4 + 0.5x_5 + x_6 = 31.58$$
$$x_1 + x_2 + x_3 + x_4 + x_5 + x_6 = 171$$
$$2.5x_1 + 5.0x_2 + 2.268x_3 + 5.67x_4 + 11.34x_5 + 8.1x_6 = 735.762.$$

Of course, in addition to these equations, we require the variables to be not only nonnegative, but nonnegative integers; but we ignore these requirements for the moment.

Example 4: Balancing chemical reactions

In a chemical reaction, two or more input chemicals combine and react to form one or more output chemicals. **Balancing this chemical reaction** deals with the problem of determining how many molecules of each input chemical participate in this reaction, and how many molecules of each output chemical are produced in the reaction. This leads to a system of linear equations for which the simplest positive integer solution is needed.

As an example, consider the well known reaction in which molecules of hydrogen (H) and oxygen (O) combine to produce molecules of water. A molecule of hydrogen (oxygen) consists of two atoms of hydrogen (oxygen) and is represented as H_2 (O_2); in the same way a molecule of water consists of two atoms of hydrogen and one atom of oxygen and is represented as H_2O. The question that arises about this reaction is: how many molecules of hydrogen combine with how many molecules of oxygen in the reaction, and how many molecules of water are produced as a result. Let

x_1, x_2, x_3 denote the number of molecules of hydrogen, oxygen, water respectively, involved in this reaction.

Then this chemical reaction is represented as:

$$x_1 H_2 + x_2 O_2 \longrightarrow x_3 H_2 O.$$

Balancing this chemical reaction means finding the simplest positive integral values for x_1, x_2, x_3 that leads to the same number of atoms of hydrogen and oxygen, the two elements appearing in this reaction, on both sides of the arrow.

Equating the atoms of each element on both sides of the arrow leads to an equation that the variables have to satisfy. In this reaction those equations are: $2x_1 = 2x_3$ for hydrogen atoms, and $2x_2 = x_3$ for oxygen atoms. This leads to the system of equations:

$$2x_1 - 2x_3 = 0$$
$$2x_2 - x_3 = 0$$

in x_1, x_2, x_3. $x_1 = 2, x_2 = 1, x_3 = 2$ leads to a simple integer solution to this system, leading to the balanced reaction $2H_2 + O_2 \longrightarrow 2H_2O$.

In the same way balancing any chemical reaction leads to a system of linear equations in which each variable is associated with either an input, or an output chemical; and each constraint corresponds to a chemical element in some of these chemicals. We require the simplest positive integer solution to this system, but for the moment we will ignore these requirements.

Example 5: Parameter estimation in curve fitting using the method of least squares

A common problem in all areas of scientific research is to obtain a mathematical functional relationship between a variable, say t, and a variable y whose value is known to depend on the value of t; from data on the values of y corresponding to various values of t, usually determined experimentally. Suppose this data is:

when $t = t_r$, the observed value of y is y_r; for $r = 1$ to k.

Using this data, we are required to come up with a mathematical expression for y as a function of t. This problem is known as a **curve fitting problem**, handling it involves the following steps.

Algorithm for Solving a Curve Fitting Problem

BEGIN

Step 1: Select a model function $f(t)$

An explicit mathematical function, let us denote it by $f(t)$, containing possibly several unknown parameters (or numerical constants with unknown values) that seems to provide a good fit for y is selected in this step. The selected function is called the **model function**. There may be theoretical considerations which suggest a good model function. Or one can create a plot of the observed data on the 2-dimensional Cartesian plane, and from the pattern of these points think of a good model function.

Step 2: Parameter estimation

Let $a_0, a_1, \ldots, a_n$ denote the parameters in the model function $f(t)$ whose numerical values are unknown.

Measuring the values of y experimentally is usually subject to unknown random errors, so even if the model function $f(t)$ is an exact fit for y, we may not be able to find any values of the parameters $a_0, a_1, \ldots, a_n$ that satisfy:

y_r = observed value of y when $t = t_r$ is $= f(t_r)$, for $r = 1$ to k.

In this step we find the best values for the parameters $a_0, \ldots, a_n$ that make $f(t_r)$ as close to y_r as possible, for $r = 1$ to k.

The **method of least squares** for parameter estimation selects the values of the parameters $a_0, \ldots, a_n$ in $f(t)$ to be those that minimize the sum of squares of deviations $(y_r - f(t_r))$ over $r = 1$ to k; i.e.,

$$\text{minimize } L_2(a_0, \ldots, a_n) = \sum_{r=1}^{k} (y_r - f(t_r))^2.$$

The vector $(\bar{a}_0, \ldots, \bar{a}_n)$ that minimizes $L_2(a_0, \ldots, a_n)$ is known as the **least squares estimator** for the parameter vector; and with the values of the parameters $(a_0, \ldots, a_n)$ fixed at $(\bar{a}_0, \ldots, \bar{a}_n)$, the function $f(t)$ is known as the **least squares fit** for y as a function of t.

From calculus we know that a necessary condition for $(\bar{a}_0, \ldots, \bar{a}_n)$ to minimize $L_2(a_0, \ldots, a_n)$ is that it must satisfy the equations obtained by setting the partial derivatives of $L_2(a_0, \ldots, a_n)$ WRT $a_0, \ldots, a_n$ equal to 0.

$$\frac{\partial L_2(a_0, \ldots, a_n)}{\partial a_j} = 0 \quad j = 0 \text{ to } n.$$

This system is known as the **system of normal equations**.

When the model function $f(t)$ satisfies certain conditions, the system of normal equations is a system of linear equations that can be solved by the methods to be discussed later on in this chapter. Also, any solution to this system is guaranteed to minimize $L_2(a_0, \ldots, a_n)$. A general class of model functions that satisfy these conditions are the polynomials. In this case,

$$f(t) = a_0 + a_1 t + a_2 t^2 + \cdots + a_n t^n$$

where the coefficients $a_0, \ldots, a_n$ are the parameters in this model function. We will discuss this case only.

If $n = 1$ (i.e., $f(t) = a_0 + a_1 t$), we are fitting a linear function of t for y. In Statistics, the problem of finding the best linear function of t that fits y (i.e., fits the given data as closely as possible) is known as the **linear regression problem**. In this special case, the estimates for the parameters obtained are called **regression coefficients**, and the best fit is called the **regression line** for y in terms of t.

If $n = 2$, we are trying to fit a quadratic function of t to y. And so on.

When the model function $f(t)$ is the polynomial $a_0 + a_1 t + \cdots + a_n t^n$, we have $L_2(a_0, \ldots, a_n) = \sum_{r=1}^{k} (y_r - a_0 - a_1 t_r - \cdots - a_n t_r^n)^2$. So,

$$\frac{\partial L_2(a_0, \ldots, a_n)}{\partial a_j}$$

$$= 2 \left(\sum_{r=1}^{k} y_r t_r^j - a_0 \sum_{r=1}^{k} t_r^j - a_1 \sum_{r=1}^{k} t_r^{j+1} - \cdots - a_n \sum_{r=1}^{k} t_r^{n+j} \right)$$

for $j = 0$ to n. Hence the normal equations are equivalent to

$$a_0 \sum_{r=1}^{k} t_r^j + a_1 \sum_{r=1}^{k} t_r^{j+1} + \cdots + a_n \sum_{r=1}^{k} t_r^{n+j} = \sum_{r=1}^{k} y_r t_r^j \quad \text{for } j = 0 \text{ to } n$$

which is a system of $n+1$ linear equations in the $n+1$ parameters $a_0, \ldots, a_n$.

If $(\bar{a}_0, \ldots, \bar{a}_n)$ is the solution of this system of linear equations, then $\bar{f}(t) = \bar{a}_0 + \bar{a}_1 t + \cdots + \bar{a}_n t^n$ is the best fit obtained for y.

Step 3: Checking how good the best fit obtained is

Let $(\bar{a}_0, \ldots, \bar{a}_n)$ be the best values obtained in Step 2 for the unknown parameters in the model function $f(t)$; and let $\bar{f}(t)$ denote the best fit for y, obtained by substituting $\bar{a}_j$ for a_j for $j = 0$ to n in $f(t)$. Here we check how well $\bar{f}(t)$ actually fits y.

The minimum value of $L_2(a_0, \ldots, a_n)$ is $L_2(\bar{a}_0, \ldots, \bar{a}_n)$, this quantity is called the **optimal residual** or the **sum of squared errors**.

The optimal residual will be 0 only if $y_r - \bar{f}(t_r) = 0$ for all $r = 1$ to k; i.e., the fit obtained is an exact fit to y at all values of t observed in the experiment.

The optimal residual will be large if some or all the deviations $y_r - \bar{f}(t_r)$ are large. Thus the value of the optimal residual can be used to judge how well $\bar{f}(t)$ fits y. The smaller the optimal residual the better the fit obtained.

If the optimal residual is considered too large, it is an indication that the model function $f(t)$ selected in Step 1 does not provide a good approximation for y in terms of t. In this case Step 1 is repeated to select a better model function, and the whole process repeated.

END

Example: Example of a curve fitting problem

As a numerical example consider the yield y from a chemical reaction, which depends on the temperature t in the reactor. The yield is measured at four different temperatures on a specially selected scale, and the data is given below.

Temperature t	Yield y
−1	7
0	9
1	12
2	10

We will now derive the normal equations for fitting a quadratic function of t to y using the method of least squares. So, the model function is $f(t) = a_0 + a_1 t + a_2 t^2$, involving three parameters a_0, a_1, a_2. In this example the sum of squared deviations

$$L_2(a_0, a_1, a_2) = (7 - a_0 + a_1 - a_2)^2 + (9 - a_0)^2$$
$$+ (12 - a_0 - a_1 - a_2)^2 + (10 - a_0 - 2a_1 - 4a_2)^2.$$

So, $\frac{\partial L_2}{\partial a_0} = -2(7 - a_0 + a_1 - a_2) - 2(9 - a_0) - 2(12 - a_0 - a_1 - a_2) - 2(10 - a_0 - 2a_1 - 4a_2)^2 = -76 + 8a_0 + 4a_1 + 12a_2$. Hence the normal equation $\frac{\partial L_2}{\partial a_0} = 0$ leads to

$$8a_0 + 4a_1 + 12a_2 = 76.$$

Deriving $\frac{\partial L_2}{\partial a_1} = 0$ and $\frac{\partial L_2}{\partial a_2} = 0$, in the same way, we have the system of normal equations given below.

$$8a_0 + 4a_1 + 12a_2 = 76$$
$$4a_0 + 12a_1 + 16a_2 = 50$$
$$12a_0 + 16a_1 + 36a_2 = 118.$$

This is a system of three linear equations with three unknowns. If $(\bar{a}_0, \bar{a}_1, \bar{a}_2)$ is its solution, $\bar{f}(t) = \bar{a}_0 + \bar{a}_1 t + \bar{a}_2 t^2$ is the quadratic fit obtained for y in terms of t.

We considered the case where the variable y depends on only one independent variable t. When y depends on two or more independent variables, parameter estimation problems are handled by the method of least squares in the same way by minimizing the sum of squared deviations.

Historical note on the method of least squares: The method of least squares is reported to have been developed by the German mathematician Carl Friedrich Gauss at the beginning of the 19th century while calculating the orbit of the asteroid Ceres based on recorded observations in tracking it earlier. It was lost from view when the astronomer Piazzi tracking it fell ill. Gauss used the method of least squares and the Gauss–Jordan method for solving systems of linear equations in this work. He had to solve systems involving 17 linear equations, which are quite large for hand computation. Gauss's accurate computations helped in relocating the asteroid in the skies in a few months time, and his reputation as a mathematician soared.

The Importance of Systems of Linear Equations

Systems of linear equations pervade all areas of scientific computation. They are the most important tool in scientific research, development and technology. Many other problems which do not involve linear equations directly in their statement are solved by iterative methods that usually involve in each iteration, one or more systems of linear equations to be solved.

The Scope of this Book, Limitations

The branch of mathematics called **Linear Algebra** initially originated for the study of systems of linear equations and for developing methods to solve them. This occurred more than 2000 years ago. Classical linear algebra concerned itself with only linear equations, and no linear inequalities. We limit the scope of this book to linear algebra related to the study of linear equations.

Extensions to Linear Inequalities, Linear Programming, Integer Programming, and Combinatorial Optimization

Historical note on linear inequalities, linear programming: Methods for solving systems of linear constraints including linear inequalities and/or nonnegativity restrictions or other bounds on the variables belong to the modern subject called **Linear Programming** which began with the development of the Simplex Method to solve them by George B. Dantzig in 1947. Most of the content of this book, with the possible exception of eigen values and eigen vectors is an essential prerequisite for the study of linear programming. All useful methods for solving linear programs are iterative methods that need to solve in each iteration one or more systems of linear equations formed with data coming from the current solution. The fundamental theoretical result concerning linear inequalities, called **Farkas' Lemma**, is the key for the development of efficient methods for linear programming. It appeared in *J. Farkas, "Über die Theorie der einfachen Ungleichungen", (in German), J. Reine Angew. Math., 124 (1902) 1–24.* Thus linear programming is entirely a 20th century subject. Linear programming is outside the scope of this book.

We noticed that some applications involved obtaining solutions to systems of linear equations in which the variables are required to be nonnegative integers. The branch of mathematics dealing with methods for obtaining integer solutions to systems of linear equations (no inequalities) is called **linear Diophantine equations**, again a classical subject that originated a long time ago. Unfortunately, linear Diophantine equations has remained a purely theoretical subject without many applications. The reason for this is that in applications which call for an integer solution to a system of linear constraints, the system usually contains nonnegativity restrictions or other bounds on the variables and/or other inequalities. Problems in which one is required to find integer solutions to systems of linear constraints including some inequalities and/or bounds on variables fall under the modern subjects known as **Integer Programming**, and **Combinatorial Optimization**. These subjects are much further extensions of linear programming, and were developed after linear programming in the 20th century. Approaches for solving integer programs usually need the solution of a sequence of linear programs. All these subjects are outside the scope of this book.

Figure 1.4 summarizes the relatioships among these subjects. The subject at the head-end of each arrow is an extension of the one at the tail-end.

Exercises

1.2.1: Data on the thickness of various US coins is given below.

Coin	Thickness in mm
Penny	1.55
Nickel	1.95
Dime	1.35
Quarter	1.75
Half dollar	2.15
Dollar	2.00

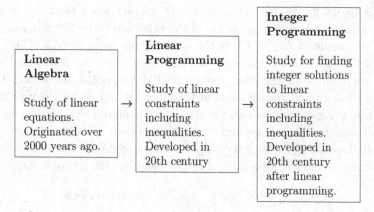

Figure 1.4

Formulate a version of the coin assembly problem in Example 3 in which the total value of the coin assembly has to be $55.93, the total number of coins in the assembly has to be 254, the total weight of the assembly has to be 1367.582 grams, and the total height of the stack when the coins in the assembly are stacked up one on top of the other has to be 448.35 mm, as a system of linear equations by ignoring the nonnegative integer requirements on the decision variables.

1.2.2: In the year 2000, the US government introduced two new one dollar coins. Data on them is given below.

Coin	Thickness in mm	Weight in grams
Sacagawea golden	1.75	5.67
Sacagawea Silver toned	2	8.1

Consider the problem stated in Exercise 1.2.1 above. The set of coins assembled has to satisfy all the constraints stated in Exercise 1.2.1. It may consist of a nonnegative integer number of pennies, nickels, dimes, quarters, half dollars, and dollars mentioned in Exercise 1.2.1; in addition, the set is required to contain exactly two Sacagawea golden dollars, and exactly 3 Sacagawea silver toned dollars. Discuss how this changes the linear equations model developed in Exercise 1.2.1.

1.2.3: Data on the composition of various US coins is given below. Consider a version of the coin assembly problem in Example 3 in which the total value of the coin assembly has to be $109.40, the total number of coins in the assembly has to be 444, the total weight of the assembly has

to be 2516.528 grams, the total height of the stack when the coins in the
assembly are stacked up one on top of the other has to be 796.65 mm; and
the total amount of Zn, Ni, Cu in the assembly has to be 268.125, 256.734,
and 1991.669 g, respectively. Also, among these coins, the penny and the
nickel are plain; while the dime, quarter, half dollar, dollar are reeded. The
number of plain and reeded coins in the assembly has to be 194 and 250,
respectively. Again, among these coins, only the quarter, half dollar, dollar
have an eagle on the reverse; and the number of such coins in the assembly
is required to be 209. Formulate this problem as a system of linear equations
by ignoring the nonnegative integer requirements on the decision variables.

Coin	Fraction in coin, by weight, of element		
	Zn	Ni	Cu
Penny	0.975	0	0.025
Nickel	0	0.25	0.75
Dime	0	0.0833	0.9167
Quarter	0	0.0833	0.9167
Half dollar	0	0.0833	0.9167
Dollar	0	0.0833	0.9167

1.2.4: A container has been filled with water from two taps in two different
experiments. In the first experiment, the first tap was left on at full level
for six minutes, and then the second tap was left on at full level for three
minutes; resulting in a total flow of 114 quarts into the container. In the
second experiment a total of 66 quarts has flown into the container when
the first tap was left on at full level for two minutes, and then the second tap
was left on at full level for three minutes. Assuming that each tap has the
property that the flow rate through it at full level is a constant independent
of time, find these rates per minute. Formulate this problem as a system
of linear equations (from Van Nostrand Reinhold *Concise Encyclopedia of
Mathematics*, 2nd edition, 1989).

1.2.5: An antifreeze has a density of 1.135 (density of water = 1). How
many quarts of antifreeze, and how many quarts of water have to be mixed
in order to obtain 100 quarts of mixture of density 1.027? (From Van Nostrand Reinhold *Concise Encyclopedia of Mathematics*, 2nd edition, 1989.)

1.2.6: A person is planning to buy special long slacks (SLS), fancy shirts
(FS) that go with them; special half-pants (SHP), and special jackets (SJ)
that go with them. Data on these items is given below.

The person wants to spend exactly $1250, and receive exactly 74 bonus coupons from the purchase. Also, the total number of items purchased should be 14. The number of SLS, FS purchased should be equal; as also the number of SHP, SJ purchased. Formulate the person's problem using a system of linear equations, ignoring any nonnegative integer requirements on the variables.

Item	Price ($/item)	Bonus coupons*
SLS	120	10
FS	70	4
SHP	90	5
SJ	80	3

*awarded per item purchased

1.2.7: A person has taken a boat ride on a river from point A to point B, 10 miles away, and back, to point A on a windless day. The river water is flowing in the direction of A to B at a constant speed and the boat always travels at a constant speed in still water. The trip from A to B took one hour, and the return trip from B to A took 5 hours. Formulate the problem of finding the speeds of the river and the boat using a system of linear equations.

1.2.8: A company has two coal mines which produce metallurgical quality coal to be used in its three steel plants. The daily production and usage rates are as follows:

Mine	Daily production
M_1	800 tons
M_2	400 tons

Plant	Daily requirement
P_1	300 tons
P_2	200 tons
P_3	700 tons

Each mine can ship its coal to each steel plant. The requirement at each plant has to be met exactly, and the production at each mine has to be shipped out daily. Define the decision variables as the amounts of coal shipped from each mine to each plant, and write down the system of linear equations they have to satisfy.

1.2.9: The yield from a chemical reaction, y, is known to vary as a third degree polynomial in $t - 105$ where t is the temperature inside the reactor in degrees centigrade, in the range $100 \leq t \leq 110$. Denoting the yield when the temperature inside the reactor is t by $y(t)$, we therefore have

$$y(t) = a_0 + a_1(t - 105) + a_2(t - 105)^2 + a_3(t - 105)^3$$

in the range $100 \leq t \leq 110$, where a_0, a_1, a_2, and a_3 are unknown parameters. To determine the values of these parameters, data on observations on the yield taken at four different temperatures is given below. Write the equations for determining the parameters a_0, a_1, a_2, a_3 using this data.

Temperature	Yield
103	87.2
105	90
107	96.8
109	112.4

1.2.10: A pizza outlet operates from noon to midnight each day. Its employees begin work at the following times: noon, 2 PM, 4 PM, 6 PM, 8 PM, and 10 PM. All those who begin work at noon, 2 PM, 4 PM, 6 PM, or 8 PM, work for a four hour period without breaks and then go home. Those who begin work at 10 PM work only for two hours and go home when the outlet closes at midnight.

On a particular day the manager has estimated that the outlet needs the following number of workers during the various two hour intervals.

Time interval	No. workers needed
12 noon − 2 PM	6
2 PM − 4 PM	10
4 PM − 6 PM	9
6 PM − 8 PM	8
8 PM − 10 PM	5
10 PM − 12 midnight	4

The manager needs to figure out how many workers should be called for work at noon, 2 PM, 4 PM, 6 PM, 8 PM, and 10 PM on that day to meet the staffing requirements given above exactly. Write the system of equations that these decision variables have to satisfy.

1.2.11: A shoe factory makes two types of shoes, J (jogging shoes) and W (walking shoes). Shoe manufacturing basically involves two major operations: cutting in the cutting shop and assembly in the assembly shop. Data on the amount of man-minutes of labor required in each shop for each type and the profit per pair made, is given below.

Type	Man-minutes needed/pair in		Profit/pair
	Cutting shop	Assembly shop	
J	10	12	10$
W	6	3	8$
Man-mts. available	130,000	135,000	

Assuming that all man-minutes of labor available in each shop are used up exactly on a particular day, and that the total profit made on that day is $140,000, develop a linear equation model to determine how many pairs of J and W are manufactured on that day.

1.2.12: A farmer has planted corn and soybeans on a 1000 acre section of the family farm, using up all the land in the section.

An acre planted with soybeans yields 40 bushels of soybeans, and an acre planted with corn produces 120 bushels of corn. Each acre of corn requires 10 manhours of labor, and each acre of soybeans needs 6 manhours of labor, over the whole season. The farmer pays $5/manhour for labor.

The farmer sells soybeans at the rate of $8/bushel, and corn at the rate of $4/bushel.

The farmer had to pay a total of $38,000 for labor over the season, and obtained a total of $384,000 from the sale of the produce.

Develop a system of linear equations to determine how many acres are planted with corn and soybeans and how many bushels of corn and soybeans are produced.

1.2.13: In a plywood company, logs are sliced into thin sheets called green veneer. To make plywood sheets, the veneer sheets are dried, glued, pressed in a hot press, cut to an exact size, and then polished.

There are different grades of veneer and different grades of plywood sheets. For our problem assume that there are two grades of veneer (A and B), and that the company makes two grades of plywood sheets (regular and premium). Data on inputs needed to make these sheets and the availability of these inputs is given below.

Input	Needed for 1 sheet of		Available/day
	Regular	Premium	
A-veneer sheets	2	4	100,000
B-veneer sheets	4	3	150,000
Pressing mc. minutes	1	2	50,000
Polishing mc. minutes	2	4	90,000
Selling price($/sheet)	15	20	

On a particular day, the total sales revenue was $575,000. 10,000 A-veneer sheets and 20,000 B-veneer sheets of that day's supply were unused. The polishing machine time available was fully used up, but 5,000 minutes of available pressing time was not used. Use this information and develop a system of all linear equations that you can write for determining the number of regular and premium plywood sheets manufactured that day.

1.2.14: A couple, both of whom are ardent nature lovers, have decided not to have any children of their own, because of concerns that the already large human population is decimating nature all over the world. One is a practicing lawyer, and the other a practicing physician. The physician's annual income is four times that of the lawyer.

They have a fervent desire to complete two projects. One is to buy a 1000 acre tract of undeveloped land and plant it with native trees, shrubs and other flora, and provide protection so that they can grow and flourish without being trampled by humanity. This is estimated to require 10 years of the wife's income plus 20 years of the husband's income.

The other project is to fund a family planning clinic in their neighborhood, which is exactly half as expensive as the first project, and requires 6 years of their combined income.

They read in a magazine article that Bill Gate's official annual salary at his company is half-a-million dollars, and thought "Oh!, that is our combined annual income".

Express as linear equations, all the constraints satisfied by their annual incomes described above.

1.2.15: A university professor gives two midterm exams and a final exam in each of his courses. After teaching many courses over several semesters he has developed a theory that the score of any student in the final exam of one of his courses, say x_3, can be predicted from that student's midterm exam scores, say x_1, x_2; by a function of the form $a_0 + a_1 x_1 + a_2 x_2$ where a_0, a_1, a_2 are unknown parameters.

In the following table, we give data on the scores obtained by 6 students in one of his courses in all the exams. Using it, it is required to determine the best values for a_0, a_1, a_2 that will make x_3 as close as possible to $a_0 + a_1x_1 + a_2x_2$ by the method of least squares. Develop the system of normal equations from which these values for a_0, a_1, a_2 can be determined.

Student no.	Score obtained in		
	Midterm 1	Midterm 2	Final
1	58	38	57
2	86	76	73
3	89	70	58
4	79	83	68
5	77	74	77
6	96	90	84

1.2.16: The following table gives data on $x_1 =$ height in inches, $x_2 =$ age in years, and $x_3 =$ weight in lbs of six male adults from a small ethnic group in a village. It is believed that a relationship of the form: $x_3 = a_0 + a_1x_1 + a_2x_2$ holds between these variables, where a_0, a_1, a_2 are unknown parameters. Develop the normal equations to determine the best values for a_0, a_1, a_2 that make this fit as close as possible using the method of least squares.

Person	Height	Age	Weight
1	62	40	152
2	66	45	171
3	72	50	184
4	60	35	135
5	69	54	168
6	71	38	169

1.2.17: They have conducted experiments at an agricultural research station to determine the effect of the application of the three principal fertilizer components, nitrogen (N), phosphorous (P), and potassium (K), on yield in the cultivation of sugar beets. Various quantities of N, P, K are applied on seven research plots and the yield measured (all quantities are measured in their own coded units). Data is given in the following table.

Let N, P, K, and Y denote the amounts of N, P, K applied and the yield respectively. It is required to find the best fit for Y of the form $a_0 + a_1N + a_2P + a_3K$, where a_0, a_1, a_2, a_3 are unknown parameters. Develop the normal equations for finding the best values for these parameters that give the closest fit, by the method of least squares.

Plot no.	Quantity applied			Yield
	N	P	K	
1	4	4	4	90
2	5	4	3	82
3	6	2	4	87
4	7	3	5	100
5	6	8	3	115
6	4	7	7	110
7	3	5	3	80

1.2.18: Following table gives information on the box-office revenue (in million dollar units) that some popular movies released in 1998 earned during each weekend in theaters in USA. The 1st column is the weekend number (1 being the first weekend) after the release of the movie, and data is provided for all weekends that the movie remained among the top-ten grossing films.

Weekend	Armageddon	Rush hour	The wedding singer
1	36.1	33.0	18.9
2	23.6	21.2	12.2
3	16.6	14.5	8.7
4	11.2	11.1	6.2
5	7.6	8.2	4.7
6	5.3	5.9	3.2
7	4.1	3.8	
8		3.3	

For each movie, it is believed that the earnings $E(w)$ during weekend w can be represented approximately by the mathematical formula

$$E(w) = ab^w$$

where a, b are positive parameters that depend on the movie.

For each movie separately, assuming that the formula is correct, write a system of linear equations from which a, b can be determined (you have to make a suitable transformation of the data in order to model this using linear equations).

Also, if this system of linear equations has no exact solution, discuss how the least squares method for parameter estimation discussed in Example 5 can be used to get the best values for the parameters a, b that minimizes the sum of squared deviations in the above system of linear equations. Using the procedure discussed in Example 5 derive another system of linear equations from which these best estimates for a, b can be obtained.

(Data obtained from the paper: R. A. Powers, "Big Box Office Bucks", *Mathematics Teacher*, 94, no. 2 (112–120) Feb. 2001.)

1.2.19: Minnesota's shrinking farm population: Given below is the yearly data on the number of farms in the State of Minnesota between 1983 (counted as year 0) to 1998 from the Minnesota Dept. of Agriculture. Here a farm is defined to be an agricultural establishment with \$1000 or more of agricultural sales annually. Let

t = year number (1983 corresponds to $t = 0$)

$y(t)$ = number of farms in Minnesota in year t in thousands.

t	0	1	2	3	4	5	6	7
$y(t)$	102	97	96	93	92	92	90	89

t	8	9	10	11	12	13	14	15
$y(t)$	88	88	86	85	83	82	81	80

Assume that $y(t) = a + bt$ where a, b are unknown real parameters. Write all the linear equations that a, b have to satisfy if the assumption is correct.

This system of equations may not have an exact solution in a, b. In this case the method of least squares takes the best values for a, b to be those that minimize the sum of square deviations $L(a, b) = \sum_{t=0}^{15}(y(t) - a - bt)^2$. From calculus we know that these values are given by the system of equations:

$$\frac{\partial L(a,b)}{\partial a} = 0, \quad \frac{\partial L(a,b)}{\partial b} = 0.$$

Derive this system of equations and show that it is a system of two linear equations in the two unknowns a, b.

1.2.20: Archimedes' Cattle Problem, also known as **Archimedes' Problema Bovinum**: The Greek mathematician Archimedes composed this problem around 251 BC in the form of a poem consisting of 22 elegiac couplets, the first of which says "If thou art diligent and wise, O stranger, compute the number of cattle of the Sun, who once upon a time grazed on the fields of the Thrinacian isle of Sicily." The Sun god's herd consisting of bulls and cows, is divided into white, black, spotted, and brown cattle. The requirement is to determine the composition of this herd, i.e., the number of cattle of each sex in each of these color groups. Ignoring the nonnegative integer requirements, formulate this problem as a system of linear equations, based on the following information given about the herd.

Among the bulls:

- the number of white ones is greater than the brown ones by (one half plus one third) the number of black ones;
- the number of black ones is greater than the brown ones by (one quarter plus one fifth) the number of the spotted ones;
- the number of spotted ones is greater than the brown ones by (one sixth plus one seventh) the number of white ones.

Among the cows:

- the number of white ones is (one third plus one quarter) of the total black cattle;
- the number of black ones is (one quarter plus one fifth) the total of the spotted cattle;
- the number of spotted is (one fifth plus one sixth) the total of the brown cattle;
- and the number of brown ones is (one sixth plus one seventh) the total of the white cattle.

1.2.21: Monkey and Coconut Problem: There are 3 monkeys, and a pile of N coconuts.

A traveling group of six men come to this scene one after the other. Each man divides the number of coconuts in the remaining pile by 6, and finds that it leaves a remainder of 3. He takes one sixth of that remaining pile himself, gives one coconut each to the three monkeys, leaves the others in the pile, and exits from the scene. The monkeys immediately transfer the coconuts they received to their private treasure piles.

After all the men have left, a group of 5 female tourists show up on the scene together. They find that if the number of remaining coconuts is divided by 5, it leaves a remainder of 3. Each one of them take one fifth of the remaining pile, and give one coconut to each of the three monkeys.

Write the system of equations satisfied by N and the number of coconuts taken by each man, and each female tourist.

1.2.22: The Classic Monkey and the Bananas Problem: Three sailors were shipwrecked on a lush tropical island. They spent the day gathering bananas, which they put in a large pile. It was getting dark and they were very tired, so they slept, deciding to divide the bananas equally among themselves next morning. In the middle of the night one of the sailors awoke and divided the pile of bananas into 3 equal shares. When he did so, he had one banana left over, he fed it to a monkey. He then hid his share in a secure place and went back to sleep. Later a 2nd sailor awoke and divided the remaining bananas into 3 equal shares. He too found one banana left over which he fed to the monkey. He then hid his share and went back to sleep. The 3rd sailor awoke later on, and divided the remaining bananas into 3 equal shares. Again one banana was left over, he fed it to the monkey; hid his share and went back to sleep.

When all the sailors got up next morning, the pile was noticeably smaller, but since they all felt guilty, none of them said anything; and they agreed to divide the bananas equally. When they did so, one banana was left over, which they gave to the monkey.

Write the system of linear equations satisfied by the number of bananas in the original pile, and the number of bananas taken by each of the sailors in the two stages. Find the positive integer solution to this system of equations that has the smallest possible value for the number of bananas in the original pile.

1.2.23: (From C. Ray Wylie, *Advanced Engineering Mathematics*, 6th ed., McGraw-Hill, 1995) Commercial fertilizers are usually mixtures containing specified amounts of 3 components: potassium (K), phosphorus (P), and nitrogen (N). A garden store carries three standard blends M_1, M_2, M_3 with the following compositions.

Blend	% in blend, of		
	K	P	N
M_1	10	10	30
M_2	30	10	20
M_3	20	20	30

(i): A customer has a special need for a fertilizer with % of K, P, N equal to 15, 5, 25 respectively. Discuss how the store should mix M_1, M_2, M_3 to produce a mixture with composition required by the customer. Formulate this problem as a system of linear equations.

(ii): Assume that a customer has a special need for a fertilizer with % of K, P, and N equal to k, p, and n respectively. Find out the conditions that k, p, n have to satisfy so that the store can produce a mixture of M_1, M_2, M_3 with this specific composition.

1.2.24: Balancing chemical reactions: Develop the system of linear equations to balance the following chemical reactions, and if possible find the simplest positive integer solution for each of these systems and express the chemical reaction using it. Besides hydrogen (H), oxygen (O) mentioned above, the elements involved in these reactions are: lead (Pb), carbon (C), sulpher (S) nitrogen (N), phosphorous (P), chlorine (Cl), sodium (Na), cerium (Ce), molybdenum (Mo), tantalum (Ta), Chromium (Cr), arsenic (As), and manganese (Mn).

(i) $C_n H_{2n+2}$ is the formula for a basic hydrocarbon for positive integer values of n. $n = 8, 10, 12$ correspond to common liquid hydrocarbons. Each of these hydrocarbons combines with oxygen in a chemical reaction (*combustion*) to produce carbon dioxide (CO_2) and water (H_2O). Balance this reaction for $n = 8, 10, 12$.

(ii) CH_4 is the formula for methane which is the main component in natural gas. Balance the chemical reaction in which methane combines with oxygen to produce carbon dioxide and water.

(iii) $C_2 H_{22} O_{11}$ is the formula for sucrose (sugar). Balance the reaction in which sucrose combines with oxygen to produce carbon dioxide and water.

(iv)   Balance the chemical reaction in which NO_2 and H_2O combine to form HNO_2 and HNO_3.

(v)   Balance the reaction in which the chemical compounds PbN_6 and $CrMn_2O_8$ combine to produce Pb_3O_4, CrO_3, MnO_2 and NO.

(vi)   Balance the reaction in which the chemical compounds MnS, $As_2 Cr_{10} O_{35}$, and $H_2 SO_4$ combine to produce $HMnO_4$, AsH_3, $CrS_3 O_{12}$ and H_2O.

(vii)   Balance the reaction in which the chemical compounds $Ta_2(NH_2)_4(PH_3)_2 H_2$ and O_2 combine to produce $Ta_2 O_5$, H_2O, $P_2 O_5$, and NO_2.

(viii) Balance the reaction in which the chemical compounds $Ta_2(NH_2)_4(PH_3)_2H_2$ and Cl_2 combine to produce $TaCl_5$, PCl_5, HCl, and NCl_3.

(ix) Balance the reaction in which the chemical compounds $Na_8CeMo_{12}O_{42}$ and Cl_2 combine to produce $NaCl$, $CeCl_4$, $MoCl_5$, and Cl_2O.

The reactions (vii), (viii), (ix) are given to me by K. Rengan.

1.2.25: Two couples C_1, C_2 who are very close friends, had their wedding anniversaries on the same date. They decided to celebrate it together.

First they had dinner at a nice restaurant. To cover the restaurant bill couple C_2 paid twice as much as C_1. Then they went to a theater show. For the tickets couple C_1 paid twice as much as C_2. Together, the theater tickets cost 50% more than the restaurant bill.

Then they went for dancing in a nightclub. For the entrance fees here couple C_1 paid as much as C_2 paid at the restaurant and theater together; and C_2 paid as much as C_1 paid at the restaurant and theater together.

After it was all over, they realized that the evenings program cost them together a total of \$750. Formulate a system of linear equations to determine who spent how much at each place, and who spent more overall.

1.2.26: Two couples $(W_1, H_1), (W_2, H_2)$ met at a weight loss clinic. They found out that all of them together weigh 700 lbs, that the two wives have the same weight which is the average weight of the two husbands, and that H_2 is one-third heavier than H_1. Formulate a system of linear equations to determine their weights individually.

1.2.27: In an enclosed area in a cattle ranch there are cows grazing and cowboys riding on horseback. The total number of legs (heads) of all the cows, horses, and cowboys put together is 2030 (510). Formulate a system of linear equations to determine the number of cows and cowboys in the area.

1.2.28: A college student went into a jewellery store in which all the items are on sale for \$40, with \$1000 in her purse. She bought some ear ring pairs (sold only in pairs), necklaces, single bangles, and ordinary rings; spending 60% of her money.

The number of ear ring pairs and ordinary rings that she purchased is twice the number of necklaces. The number of single bangles and ordinary rings that she purchased is the total number of ear rings in all the pairs she purchased. The number of necklaces and single bangles that she purchased is 50% more than the number of ear ring pairs and ordinary rings. The number of necklaces and ordinary rings that she purchased is one-half the number of ear ring pairs and single bangles. Formulate a system of linear equations to determine how she spent her money.

1.2.29: A graduate who just got a new job went to the department store with $1000 in his pocket. He purchased some $30 shirts, $60 trousers, $80 sweaters, and $100 shoe pairs and stopped when he realized that he will have only $40 left in his pocket after paying for all of them. At that time he had purchased a total of 21 items.

The amount spent on shirts is equal to the amount spent on trousers. The number of sweaters purchased is equal to the number of shoe pairs purchased. Formulate a system of linear equations to determine what things were purchased by him.

1.2.30: (From Mahaaviiraa's *Ganiita saara sangraha*, 850 AD) The total price of 9 citrons and 7 wood-apples is 107 rupees; and that of 7 citrons and 9 wood-apples is 101 rupees. Write the system of linear equations to determine the price of each fruit.

1.2.31: (Adopted from Kordemsky's *The Moscow Puzzles*, Charles Scribner's Sons, NY 1972).

(i) A boy has as many sisters as brothers, but each sister has only half as many sisters as brothers. Formulate a system of linear equations to determine how many brothers and sisters are in the family.

(ii) There are three tractor stations in the city. Tractors are in short supply, and they lend each other equipment as needed. The first station lent the 2nd and 3rd as many tractors as they each already had. A few months later the 2nd station lent the 1st and the 3rd as many as they each had at that time. Still later, the 3rd lent the 1st and the 2nd as many as they each had at that time. Each station now had 24 tractors. Formulate a linear equation model to determine how many tractors each station had originally.

(iii) Three brothers shared 24 apples, each getting a number equal to his age three years before. Then the youngest proposed the following swap "I will keep only half the apples I got and divide the rest between you two equally. Then the middle brother, keeping half his accumulated apples, must divide the rest equally between the oldest brother and me; and then the oldest brother must do the same."

Everyone agreed, the result was that each ended with 8 apples. Formulate a linear equation model to determine how old the brothers are.

(iv) A sister S and brothers B_1 to B_4 went mushroom hunting. S collected 45 mushrooms, but B_1 to B_4 all returned empty handed. Out of sympathy she distributed all her mushrooms among her brothers.

On the way back B_1 picked up 2 new mushrooms, B_2 picked up new mushrooms to double what he had, B_3 lost two mushrooms, and B_4 lost half his mushrooms. At the end all the brothers had the same number of mushrooms. Formulate a linear equation model to determine how many mushrooms S gave each boy originally.

1.2.32: Magic squares

(i) **Lo Shu, the most famous magic square:** Associate a variable with each cell of a 3×3 array as in the following:

x_1	x_2	x_3
x_4	x_5	x_6
x_7	x_8	x_9

It is required to find values for all the variables satisfying the magical property that the sum of the variables in every row, column, and the two diagonals is the same, here 15. Write this system of linear equations.

Find a solution to this system, in which x_1 to x_9 assume distinct integer values from 1 to 9 (this yields *Lo Shu* (Shu means writing or document in Chinese) which is said to have first been revealed on the shell of a sacred turtle which appeared to the mythical emperor Yu from the waters of the Lo river in the 23rd century BC).

(ii) **Magic square from India** (more than 2000 year old): Same as the above, except that we now have 16 variables x_1 to x_{16} arranged in a 4×4 array; and the sum of all the variables in each row, column, and the two diagonals is required to be 34. Write this system of linear equations. In the magic square the variables are required to assume distinct integer values from 1 to 16, find it.

1.2.33: A group of 12 students from a class went to eat lunch in a restaurant, agreeing to split the bill equally. When the bill for a arrived, two of them discovered that they did not have any money with them. Others in the group made up the difference by paying an extra b each.

Then one of these two students who did not have any money, took them all to an ice cream parlor where he maintains an account. Each one in the group took ice cream worth $b, and the total bill this time came to one-fifth of the total bill at the restaurant, which the student charged to his account.

Develop a system of linear equations to determine how much each student paid.

1.2.34: A kid's piggybank has cents, nickels, dimes, quarters and half-dollars. The number of cents is the total number of nickels and dimes together. The number of dimes is the total number of quarters and half-dollars together. The number of nickels is twice the number of half-dollars. The value of all the dimes is the same as the value of all the quarters. The total value in the bank is $29.10. Develop a system of linear equations to find the total number of coins in the piggy bank.

1.2.35: A group of 12 people consists of doctors, lawyers, and engineers. The avarage ages of doctors, lawyers, engineers in the group is 60, 30, 40 years respectively. The average age of all the group members is 40 years. The total ages of the doctors is the same as the total ages of the lawyers, and this number is one half of the total ages of the engineers. Write a system of linear equations to determine the number of doctors, lawyers, and engineers in the group.

1.2.36: The sum of the first 5 terms in an arithmetic sequence is 40, and the sum of the first 10 terms is 155. Develop a system of linear equations to find the sum of the first 20 terms.

1.2.37: A fundraiser for storm victims was organized in a theater with tickets priced at $10, $50, $100; and a total of 1600 tickets were sold. The number of $50 tickets sold was the sum of two times the number of $100 tickets plus the number of $10 tickets sold. The sum of the numbers of $10, $50 tickets sold was three times the number of $100 tickets sold. The sum of the numbers of $50, $100 tickets sold was seven times the number of $10 tickets sold. Formulate a system of linear equations to determine the amount of money raised.

1.2.38: A batch of 30 lbs of blended coffee was made by blending two varieties of coffee beans valued at $2/lb, and $5/lb respectively. The mixture has a value of $3/lb. Develop a system of linear equations to determine how many lbs of each variety were used in the blend.

1.2.39: A lady has a total of $35,000 invested in three separate investments I1, I2, I3 that yield annual interest of 5%, 10%, 15% respectively. If she can rollover the amount invested in I1 completely into the investment I2 (I3) her net annual yield will change to 12.86% (13.57%). If she can roll

over the amount invested in I2 completely into the investment I3 her net
annual yield will change to 13.57%. Formulate a system of linear equations
to determine her present net annual yield.

1.2.40: A husband and wife team went on a car trip to a destination
800 miles away. They started at 6 AM. At the beginning the husband
drove the car nonstop at an average speed of 70 miles/hour for some time.
Then they stopped and did some touring and sightseeing on foot. Then the
wife took the wheel and drove nonstop at an averag speed of 60 miles/hour
to the destination. The time spent sightseeing in the middle was one-sixth
of the total time spent driving. The distance driven by the wife was 30%
of the total distance. Formulate a system of linear equations to determine
when they reached their destination.

1.2.41: Problems dealing with snow shoveling rates of people: In
these problems assume that each person always shovels snow at a constant
rate independently of any others who may also be shoveling snow on the
same driveway.

(i) Husband and wife working together can clear their driveway in
36 minutes. Working alone, the lady requires half an hour more than her
husband to complete the job. Develop a system of linear equations to determine how long it would take each of them to complete the job working alone.

(ii) A family of father, mother, son have to clear their driveway. Father
and son working together can finish it in 4 hours. If son and mother work
on it they will finish in 3 hours. If father and mother work on it they will
need 2.4 hours. Formulate a system of linear equations to determine how
long it would take each of them to complete the job working alone.

1.2.42: Problems dealing with ages of people

(i) Persons P_1, P_2, P_3 all have birthdays on 9 September. P_1's present age
is 5 years less than the sum of the present ages of P_2, P_3. In 10 years P_1 will
be two and a half times as old as P_2 will be then. 5 years ago P_2 was one-
sixth as old as P_3 was. Develop a system of linear equations to determine
their present ages.

(ii) (From Metrodorus, *Greek Anthology*, about 500 AD) A person named
Demochares lived one-fourth of his life as a boy, one-fifth as a youth, one-
third as a man, and the remaining 13 years in his dotage before he passed
away. Develop a linear equation model to determine how old he was when
he died.

(iii) In 10 years a father will be twice as old as his son will be then. 5 years ago the father was 5 times as old as his son was then. Develop a system of linear equations to determine their present ages.

(iv) 6 years ago Rod was 4 times as old as Chandra. 2 years ago Rod was twice as old as Chandra. Develop a system of linear equations to determine their present ages.

(v) Brother B and sisters S_1, S_2 have discovered that when B was half the age of S_1, S_2 was 30 years old. When S_1 was half as old as S_2, B was just born. The sum of the ages of B and S_2 is twice the age of S_1. Develop a system of linear equations to find all their ages.

(vi) A teenage girl fell in love with an older man twice her age. The sum of the digits in both their ages are equal. Also, one year ago the man was one year older than twice her age. Develop a system of linear equations to determine their ages.

(vii) B_1, B_2, B_3 are brothers, and B_1's age is the sum of the ages of B_2, B_3 two years ago. The sum of the ages of B_1, B_3 is one more than twice the age of B_2. Also, the difference between the ages of B_1 and B_2 is one more than half the age of B_2. Formulate a system of linear equations to determine their ages.

(viii) Three brothers all of whose ages are of two digits, have the interesting property that the sum of the ages of any two of them is the number obtained by writing the age of the third in reverse order. Formulate a system of linear equations to determine their ages.

(ix) Three people P_1, P_2, P_3 discovered that their present ages total 90. When P_1 was twice as old as P_2, P_3 was 5 years old. When P_2 was twice as old as P_3, P_1 was 50. Formulate a system of linear equations to find their ages.

1.2.43: A clothing store received a truckload consisting of $20 shirts, $30 slacks, and $60 sweaters. The total value of the slacks received was 65.22% of the sum of the values of all the shirts and sweaters received. The total value of the shirts received was 111.11% of the sum of the values of all the slacks and sweaters received. If each oh the slacks had been valued $10 more, then the total value of the slacks received would have been equal to the total value of the shirts received. Formulate a system of linear equations to determine the total value of the truckload received.

1.2.44: A carpenter and her assistant did some work at a home. The assistant arrived first and started the work. 4 hours later the carpenter arrived and both of them worked for 3 more hours to finish the work. The homeowner paid both of them together $360 for the work. If each of them had worked 2 hours longer, the total bill would have been $520. Formulate a system of linear equations to determine their hourly rate of pay.

1.2.45: A girl's piggybank consists of a total of 220 coins consisting of nickels, dimes, and quarters, and has a value of $35.

If the numbers of nickels and dimes are doubled, the value would go up by 28.57% of the present value.

If the number of quarters is tripled, and the number of nickels is changed to 11 times their present number the value will be tripled.

If the number of dimes is tripled, and the number of nickels increased by 75%, the value will go up by 50%.

Formulate a system of linear equations to determine the number of coins of each type in the piggybank.

1.2.46: A person receives yearly income of $13,600 from investments he made during his working years in bonds bearing 5%, 7%, 8%, and 10% interest annually. If the amounts invested in the 5%, 7% bonds are interchanged, he will earn $400 less annually. If the amounts invested in the 5%, 8% bonds are interchanged, he will earn $1200 less annually. If the amounts invested in the 5%, 10% bonds are interchanged, he will earn $1500 less annually. If he could transfer the amounts invested in the 5% and 7% bonds to the 10% bond, he will earn $2200 more annually. Develop a system of linear equations to determine the total amount of money invested in all the bonds.

1.2.47: The cupola already contains 100 lbs of an alloy consisting of 20% Cu (copper), 5% Sn (tin), 50% Al (aluminium), and 25% Zn (zinc) by weight. Two other alloys: R1, R2 with % composition by weight of (Cu, Sn, Al, Zn) equal to (10%, 10%, 60%, 20%), (10%, 20%, 40%, 30%) respectively are available to add to the cupola; as also pure Cu and Zn. Formulate a system of linear equations to determine how much R1, R2, Cu, Zn to add to the cupola to make a batch of 400 lbs with % composition by weight of (Cu, Sn, Al, Zn) equal to (16.25%, 8.75%, 47.5%, 27.5%).

1.2.48: 4 lbs of coffee, 10 lbs of sugar and 2 lbs of butter in 2000 cost $41. By 2001 coffee is 10% more expensive, sugar is 20% more expensive, and butter is 25% more expensive; and the same order costs $43.64. Also, in 2000, 10 lbs each of coffee, sugar, and butter together cost $95. Formulate a system of linear equations to determine the 2001 prices of coffee, sugar, and butter.

1.2.49: Problems dealing with speeds of vehicles: (Adopted from M. R. Spiegel, *Theory and Problems of College Algebra*, Schaum's Outline Series, McGraw-Hill, 1956). In these problems assume that each vehicle always travels at a constant speed, but the speeds of different vehicles may be different. Formulate a system of linear equations to determine the speeds of the vehicles discussed from the information given.

(i)   Two race cars racing around a circular mile track starting from the same point at the same time, pass each other every 18 seconds when traveling in opposite directions; and every 90 seconds when traveling in the same direction.

(ii)   A passenger at the front of a train A notices that he passes the complete length of 330 feet of train B in 33 seconds when traveling in the same direction as B, and in 3 seconds when traveling in opposite directions.

1.2.50:   (Adopted from the issues of *Mathematics Teacher*).

(i)   In a school each student earns between 0 to 100 points on each test. A student has already taken $n + 2$ tests.

At the end of the nth test his average score per test was m.

On the $(n + 1)$th test he got a grade of 98 and his average points per test became $m + 1$.

On the last test he got a grade of 70 and his average points per test became $m - 1$.

Develop a system of linear equations to determine the number of tests taken so far, and the average points per test at this stage.

(ii)   When a pile of dimes is distributed equally into 100 red boxes, 50 dimes were left over. When the same pile was distributed equally into 110 blue boxes none was left over, but each blue box contained 5 fewer dimes than the red ones before. Develop a system of linear equations to determine the number of dimes in the pile.

(iii)   A train traveling at constant speed takes 20 seconds from the time it first enters a 300 meter long tunnel until it completely emerges from the tunnel at the other end. A stationary ceiling light in the tunnel was directly above the train for 10 seconds. Find the length of the train and its traveling speed using a linear equation model.

1.2.51:   Three boys have some $ cash in their pockets. If each of them put $2 more in their pocket, boy 1 will have twice the amount of boys 2, 3 combined. If boy 1 transfers $6 from his pocket to the pockets of each of boys 2, 3; he will have half as much as they both have. If boy 3 transfers $1

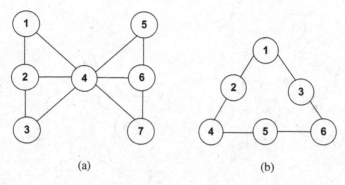

Figure 1.5

from his pocket to the pockets of each of the others, he will have one-tenth of the money in all the pockets. If $2 is taken away from the pocket of each boy, boy 1 will have 5 times as much money as the other two combined. Formulate a system of linear equations to determine how much money each boy posesses.

1.2.52: To encourage energy conservation, California has adopted the following electricity rate structure: each household consuming between 0 to 500 kwh in a month will pay at the rate of a cents/kwh. If the consumption in a month is between 500 kwh to 1500 kwh, the first 500 kwh is charged at a cents/kwh, and any amount beyond that at the greater rate of b cents/kwh. If the consumption in a month exceeds 1500 kwh the first 1500 kwh are charged as mentioned before, and any amount above 1500 kwh is charged at the highest rate of cents c/kwh. Given that the monthly electricity bill when the consumption is 300, 600, 1600 kwh; is $30, $65, $220 respectively; use a linear equation model to find the various rates.

1.2.53: Problems based on network diagrams

(i) Given in Figures 1.5, 1.6, 1.7 are several network diagrams, each consisting of nodes represented by little circles, and edges joining some pairs of nodes. Each diagram leads to a separate exercise, in each do the following. For each node j, associate a decision variable x_j. You are required to find values for the variables so that the variables along any n-node straight path is b mentioned there. Write these conditions on the variables as a system of linear equations. If possible try to find a solution to this system in which the values of the variables form a permutation of the node numbers.

For (a) $n = 3$ and $b = 12$.
For (b) $n = 3$ and do separately for $b = 9$, 10, 11, 12 respectively.
For (c) $n = 4$ and $b = 34$.

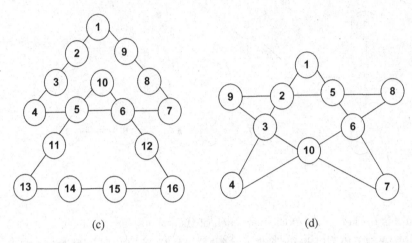

(c)   (d)

Figure 1.6

(ii) Same as in (i) with $n = 4$ for the pentagram in figure (d), but here b is not given; all you are required to do is to find the values for the variables so that the sums of all the variables along every 4-node straight path is the same. Write this as a system of linear equations.

(iii) **Magic circles:** (Adopted from J. S. Madachy, *Madachy's Mathematical Recreations*) Associate a variable x_j with point j in figure (e). Find the values for the variables such that the sum over all intersections is the same constant for all the circles.

1.2.54: Finding new girlfriend's house number: While thinking fondly about the four-digit house number of his new girlfriend Rama, Venu discovered some of its amazing properties. For example, to divide it by 7, you only have to delete its second digit from left. The sum of its two rightmost digits is five times the sum of its two leftmost digits. To multiply by three, all you have to do is to add 21 to the number consisting of its two leftmost digits. If you interchange the two rightmost digits, its value goes down by 4.29%. Develop a system of linear equations to determine Rama's house number.

1.2.55: Is this a linear system?: On returning from the department store John explained to his wife Mary that the prices of one type of shirts and one type of slacks that he liked were integer numbers in \$. He purchased as many shirts as the sum of the prices of a shirt and a slack; and as many slacks as the sum of the price of a shirt and one-half the price of a slack.

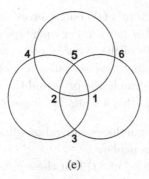

(e)

Figure 1.7

He purchased 50 items in all, and spent \$700. Develop a system of equations to determine how many shirts and slacks he purchased. Is this a system of linear equations? Explain.

1.2.56: Squashes galore!: On a hot summer day they were selling yellow, black, white, and big-8 summer squashes at 20, 25, 30, 35 cents each respectively in a Farmer's market. A restaurant owner purchased all four types and paid exactly the same total amount for each type, getting 319 squashes in all. Develop a system of linear equations to determine how many of each type she purchased.

1.2.57: Problem from ancient Chinese text Chiu-chang Suan-shu (Nine Chapters on the Mathematical Art): The yield from one sheaf of inferior grain, two sheaves of medium grain, and three sheaves of superior grain is 39 tou. The yield from one sheaf of inferior grain, three sheaves of medium grain, and two sheaves of superior grain is 34 tou. The yield from three sheaf of inferior grain, two sheaves of medium grain, and one sheaf of superior grain is 26 tou. What is the yield of inferior, medium, and superior grains? Formulate the system of linear equations to determine these. (This Chinese text is estimated to be over 2000 years old.)

1.2.58: A farmer grows hogs, geese, sheep, and ostrichs on his small farm. The number of birds in his farm is 50 plus two times the number of animals. His birds and animals together have a total of 350 heads and 900 legs. The number of sheep and ostrichs is 10 less than half the number of geese and hogs. Develop a system of linear equations to determine how many of each type he has on his farm.

1.2.59: Sita purchased a total of 90 bulbs consisting of tulip, daffodil, and gladiola bulbs to plant in her garden. For each type she purchased as many bulbs as the price per bulb of that type in cents, and paid a total of $29. She purchased twice as many gladiola bulbs as daffodil bulbs. Develop a system of equations to determine the price and the number of each type of bulb that she purchased. Is this a system of linear equations? Explain.

1.2.60: Each year Rabi purchases individual gifts for his sister and her two children, a niece and a nephew.

Last year he spent a total of $240 for these gifts. That year his sister's gift and nephew's gift together cost two times his niece's gift. His nephew's gift plus three times his niece's gift cost three times his sister's gift.

This year his sister's gift will cost twice as much as his niece's gift last year. His niece's gift will be twice as expensive as his nephew's gift last year. And his nephew's gift will cost only half as much as his sister's gift last year.

Develop a system of linear equations to determine his total gift expense this year.

1.2.61: Cities 1, 2, 3 form a triangle.

From 1 to 3 via 2 is three times the distance by direct route.
From 2 to 1 via 3 is twice the distance by direct route.
From 3 to 2 via 1 is 20 miles more than the distance by direct route.

Formulate a system of linear equations to determine the lengths of direct routes between pairs of cities.

1.2.62: Three students S_1, S_2, S_3 took a test which had 50 questions, some easy, others tough. S_1 answered three times as many questions as the sum that S_2, S_3 missed. S_2, S_3 answered the same number of questions, each of them missing a fourth of what S_1 missed. Formulate a system of linear equations to determine how many each of them answered.

1.2.63: A middle school teaches grades 7, 8, 9, 10 and has a total enrollment of 330 students. They receive annual funding at the rate of $4000 per 7th grade student, $4300 per 8th grade student, $4500 per 9th garde student, and $4700 per 10th grade student; and received a total funding of $1,457,000 for the year. 30% of students in 7th grade, 25% in 8th grade, and 20% in grades 9 and 10 take music classes; this year they have a total of 76 students enrolled in music classes. Also, this year they have a total of 31 students enrolled in AP courses, consisting of 15% of students of 8th grade and 10% of students in each of grades 9 and 10. Formulate a system of linear equations to determine the number of students in each grade.

1.2.64: The automobile route to a scenic mountainous area is 50 miles each way with only 10 miles of it level road each way, the rest is either uphill or downhill. A person maintained an average speed of 20 miles/hour uphill, 50 miles/hour downhill and 40 miles/hour on the level road on a car trip to this region. The return trip took 44.44% more than the riding time to get there (not counting stops). Formulate a system of linear equations to determine the total driving time for the trip.

1.2.65: A family purchased cantaloupes (50 cents each), honeydews ($1 each), oranges (20 cents each), grape fruit (15 cents each), and mangoes (75 cents each) at a road side fruit stall on their way back from vacation, and paid $26.30 for all of them, obtaining 80 fruit in all. They purchased 6 times as many citrus fruit as melons; as many mangoes as one-fourth of oranges and cantaloupes put together; and as many oranges as 20% more than the honeydews and grapefruit put together. Formulate a syatem of linear equations to determine how many fruit of each type they purchased.

1.2.66: (Adopted from J. A. H. Hunter, *Entertaining Mathematical Teasers and How to Solve Them*, Dover, NY, 1983).

(i) A kid's mother organized a birthday party for him. She purchased enough apples at 50 cents/apple to give two to each child invited, but a quarter of the invited kids did not show up. So she distributed two apples to each of the 16 boys who showed up, and then had sufficient apples to give three to each of the girls who showed up. Formulate a system of linear equations to determine how much money she spent on the apples.

(ii) A dealer of speciality teas imports pure oolong tea at $3.29/lb, and lapsang souchong tea at $2.38/lb from China; blends them and sells the blend at $4.48/lb which gives him a 60% markup. Formulate a system of linear equations to determine the proportions in the blend.

(iii) In an office there are both men and women working as clerks. If they had twice as many girls' or three times as many men, they would have 30 clerks. Formulate a system of linear equations to determine how many clerks they have.

1.2.67: In an ordered list of six numbers every one (except the end ones) is equal to the sum of the two adjacent numbers in the list. The sum of all the numbers is the sum of the first three. The third number is -6. Formulate a system of linear equations to find all the numbers in the list.

1.2.68: There are 60 girls in an undergraduate class. Each is either a blonde or brunette, and has either blue or brown eyes. Among them 40 are blondes and 22 have blue eyes. Formulate a system of linear equations to determine the number of girls in this class in each of the four hair and eye color combinations.

1.2.69: A four digit integer number has the property that multiplying it by 9 yields the four digit number obtained by writing the digits in the original number in reverse order. Formulate a system of linear equations to determine the original number.

1.2.70: A school principal noticed that the number of students in her school has the property that when it is divided by k, a reminder of $k - 1$ is left for $k = 2, 3, \ldots, 10$. Formulate a system of linear equations to find the number of students in the school.

1.2.71: Three numbers are in arithmetic progression, three other numbers are in geometric progression. Adding coorresponding terms of these two progressions we obtain 9, 16, and 28 respectively. Adding all three terms of the arithmetic progression we obtain 18. Formulate a system of equations to find the terms of both progressions. Is it a system of linear equations?

1.2.72: (Adopted from G. Polya, J. Kilpatrick, *The Stanford Mathematics Problem Book With Hints and Solutions*, Teachers College Press, Columbia University, NY, 1974)

(i) Unlimited numbers of coins- cents, nickels (5 cent coins), dimes (10 cent coins), quarters (25 cent coins), and half-dollars are available. It is required to determine in how many different ways you can assemble a collection of these coins so that the total value of this collection is n cents for a positive integer n, say, for n varying between 0 to 50. Formulate a system of linear equations to solve this problem, and show that for $n = 50$, the number of different ways is 50.

(ii) 10 people are sitting around a round table. $20 is to be distributed among them satisfying the rule that each person receives exactly one-half of the sum of what his (her) two neighbors receive. Formulate a system of linear equations to determine how to distribute the money. Using the special properties of this system (or the results you will learn in Chapters up to 4) argue that there is only one way to distribute the money here.

1.2.73: A magical card trick: As a child, I was enthralled by a card trick performed by the great magician Sirkar in India. It involves a deck of playing cards. The deck has 52 cards divided into 4 suits of 13 cards each. As they call them in India, these are kalaavaru and ispeetu (the black suits), and aatiinu and diamond (the red suits).

Sirkar gave me the deck and said "shuffle it as much as you want, then divide it into two piles of 26 cards each and place the piles on the table". He then covered the two piles with his hands and uttered the magical chant "Om, bhurbhuvasvuh" three times saying that he is using his power of magic to make sure that the number of black cards in the first pile becomes equal to the number of red cards in the second pile. He then asked me to count and verify the truth of his statement, which I did.

Define four decision variables; the number of black, red cards in piles 1, 2. Formulate the system of linear equations that these variables have to satisfy. Using this system, unravel the magic in this trick.

1.2.74: The sports department in a middleschool offers classes in bicyling, swimmimg, skating, and tennis. A class in this school has 25 pupils. In this class 17 pupils are enrolled in the bicycling class, 13 in the swimming class, 8 in the skating class, and 6 in the tennis class. None of the pupils enrolled in any of bicycling, swimming, or skating classes is enrolled in the tennis class. It is required to determine how many pupils of this class are enrolled in various combinations of sports classes. Formulate this using a system of linear equations.

1.2.75: People can be divided into four distinct blood groups: O, A, B, and AB. Blood transfusion with donated blood will be without danger to the receiver only under the following criteria: Every one can accept blood for transfusion from a person of the O blood group, or from a person in his/her own blood group. Also, a person in the AB group can accept blood from any person.

Among the members in a club willing to donate blood, there are 47 (64) [67] {100} members from whom an O (A) $[B]$ $\{AB\}$ can accept blood for transfusion. Formulate a system of linear equations to determine how many of these members belong to the various blood groups.

1.2.76: At a club meeting they served: coffee, a chacolate drink (for drinks); pie, a cacolate dessert (for dessert); and a nonvegetarian dish with meat, a vegetarian dish (for main dish); and members select what they like. There was a bill for each item separately for the whole club. The drink items together cost $35. The dessert items together cost $50. The cacolate items together cost $55. The main dishes together cost $110. All the items which have no meat, together cost $135. The tip which is calculated at 10%

of the total cost came to \$19.50. Formulate a system of linear equations to determine the cost of each item separately.

1.2.77: Newspaper buying habits of households in Ann Arbor: In any quarter of the year, the households in Ann Arbor can be classified into four classes: C_1 = those who do not subscribe to any daily newspaper, C_2 = those who subscribe to the Detroit Free Press only, C_3 = those who subscribe to Ann Arbor News only, and C_4 = those who subscribe to both the Detroit Free Press and the Ann Arbor News.

Consumer research has shown that at the end of every quarter, households move from the class in which they currently belong, to another class, in the following proportions.

Table: Fraction of households belonging to the class defined by the row, who stay in same class or move to class defined by column, at end of a quarter.

	C_1	C_2	C_3	C_4
C_1	0.6	0.2	0.1	0.1
C_2	0.5	0.3	0.2	0
C_3	0.5	0.2	0.1	0.2
C_4	0.2	0.7	0.1	0

These transition fractions have remained very stable at these values for a very long time.

Let p_1, p_2, p_3, p_4 denote the proportion of households in Ann Arbor belonging to the classes C_1 to C_4 respectively, in a quarter. Typically we would except the values of p_1 to p_4 to change from one quarter to the next, but surprisingly their values have remained essentially unchanged for a long time in spite of the transitions of households taking place from one quarter to the next mentioned above. Use this information to construct a system of linear equations satisfied by these proportions p_1 to p_4.

1.2.78: Figure 1.8 shows a rectangle with sides a, b cut into squares with sides x_1 to x_6. Write the system of linear equations that $x_1, \ldots, x_7$ have to satisfy.

1.2.79: An avid gardener has created nine mounds for planting different types of zucchini plants in his garden. He planted green zucchini seeds on some mounds, yellow zucchini seeds on some, and white zucchini seeds on the rest. On each mound as many plants sprouted as the number of mounds allotted to zucchini of that color; and all the sprouted plants started growing

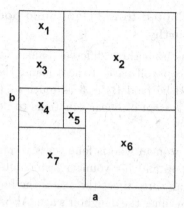

Figure 1.8

vigorously. On a certian morning each zucchini plant had as many squashes as the total number of plants of that zucchini color, and he counted a total of 353 squashes in the garden. He noticed that only half the yellow squashes, one third of the white squashes, and one eighth of the green squashes are of harvestable size and he harvested all of them (67 total) that morning.

Develop a system of equations to determine how many mounds he allocated to zucchinis of each color. Is this a system of linear equations?

1.2.80: A small farmer grows three types of chicken, each in its own pen. He has noticed that the chicken in pens 1, 2, 3 lay 24, 22, 19 eggs per chicken per month; and he has collected 5110 eggs in all one month.

All his chicken are either white or brown, and the proportions of brown chicken in pens 1, 2, 3 are 0.5, 0.25, 0.4 respectively. He counted a total of 90 brown chickens in all the pens together.

He estimates that the chicken in pens 1, 2, 3 consume on an average 4, 3, 2 units of feed per chicken per day respectively. He has been using 740 units of feed daily.

Develop a system of linear equations to determine how many chickens are housed in each pen.

1.2.81: On a holiday, players 1, 2, 3, 4 played singles tennis all day.

The total number of games in which player 1, 2, 3, 4 participated was 13, 10, 15, 14 respectively. The total number of games in which player 1, 2, 3, 4 was not one of the players was 13, 16, 11, 12 respectively.

Develop a system of linear equations to determine how many games were played by each pair of players.

1.2.82: Problems adopted from "Ganitamlo podupu kadhalu" in Telugu by CSRC Murthy

(i)   A person has loans from three different banks. One month he saved $1000 which he used to payoff part of these loans. He paid Bank 2 $100 more than Bank 1, and he paid Bank 3 as much as the sum he paid to Banks 1, 2. Develop a system of linear equations to determine how much he paid each bank.

(ii)   A man married a woman who is four years younger than him. They have two children, a boy and his younger sister. On a certain day, the father's age is 1 less than three times the age of the boy; and the mother's age is 1 more than three times the daughter's age. Also, the sum of the ages of the children on that day is 24.

Develop a system of linear equations to determine the ages of members of this family.

(iii)   At a party people had hot coffee or tea, or cold OJ (orange juice) for drinks, a total of 15 drinks. The number of coffees ordered was five more than the sum of teas and OJ ordered. Also, the number of hot drinks ordered was four times the number of cold drinks ordered. Develop a system of linear equations to determine how many drinks of each type were ordered.

1.3   Solving a System of Linear Equations is a Fundamental Computational Tool

In the previous section, we have seen a variety of applications in which a system of linear equations to be solved appears directly. Actually, solving a system of linear equations is a very fundamental computational tool used in almost all areas of scientific computation. Many problems which are not themselves systems of linear equations are solved either

- by approximating them through a system of linear equations, or
- by algorithms using iterative steps, each of which involves solving one or more systems of linear equations.

Here we provide an example of the first type of problem in differential equations, which are equations that involve functions and their derivatives. They appear in many studies in engineering, physics, chemistry, and biology. Most differential equations appearing in these areas are very hard to solve through exact mathematical techniques, so they are usually handled numerically by approximating them through a system of linear equations.

For a specific example, consider a real valued differentiable function $f(t)$ of one variable t, defined over the interval $a \leq t \leq b$. We denote this interval by $[a, b]$. Let $f'(t), f''(t)$ denote the first and second derivatives of $f(t)$. Suppose we do not know the function $f(t)$ explicitly, but know that it and its derivatives satisfy the following equations

$$u(t)f''(t) + v(t)f'(t) + w(t)f(t) = z(t) \quad \text{for all } t \text{ in } [a, b]$$
$$f(a) = \alpha, \quad f(b) = \beta$$

where $u(t), v(t), w(t), z(t)$ are known real valued functions, and α, β are given constants.

Since the values of the unknown function $f(t)$ at the two boundary points a, b of the interval $[a, b]$ are given, the problem of finding the function $f(t)$ satisfying the above equations is known as a **two-point boundary value problem**. It is an example of a differential equation that appears in many applications.

To handle such problems, numerical methods are employed to find approximate values of $f(t)$ at a set of discrete points inside the interval $[a, b]$. The approximating technique is called **discretization**, and proceeds as follows. Divide the interval $[a, b]$ into $n + 1$ subintervals of equal length $h = (b - a)/(n + 1)$. h is called the step size. The boundary points of the subintervals, called the **grid points** are

$$t_0 = a, \quad t_1 = a + h, \quad t_2 = a + 2h, \ldots, t_n = a + nh,$$
$$t_{n+1} = a + (n + 1)h = b.$$

The aim now is to find the values of $f(t)$ at the grid points, i.e., $f(t_1), \ldots, f(t_n)$, which are the unknowns. Discretization converts the differential equations corresponding to the grid points into a system of linear equations, using approximations for the derivatives at the grid points in terms of the unknowns. Derivative approximations called **central difference approximations**, given below for $i = 1$ to n, are used.

$$f'(t_i) \approx (f(t_{i+1}) - f(t_{i-1}))/2h$$
$$f''(t_i) \approx (f(t_{i-1}) - 2f(t_i) + f(t_{i+1}))/h^2.$$

Substituting these expressions for $f'(t_i), f''(t_i)$ in

$$u(t_i)f''(t_i) + v(t_i)f'(t_i) + w(t_i)f(t_i) = z(t_i)$$

for $i = 1$ to n, we get a system of n linear equations in the n unknowns $f(t_1), \ldots, f(t_n)$. Solving this system of linear equations gives approximations to the values of the function $f(t)$ at the grid points. Once approximations to $f(t_1), \ldots, f(t_n)$ are obtained, an approximation to the value of $f(t)$ at any point t which is not a grid point, can be obtained using standard interpolation techniques.

For better derivative approximations, we need to make the step size h small; this makes n large and leads to a large system of linear equations.

1.4   The Detached Coefficient Tableau Form

The **detached coefficient tableau form** is a neat way of organizing the data in a system of linear equations so that:

- the coefficients of all the variables in each constraint appear in a row of the tableau,
- the coefficients of each variable in all the constraints appear in a column of the tableau.

To represent a system of m linear equations in n variables (say, $x_1, \ldots, x_n$) in detached coefficient tableau form:

- Set up an $m \times (n + 1)$ two dimensional array.
- Arrange the variables in the system in some order, say x_1 to x_n in the natural order of increasing subscripts, and allocate one variable to each of columns 1 to n in that order, and record this by writing that variable in that column in a top row of the array. Then allocate the last column on the right, column $n + 1$, to the RHS (right hand side) constants.
- Take the constraints in any order, say constraint 1 to constraint m in natural order, and record the coefficients of the variables and the RHS constant in that constraint, in the order selected above, in a row of the tableau, one after the other.

Example 1:

Consider the 5 constraint, 4 variable system of linear equations derived in Example 1, Section 1.2, which is reproduced below.

$$5x_1 + 7x_2 + 2x_3 + x_4 = 4.43$$
$$3x_1 + 6x_2 + x_3 + 2x_4 = 3.22$$
$$4x_1 + 5x_2 + 3x_3 + x_4 = 3.89$$
$$88x_1 + 82x_2 + 94x_3 + 96x_4 = 88.46$$
$$x_1 + x_2 + x_3 + x_4 = 1$$

By taking the variables and constraints in natural order, the detached coefficient tableau form for this system is:

x_1	x_2	x_3	x_4	
5	7	2	1	4.43
3	6	1	2	3.22
4	5	3	1	3.89
88	72	94	96	88.46
1	1	1	1	1

Example 2:

Consider the following system of linear equations.

$$3x_4 - 7x_2 + 8x_1 = -13$$
$$-x_1 + 2x_3 = -2$$
$$x_2 - 13x_5 - 7x_4 + 4x_3 = 18$$
$$-2x_3 + 7x_1 = 12$$
$$20x_5 - 19x_3 = -1$$

Recording the variables and constraints in their natural order, this system in detached coefficient form is:

x_1	x_2	x_3	x_4	x_5	
8	-7	0	3	0	-13
-1	0	2	0	0	-2
0	1	4	-7	-13	18
7	0	-2	0	0	12
0	0	-19	0	20	-1

Discussion: Rows Correspond to Constraints, Columns to Variables

Notice that in the detached coefficient tableau form, the "=" symbol for each constraint is not explicitly recorded, but is implicitly understood.

Clearly, the ith constraint in the system can be read directly from the ith row in the tableau by multiplying the coefficients in this row by the corresponding variables and forming the equation. For example, from the fourth row of the above tableau we read the equation $7x_1 - 2x_3 = 12$, which is the same as the fourth equation in the system in this example. Hence, rows in the tableau correspond to constraints in the system. That's why people sometimes refer to constraints in a system as "rows" in informal conversations.

Similarly, the column in the tableau under a variable is the column vector of coefficients of that variable in the various constraints of the system. Thus each column in the tableau corresponds to a variable in the system.

Let the symbol A denote the detached coefficient tableau for a system of constraints, in which the variables are listed in the order x_1 to x_n. The ith row in the tableau (without the RHS constant in this row) forms a row vector which we denote by $A_{i.}$ (the dot following the i in the subscript indicates that it is the ith row vector in the tableau A). $A_{i.}$ is the row vector of the coefficients of the variables in the order x_1 to x_n in the ith constraint of the system.

Since each entry in $A_{i.}$ corresponds to a specific variable, if we change the order of the entries in $A_{i.}$, it may lead to a different constraint. For example, in the above tableau, $A_{4.} = (7, 0, -2, 0, 0)$. If we interchange the first two entries here, we get the row vector $(0, 7, -2, 0, 0)$, which leads to the constraint $7x_2 - 2x_3 = 12$, not the same as the fourth constraint in the system. That's why the order of entries in a vector is very important.

The jth column in the tableau forms a column vector which we denote by $A_{.j}$ (the dot before the j in the subscript indicates it is the jth column vector in the tableau A), it is the column vector of coefficients of the variable x_j in the various constraints of the system.

The Column Vector of Every Variable in the Tableau Must Contain Atleast One Nonzero Entry

A variable appears in a system of linear equations only if it has a nonzero coefficient in at least one equation in the system. Hence an important property of the detached coefficient form is that the column vector corresponding to every variable in the system must have at least one nonzero entry.

Of course the rightmost column in the tableau, the column of the RHS constants, may consist of all zero entries.

Exercises

1.4.1: Write the following systems of equations in detached coefficient form.

(i) $-2x_6 - 3x_1 + x_5 = -12$
      $x_1 - 7x_4 + 2x_3 = 17$
      $-5x_2 + 8x_3 = 13$

(ii)   Here variables have two subscripts (x_{11} etc.)

$$x_{11} + x_{12} + x_{13} = 800$$
$$x_{21} + x_{22} + x_{23} = 300$$
$$x_{11} + x_{21} = 400$$
$$x_{12} + x_{22} = 500$$
$$x_{13} + x_{23} = 200$$

(iii) $-3x_4 + 7x_2 = -10 + 6x_5 - 17x_1$
$$8x_6 - 3x_1 = 15 - 13x_2 + 12x_4 - 9x_3$$
$$-22 + 7x_5 - 4x_2 + 14x_6 = 33 + 2x_1 + 6x_3$$
$$88x_2 + 65x_7 - 5x_2 = 29 - 4x_3 - 3x_4$$
$$x_5 - 7x_3 = -27 + 6x_7 - 8x_5$$
$$39 + 4x_7 - 13x_4 + 11x_2 = 0$$
$$0 = 13x_7 - 5x_3 + 25x_1 - 9x_5$$

(iv) $x_1 + x_2 = 110$
$$x_1 = x_3 + x_4$$
$$x_2 = x_5 + x_8$$
$$x_6 = x_3 + x_5$$
$$x_7 = x_4 + x_8$$
$$x_7 + x_6 = 110$$

1.4.2: Write all the constraints in the system corresponding to the following detached coefficient forms:

x_1	x_2	x_3	x_4	x_5	x_6	
-5	0	0	17	-13	8	-23
0	0	21	-13	0	0	37
-97	-15	0	0	0	33	-56
0	11	0	-2	0	8	0
7	0	-6	0	26	0	29

x_1	x_2	x_3	x_4	x_5	x_6	x_7	x_8	x_9	
1	1	1	0	0	0	0	0	0	1
0	0	0	1	1	1	0	0	0	1
0	0	0	0	0	0	1	1	1	1
1	0	0	1	0	0	1	0	0	1
0	1	0	0	1	0	0	1	0	1
0	0	1	0	0	1	0	0	1	1

1.5   Operations on Vectors, the Zero and the Unit Vectors, Subspaces and Coordinate Subspaces of R^n

Transposition of Vectors

A row vector can be transformed into a column vector by writing the entries in it one below the other instead of side by side and vice versa. This operation is called **transposition**. Thus the transpose of a row vector is a column vector and vice versa. The transpose of a vector u is usually denoted by the symbol u^T. For example:

$$(3, 8, 4)^T = \begin{pmatrix} 3 \\ 8 \\ 4 \end{pmatrix} ; \quad \begin{pmatrix} 7 \\ -1 \\ 0 \\ 3 \end{pmatrix}^T = (7, -1, 0, 3).$$

In particular, we will find it very convenient to write a column vector, for example $\begin{pmatrix} 3 \\ 8 \\ 4 \end{pmatrix}$, as $(3, 8, 4)^T$ to conserve vertical space in the text.

Dot (Inner) Product of Two Vectors

Now consider the detached coefficient tableau for the system in Example 2, Section 1.4, which is reproduced below.

x_1	x_2	x_3	x_4	x_5	
8	−7	0	3	0	−13
−1	0	2	0	0	−2
0	1	4	−7	−13	18
7	0	−2	0	0	12
0	0	−19	0	20	−1

Let us denote by $x = (x_1, x_2, x_3, x_4, x_5)^T$ the vector of variables in this system written as a column vector. Also, let $b = (b_1, b_2, b_3, b_4, b_5)^T = (-13, -2, 18, 12, 1)^T$ denote the column vector of RHS constants in this tableau.

Denoting this tableau by A, the first row vector of coefficients in it, $A_1. = (8, -7, 0, 3, 0)$, and it corresponds to the first constraint in the system

$$8x_1 - 7x_2 + 0x_3 + 3x_4 + 0x_5 = b_1 = -13.$$

The quantity on the left of this constraint, $8x_1 - 7x_2 + 0x_3 + 3x_4 + 0x_5$ is called the **dot product** or **inner product** of the row vector $A_1.$ with the column vector x. It is written as $A_{1.}x$ in this specific order. Using this notation, we can write the ith constraint in the system as

$$A_{i.}x = b_i.$$

The dot product or inner product of two vectors is defined only if both the vectors have the same dimension (i.e., the same number of entries). Then it is the sum of the products of the corresponding entries. Thus if $u \in R^n$ with entries $u_1, \ldots, u_n$ in that order, and $v \in R^n$ with entries $v_1, \ldots, v_n$ in that order, then the dot (inner) product of u and v is the number $u_1 v_1 + \cdots + u_n v_n$. It is denoted in mathematical literature in two ways:

- as uv where the left vector u is written as a row vector, and the right vector v is written as a column vector, or
- as $\langle u, v \rangle$ or $\langle v, u \rangle$ where either of u or v may be row or column vectors.

We caution the reader that the product written as uv where $u \in R^n$ is a column vector and $v \in R^n$ is a row vector has a totally different meaning to be explained in the next chapter.

Scalar Multiplication of Vectors

Let u be a (row or column) vector in R^n with entries $u_1, \ldots, u_n$. Let α be any real number. Multiplying the vector u by α is called **scalar multiplication of u by the scalar** α. It has the effect of multiplying each entry in u by α. Thus:

$$\alpha(u_1, \ldots, u_n) = (\alpha u_1, \ldots, \alpha u_n)$$

$$\alpha \begin{pmatrix} u_1 \\ \vdots \\ u_n \end{pmatrix} = \begin{pmatrix} \alpha u_1 \\ \vdots \\ \alpha u_n \end{pmatrix}$$

For example:

$$10(3, 0, -4) = (30, 0, -40)$$

$$-15 \begin{pmatrix} 3 \\ 0 \\ -4 \end{pmatrix} = \begin{pmatrix} -45 \\ 0 \\ 60 \end{pmatrix}$$

$$0(3, 0, -4) = (0, 0, 0)$$

Addition of Vectors, Linear Combination of Vectors

Let u, v be two vectors. The operations

$$u + v \quad \text{or} \quad u - v$$

are only defined if either both u and v are row vectors of the same dimension, or both are column vectors of the same dimension. The result is a vector of the same type obtained by performing the operation component by component. Thus:

$$(u_1, \ldots, u_n) + (v_1, \ldots, v_n) = (u_1 + v_1, \ldots, u_n + v_n)$$

$$(u_1, \ldots, u_n) - (v_1, \ldots, v_n) = (u_1 - v_1, \ldots, u_n - v_n)$$

$$\begin{pmatrix} u_1 \\ \vdots \\ u_n \end{pmatrix} + \begin{pmatrix} v_1 \\ \vdots \\ v_n \end{pmatrix} = \begin{pmatrix} u_1 + v_1 \\ \vdots \\ u_n + v_n \end{pmatrix}, \quad \begin{pmatrix} u_1 \\ \vdots \\ u_n \end{pmatrix} - \begin{pmatrix} v_1 \\ \vdots \\ v_n \end{pmatrix} = \begin{pmatrix} u_1 - v_1 \\ \vdots \\ u_n - v_n \end{pmatrix}.$$

Thus the operation

$$(u_1, \ldots, u_m) + (v_1, \ldots, v_n)$$

is not defined if $m \neq n$. And the operation

$$(u_1, \ldots, u_n) + \begin{pmatrix} v_1 \\ \vdots \\ v_n \end{pmatrix}$$

is not defined.

In general, when α, β are scalars, the general linear combination of two vectors is defined by:

$$\alpha(u_1, \ldots, u_n) + \beta(v_1, \ldots, v_n) = (\alpha u_1 + \beta v_1, \ldots, \alpha u_n + \beta v_n)$$

$$\alpha \begin{pmatrix} u_1 \\ \vdots \\ u_n \end{pmatrix} + \beta \begin{pmatrix} v_1 \\ \vdots \\ v_n \end{pmatrix} = \begin{pmatrix} \alpha u_1 + \beta v_1 \\ \vdots \\ \alpha u_n + \beta v_n \end{pmatrix}$$

Examples:

$$\begin{pmatrix} -3 \\ 7 \\ 0 \\ 8 \end{pmatrix} + \begin{pmatrix} 8 \\ -17 \\ 3 \\ -20 \end{pmatrix} = \begin{pmatrix} 5 \\ -10 \\ 3 \\ -12 \end{pmatrix}, \quad \begin{pmatrix} -3 \\ 7 \\ 0 \\ 8 \end{pmatrix} - \begin{pmatrix} 8 \\ -17 \\ -3 \\ 20 \end{pmatrix} = \begin{pmatrix} -11 \\ 24 \\ 3 \\ -12 \end{pmatrix}$$

$$2(4, -6, 8, 2) - 10(6, -5, 0, 8) = (-52, 38, 16, -76).$$

Some Properties of the Addition Operation on Vectors

In summing two vectors u and v, the order in which you write them in the sum is immaterial, i.e., $u + v = v + u$

When you want to find the sum of three vectors u, v, w, you can either add w to the sum of u, v; or u to the sum of v, w; both will give the same result; i.e., $(u + v) + w = u + (v + w)$. This property is known as the **associative law of addition**.

Also, for all vectors u, we have $u + 0 = u$ where 0 is the zero vector of the same type as u and in the same space.

In the same way, for $r = 1$ to k, let

$$x^r = \begin{pmatrix} x_1^r \\ \vdots \\ x_n^r \end{pmatrix}$$

be a column vector in R^n. The general linear combination of these k column vectors $\{x^1, \ldots, x^k\}$ with scalar multipliers $\alpha_1, \ldots, \alpha_k$, each of which can be an arbitrary real number, is

$$\sum_{r=1}^{k} \alpha_r x^r = \alpha_1 x^1 + \cdots + \alpha_k x^k = \begin{pmatrix} \sum_{r=1}^{k} \alpha_r x_1^r \\ \sum_{r=1}^{k} \alpha_r x_2^r \\ \vdots \\ \sum_{r=1}^{k} \alpha_r x_n^r \end{pmatrix}$$

and the set of linear combinations of a set of row vectors in R^n is defined in a similar way.

The set of all possible linear combinations of the set of vectors $\{x^1, \ldots, x^k\}$, $\{\alpha_1 x^1 + \cdots + \alpha_k x^k :$ where $\alpha_1, \ldots, \alpha_k$ each take arbitrary real values$\}$ is called the **linear hull** or **linear span** of $\{x^1, \ldots, x^k\}$.

For example, let

$$x^1 = \begin{pmatrix} 1 \\ 1 \\ -1 \end{pmatrix}, \quad x^2 = \begin{pmatrix} 1 \\ -1 \\ 1 \end{pmatrix}, \quad x^3 = \begin{pmatrix} -1 \\ 1 \\ 1 \end{pmatrix}.$$

Then

$$3x^1 + 4x^2 - 5x^3 = \begin{pmatrix} 3+4-(-5) \\ 3+4(-1)+5 \\ 3(-1)+4+5 \end{pmatrix} = \begin{pmatrix} 12 \\ 4 \\ 6 \end{pmatrix}$$

and so the vector $(12, 4, 6)^T$ is contained in the linear hull of $\{x^1, x^2, x^3\}$.

Given a set of vectors $\{x^1, \ldots, x^k\}$ (either all row vectors, or all column vectors) from R^n, the linear combination $\alpha_1 x^1 + \cdots + \alpha_k x^k$ is called an **affine combination** if the scalar multipliers in it satisfy the condition

$$\alpha_1 + \cdots + \alpha_k = 1.$$

The set of all possible affine combinations of $\{x^1, \ldots, x^k\}$ is called the **affine hull** or **affine span** of $\{x^1, \ldots, x^k\}$.

For the set of vectors $\{x^1, x^2, x^3\}$ from R^3 given above, the linear combination $3x^1 + 4x^2 - 5x^3$ is not an affine combination because $3 + 4 - 5 = 2$ is different from 1. But

$$-8x^1 - 4x^2 + 13x^3 = \begin{pmatrix} -8-4+13(-1) \\ -8-4(-1)+13 \\ -8(-1)-4+13 \end{pmatrix} = \begin{pmatrix} -25 \\ 9 \\ 17 \end{pmatrix}$$

is an affine combination of $\{x^1, x^2, x^3\}$ because $-8 - 4 + 13 = 1$. Hence $(-25, 9, 17)^T$ is in the affine hull of $\{x^1, x^2, x^3\}$.

Exercises

1.5.1: Find the following if they are defined:

$$(6, -2, 8) + (0, 10)$$

$$(3, 3, 3) - \begin{pmatrix} 6 \\ 8 \\ 11 \end{pmatrix}$$

$$15(1, -1, 1, 0, 1) - 21(0, 1, 5, 2, 3) + 3(0, 0, 0, 0, 1)$$

$$\left\langle \begin{pmatrix} 3 \\ -8 \\ 4 \end{pmatrix}, (1, 2, 3) \right\rangle$$

$$(1,2,3) \begin{pmatrix} 2 \\ -3 \\ 4 \end{pmatrix}$$

$$\left\langle (1,2,3,0), \begin{pmatrix} 8 \\ -17 \\ 3 \end{pmatrix} \right\rangle.$$

The Zero Vector and the Unit Vectors

We now define some important special vectors in R^n.

The vector in R^n in which all the entries are zero is called the **zero vector** and is itself denoted by the symbol "0", whether it is a row or a column vector. From context, one can determine whether the symbol 0 is denoting the row or the column zero vectors.

A vector in R^n in which the ith entry is 1 and all the other entries are 0 is called the ith **unit vector** in R^n. Thus there are n different unit vectors in R^n. For example:

$(0,0,1)$ is the third unit row vector in R^3.

$(1,0,0)^T = \begin{pmatrix} 1 \\ 0 \\ 0 \end{pmatrix}$ is first unit column vector in R^3.

Caution: For $n \geq 2$, the vector in R^n in which all the entries are 1, for example $(1,1,1) \in R^3$ is not a unit vector.

Spanning Property of the Unit Vectors

The set of n unit vectors in R^n have the special property that their linear hull is R^n itself, i.e., every vector in R^n is a linear combination of the set of n unit vectors in R^n.

For instance, let $n = 3$. Consider the column vector $(3, -69, 75)^T$. We have

$$\begin{pmatrix} 3 \\ -69 \\ 75 \end{pmatrix} = 3 \begin{pmatrix} 1 \\ 0 \\ 0 \end{pmatrix} - 69 \begin{pmatrix} 0 \\ 1 \\ 0 \end{pmatrix} + 75 \begin{pmatrix} 0 \\ 0 \\ 1 \end{pmatrix}$$

and hence $(3, -69, 75)^T$ is contained in the linear hull of the set of three unit vectors in R^3.

Similarly, let $\{I_{.1}, \ldots, I_{.n}\}$ denote the set of n unit column vectors in R^n. Consider the general column vector $x = (x_1, \ldots, x_n)^T \in R^n$. We have

$$x = \begin{pmatrix} x_1 \\ \vdots \\ x_n \end{pmatrix} = x_1 I_{.1} + \cdots + x_n I_{.n}$$

and hence x is in the linear hull of $\{I_{.1}, \ldots, I_{.n}\}$.

Subspaces of R^n

A subset S of n-dimensional vectors (either all row vectors or all column vectors) is said to be a **subspace of the n-dimensional vector space** R^n, if it satisfies the following property:

Subspace Property: for every pair of vectors $x, y \in S$, every linear combination $\alpha x + \beta y$ for any real numbers α, β is also contained in the set.

Here are some examples of subspaces of R^n.

- **The singleton set** $\{0\}$ containing the 0-vector only, is a subspace of **dimension** 0. This is the only subspace containing only one vector in it. All other subspaces contain an infinite number of vectors.
- **The j-th coordinate subspace of** R^n. For any $j = 1$ to n, this is the set of all vectors $\{x = (x_1, \ldots, x_{j-1}, x_j = 0, x_{j+1}, \ldots, x_n)$: in which the component $x_j = 0$, while all the other components may take any real values$\}$. This subspace has dimension $n - 1$.
- **Lower dimensional coordinate subspaces of** R^n of dimension $n-r <$ $n-1$. Select any subset T of r components among $x_1, \ldots, x_n$, where $r \geq 2$. Then the set of all vectors $\{x = (x_1, \ldots, x_n)$: each component $x_j \in T$ is 0$\}$ is an $(n - r)$-dimensional coordinate subspace of R^n defined by the subset of r components T. For example the set of vectors $\{x = (x_1 = 0, x_2 = 0, \ldots, x_r = 0, x_{r+1}, \ldots, x_n)\}$ is the $n - r$-dimensional subspace in which all the components $\{x_1, \ldots, x_r\}$ are fixed equal to 0.
- **Set of all linear combinations of a given set of vectors** $\Delta \subset R^n$ is a subspace of R^n containing all the vectors in Δ.

1.6   Notation to Denote the General System of Linear Equations

We now describe the algebraic notation that we will use for denoting a general system of m linear equations in n variables. (Here, $m =$ the number of equations; may be $<$, or $=$, or $> n =$ the number of variables.)

Let $x = (x_1, \ldots, x_n)^T$ denote the column vector of variables in the system.

Let a_{ij} denote the coefficient of x_j in the ith constraint, for $i = 1$ to m, $j = 1$ to n.

Let $b = (b_1, \ldots, b_m)^T$ be the column vector of the RHS constants in the constraints 1 to m. Then the system of constraints is:

$$a_{11}x_1 + a_{12}x_2 + \cdots + a_{1n}x_n = b_1$$
$$a_{21}x_1 + a_{22}x_2 + \cdots + a_{2n}x_n = b_2$$
$$\vdots \quad \vdots$$
$$a_{m1}x_1 + a_{m2}x_2 + \cdots + a_{mn}x_n = b_m.$$

In detached coefficient tableau form, it is:

x_1	$\cdots$	x_j	$\cdots$	x_n	RHS constants
a_{11}	$\cdots$	a_{1j}	$\cdots$	a_{1n}	b_1
$\vdots$		$\vdots$		$\vdots$	$\vdots$
a_{i1}	$\cdots$	a_{ij}	$\cdots$	a_{in}	b_i
$\vdots$		$\vdots$		$\vdots$	$\vdots$
a_{m1}	$\cdots$	a_{mj}	$\cdots$	a_{mn}	b_m

Denoting this tableau by A, in this system, the row vector of coefficients of the variables in the ith constraint is: $A_{i.} = (a_{i1}, \ldots, a_{in})$, $i = 1$ to m.

The column vector corresponding to the variable x_j in this tableau is: $A_{.j} = (a_{1j}, \ldots, a_{mj})^T$, $j = 1$ to n.

Denoting the System in Terms of the Rows of the Tableau

So, the ith constraint in the system can be written as: $A_{i.}x = b_i$, $i = 1$ to m, using our notation. So, the above system of equations can also be written as:

$$A_{i.}x = b_i \quad i = 1 \text{ to } m.$$

Denoting the System in Terms of the Column Vectors Corresponding to the Variables

The column vectors corresponding to the variables x_1 to x_n are $A_{.1}$ to $A_{.n}$. It can be verified that the left hand side of the general system of equations given above is the linear combination of the column vectors $A_{.1}, \ldots, A_{.n}$ with $x_1, \ldots, x_n$ as the coefficients. Thus this system of equations can also be written as a **vector equation** as:

$$x_1 A_{.1} + \cdots + x_j A_{.j} + \cdots + x_n A_{.n} = b.$$

Thus the values of the variables x_1 to x_n in any solution of the system, are the coefficients in an expression of the RHS constants column vector b as a linear combination of the column vectors corresponding to the variables, $A_{.1}$ to $A_{.n}$ in the system.

Exercises

1.6.1: Write the systems of constraints in the following detached coefficient tableaus as vector equations, expressing a column vector as a linear combination of other column vectors.

x_1	x_2	x_3	x_4	x_5	
0	-7	0	0	14	-36
-2	0	12	3	-3	22
13	17	-11	8	0	0
0	3	0	5	6	-12

x_1	x_2	x_3	x_4	x_5	
1	2	1	2	-1	11
2	1	2	1	-4	10

1.7   Constraint Functions, Linearity Assumptions, Linear and Affine Functions

Consider the following general system of m linear equations in n variables $x = (x_1, \ldots, x_n)^T$.

$$a_{i1} x_1 + \cdots + a_{in} x_n = b_i, \quad i = 1, \ldots, m.$$

The left hand side expression in the ith constraint, $a_{i1} x_1 + \cdots + a_{in} x_n$ is a real valued function of the variables called the **constraint function** for this constraint. Denoting it by $g_i(x)$, the constraints in this system can be written as

$$g_i(x) = b_i, \quad i = 1, \ldots, m.$$

Hence each constraint in the system requires that its constraint function should have its value equal to the RHS constant in it.

As an example, consider the system of 5 equations in 4 variables in Example 1 of Section 1.2, which is reproduced below.

$$5x_1 + 7x_2 + 2x_3 + x_4 = 4.43$$
$$3x_1 + 6x_2 + x_3 + 2x_4 = 3.22$$
$$4x_1 + 5x_2 + 3x_3 + x_4 = 3.89$$
$$88x_1 + 82x_2 + 94x_3 + 96x_4 = 88.46$$
$$x_1 + x_2 + x_3 + x_4 = 1.$$

The left hand side expression in the first equation, $5x_1+7x_2+2x_3+x_4 = g_1(x)$ measures the percentage by weight of Al in the mixture represented by the vector x, and this is required to be $= 4.43$. Similarly, the left hand side expressions in equations 2, 3, 4 in the system, $g_2(x) = 3x_1 + 6x_2 + x_3 + 2x_4$, $g_3(x) = 4x_1+5x_2+3x_3+x_4$, $g_4(x) = 88x_1+82x_2+94x_3+96x_4$, respectively measure the percentage by weight of Si, C, Fe respectively in the mixture represented by the vector x. $g_5(x) = x_1+x_2+x_3+x_4$ from the fifth equation is the sum of the proportions of SM-1 to SM-4 in the mixture, and unless this is equal to 1 the vector x cannot be a solution to the problem.

With these constraint functions defined as above, we can denote this system in the form

$$\begin{pmatrix} g_1(x) \\ g_2(x) \\ g_3(x) \\ g_4(x) \\ g_5(x) \end{pmatrix} = \begin{pmatrix} 4.43 \\ 3.22 \\ 3.89 \\ 88.46 \\ 1 \end{pmatrix}.$$

In the same way, in any system of linear equations, the constraint function associated with any constraint measures some important quantity as a function of the variables, and the constraint requires it to be equal to the RHS constant in that constraint.

Linear Functions

Definition 1: A real valued function, $h(x)$, of variables $x = (x_1, \ldots, x_n)^T$ is said to be a **linear function** if it satisfies the following two assumptions, which together are called the **linearity assumptions**.

Additivity or **Separability assumption:** This assumption requires that the function $h(x)$ should be the sum of n functions, in which the jth function involves only one variable x_j, for $j = 1$ to n; i.e.,

$$h(x) = h_1(x_1) + h_2(x_2) + \cdots + h_n(x_n)$$

Proportionality assumption: A real valued function of a single variable, say x_j, satisfies this assumption if the function is equal to $c_j x_j$ for some constant c_j.

The function $h(x)$ of n variables $x = (x_1, \ldots, x_n)^T$ which satisfies the additivity assumption as in the above equation, is said to satisfy this proportionality assumption, if each of the functions $h_j(x_j)$, $j = 1$ to n satisfy it. ◁

Therefore, the real valued function $h(x)$, where $x = (x_1, \ldots, x_n)^T$, is a linear function iff there exist constants $c_1, \ldots, c_n$ such that $h(x) = c_1 x_1 + \cdots + c_n x_n$ for all $x \in R^n$. In this case $c = (c_1, \ldots, c_n)^T$ is called the coefficient vector of the variables in the linear function $h(x)$. Consider the following examples.

The function of $x = (x_1, \ldots, x_5)^T$	Comments
$3x_2 - 7x_5$	A linear function with coefficient vector $(0, 3, 0, 0, -7)$.
$x_2 - 3x_2 x_3 - 4x_4$	Not a linear function, as it does not satisfy the additivity assumption because of the term $-3x_2 x_3$ in it.
$x_1^2 + x_2^2 + x_3^2 + x_4^2$	Satisfies additivity assumption, but not the proportionality assumption, because of the square terms. Hence not a linear function.

Another equivalent definition of a linear function is the following:

Definition 2: A real valued function $h(x)$ of variables $x = (x_1, \ldots, x_n)^T$ is said to be a linear function if for every pair of vectors $y = (y_1, \ldots, y_n)^T$ and $z = (z_1, \ldots, z_n)^T$ and scalars α, β, we have

$$h(\alpha y + \beta z) = \alpha h(y) + \beta h(z)$$

or equivalently, for every set of vectors $\{x^1, \ldots, x^r\}$ in R^n, and scalars $\alpha_1, \ldots, \alpha_r$, we have

$$h(\alpha_1 x^1 + \cdots + \alpha_r x^r) = \alpha_1 h(x^1) + \cdots + \alpha_r h(x^r).$$

To show that Definition 1 implies Definition 2: Suppose $h(x)$ satisfies Definition 1. Then $h(x) = c_1 x_1 + \cdots + c_n x_n$ for some constants $c_1, \ldots, c_n$. Given $y = (y_1, \ldots, y_n)^T$ and $z = (z_1, \ldots, z_n)^T$, and scalars α, β,

$$h(\alpha y + \beta z) = \sum_{j=1}^{n} c_j(\alpha y_j + \beta z_j) = \alpha \sum_{j=1}^{n} c_j y_j + \beta \sum_{j=1}^{n} c_j z_j = \alpha h(y) + \beta h(z).$$

So, $h(x)$ satisfies Definition 2.

To show that Definition 2 implies Definition 1: Suppose $h(x)$ satisfies Definition 2. Let $I_{.j}$ be the jth unit vector in R^n for $j = 1$ to n. Then the vector x satisfies

$$x = (x_1, \ldots, x_n)^T = x_1 I_{.1} + \cdots + x_n I_{.n}.$$

Applying Definition 2 using this equation, we have

$$h(x) = x_1 h(I_{.1}) + x_2 h(I_{.2}) + \cdots + x_n h(I_{.n}).$$

So, if we denote $h(I_{.j})$ (the value of the function corresponding to the jth unit vector) by c_j, we have

$$h(x) = c_1 x_1 + \cdots + c_n x_n$$

and hence $h(x)$ satisfies Definition 1.

Affine Functions

Definition 3: A real valued function, $a(x)$, of variables $x = (x_1, \ldots, x_n)^T$ is said to be an **affine function** if it is equal to a constant + a linear function of variables x; i.e., if there are constants $c_0, c_1, \ldots, c_n$ such that

$$a(x) = c_0 + c_1 x_1 + \cdots + c_n x_n$$

for all $x \in R^n$. ◧

Hence every linear function is also an affine function (if the constant $c_0 = 0$), but the converse may not be true.

As an example, the function $-300 + 2x_2 - 3x_3$ is an affine function, but not a linear function.

Another equivalent definition of an affine function is the following:

Definition 4: A real valued function $h(x)$ of variables $x = (x_1, \ldots, x_n)^T$ is said to be an affine function if for every pair of vectors $y = (y_1, \ldots, y_n)^T$

and $z = (z_1, \ldots, z_n)^T$ and scalars α, β satisfying $\alpha + \beta = 1$, we have

$$h(\alpha y + \beta z) = \alpha h(y) + \beta h(z)$$

or equivalently, for every set of vectors $\{x^1, \ldots, x^r\}$ in R^n, and scalars $\alpha_1, \ldots, \alpha_r$ satisfying $\alpha_1 + \cdots + \alpha_r = 1$, we have

$$h(\alpha_1 x^1 + \cdots + \alpha_r x^r) = \alpha_1 h(x^1) + \cdots + \alpha_r h(x^r) \quad \bowtie$$

Notice the difference between the conditions stated in Definitions 2 and 4. Definition 2 requires the condition to hold for all linear combinations of vectors, but Definition 4 requires it to hold only for all affine combinations of vectors (i.e., only when the scalar multipliers in the linear combination sum to 1).

We leave it to the reader to verify that Definitions 3 and 4 are equivalent using arguments similar to those used above to show that Definitions 1 and 2 are equivalent. Remember that a function $a(x)$ satisfying Definition 4 is of the form $c_0 + c_1 x_1 + \cdots + c_n x_n$ where

$$c_0 = a(0) \quad \text{where } 0 = (0, \ldots, 0)^T \text{ is the zero vector in } R^n.$$

and $c_j = a(I_{.j})$ for $j = 1$ to n, where $I_{.j}$ is the jth unit vector in R^n.

Sometimes, in informal conversations, people refer to affine functions also as linear functions.

A real valued function $f(x)$ of variables $x = (x_1, \ldots, x_n)^T$ which is neither linear nor affine is known as a **nonlinear function**. An example of a nonlinear function in the variables $x = (x_1, x_2)^T$ is $f(x) = 9x_1^2 + 16x_1 x_2 + 6x_2^2$.

Hence, given a system of m linear equations in variables $x = (x_1, \ldots, x_n)^T$, if the constraint functions associated with the m constraints are $g_1(x), \ldots, g_m(x)$ and the RHS constants vector is $b = (b_1, \ldots, b_m)^T$, then the system can be written as

$$g_i(x) = b_i, \quad i = 1, \ldots, m.$$

For $i = 1$ to m, by transferring the RHS constant term in the ith equation to the left, we get the affine function $a_i(x) = g_i(x) - b_i$. In this case, the above system of equations can be written in an equivalent manner as

$$a_i(x) = g_i(x) - b_i = 0, \quad i = 1, \ldots, m.$$

This shows that each constraint in a system of linear equations requires that the constraint function associated with it should have a value equal to the RHS constant in that equation. This constraint is in the model only because of this requirement.

In constructing mathematical models for real world problems, very often the constraint functions that we have to deal with do not satisfy the linearity assumptions exactly. For instance, in Example 2 of Section 1.2, we assumed that the amount of nutrient (like a vitamin) absorbed by the body from the food consumed, is proportional to the amount of the nutrient present in the food. This is only approximately true. In reality, the percentage of vitamins in the diet absorbed by the body goes up if some protein is present in the diet.

If the constraint functions violate the linearity assumptions by huge amounts, then we cannot model the equality constraints in the problem as a system of linear equations. However, in many applications, the constraint functions, while not satisfying the linearity assumptions exactly, do satisfy them to a reasonable degree of approximation. In this case, a linear equation model for the system of equality constraints in the problem provides a reasonable approximation.

The reader should read the examples and exercises in Section 1.2 and check whether the constraint functions in them do satisfy the linearity assumptions either exactly or to a reasonable degree of approximation.

Admonition to engineers: In engineering there is an oft quoted admonition: *Be wise, linearize!*. This refers to the fact that engineers very often simplify models of problems by approximating constraint functions with judiciously chosen linear functions. ⋈

A system of equations of the form

$$f_i(x) = 0, \quad i = 1, \dots, m$$

where at least one of the constraint functions $f_1(x), \dots, f_m(x)$ is nonlinear (i.e., not linear or affine) is called a system of **nonlinear equations**. Systems of nonlinear equations are much harder to solve than systems of linear equations. Usually nonlinear equations are solved by iterative methods. These require in each iteration the solution of at least one system of linear equations based on the linear approximations of functions at the current point. Solving nonlinear equations is outside the scope of this book.

Bilinear Functions

A real valued function of two or more variables is said to be a **bilinear function** if the variables can be partitioned into two vectors u, v say, such that when u is fixed at any $\bar{u}$, the function is a linear function in the remaining variables v; and similarly when v is fixed at any $\bar{v}$, the function is a linear function in u.

Thus, if the function is $f(u, v)$, it is bilinear if

for any $\bar{u}$, $f(\bar{u}, v)$ is a linear function of v, and
for any $\bar{v}$, $f(u, \bar{v})$ is a linear function of u.

An example of a bilinear function is the dot product of two vectors. Let $u = (u_1, \ldots, u_n)$, $v = (v_1, \ldots, v_n)^T$. Let the function

$$f(u, v) = uv = \text{dot product of } u \text{ and } v.$$

$f(u, v)$ is a nonlinear function of all the variables $u_1, \ldots, u_n, v_1, \ldots, v_n$; however it is bilinear since it is a linear function of v when u is fixed at any $\bar{u}$, and a linear function of u when v is fixed at any $\bar{v}$. For instance, let $n = 3$, $\bar{u} = (3, -1, -7)$, $\bar{v} = (-2, 0, 1)^T$, $u = (u_1, u_2, u_3)$, $v = (v_1, v_2, v_3)^T$. Then the dot product of u and v, $f(u, v)$ satisfies

$$f(\bar{u}, v) = 3v_1 - v_2 - 7v_3; \quad f(u, \bar{v}) = -2u_1 + u_3$$

both of which are linear functions. We state this formally in the following result.

Result 1.7.1: The dot product of two vectors is a bilinear function of the two vectors: *The dot product of two vectors is a bilinear function, whenever any of these vectors is fixed equal to some numerical vector, the dot product is a linear function of the other vector.*

1.8 Solutions to a System, General Solution; Nonbasic (Independent), Basic (Dependent) Variables in Parametric Representation

Three Possibilities

Given a system of linear equations, there are three different possibilities, which we discuss below.

(a) System Has No Solution: In this case we say that the system in **infeasible** or **inconsistent**. An example of such a system is:

$$x_1 + x_2 = 2$$
$$-x_1 - x_2 = -1$$

If there is a solution $(x_1, x_2)^T$ satisfying both these equations, then by adding them we would get $0 = 1$, a contradiction. Hence this system is infeasible, or inconsistent.

(b) System Has a Unique Solution: System may have a solution, and that solution may be the only one, i.e., **unique solution**. An example of such a system is the following:

$$x_1 = 1$$
$$x_1 + x_2 = 3$$

By subtracting the first equation from the second, we see that this system has the unique solution $(x_1, x_2)^T = (1, 2)^T$.

We will see later that if a system of m equations in the variables $x = (x_1, \ldots, x_n)^T$ has the unique solution $x = (c_1, \ldots, c_n)^T$, then the system of equations is **equivalent to**, i.e., **can be reduced to** the simple system:

$$x_1 = c_1$$
$$\vdots$$
$$x_n = c_n$$

(c) System Has an Infinite Number of Solutions: When the system has more than one solution, it actually has an infinite number of solutions. An example of such a system is:

$$x_1 + x_3 = 10$$
$$x_2 + x_3 = 20$$

For this system we see that $x = (10 - \alpha, 20 - \alpha, \alpha)^T$ is a solution for every real value of α.

In this case the set of all solutions of the system is called the **solution set** for the system.

When the solution to the system is unique, the solution set consists of that single solution.

Two systems of linear equations in the same variables, are said to be **equivalent**, if they have the same solution set.

Representation of the General Solution
in Parametric Form

Consider a system of m equations in variables $x = (x_1, \ldots, x_n)^T$ which has more than one solution. In this case we will show that the system of equations is equivalent to another, which expresses a subset of r variables as linear functions of the other variables for some $r \leq m$; i.e., that the system of equations can be reduced to one in the following form (here for simplicity we are assuming that the r variables expressed in terms of the others are $x_1, \ldots, x_r$).

$$x_1 = \alpha_1 + \beta_{1,r+1}x_{r+1} + \cdots + \beta_{1,n}x_n$$
$$x_2 = \alpha_2 + \beta_{2,r+1}x_{r+1} + \cdots + \beta_{2,n}x_n$$
$$\vdots \quad \vdots$$
$$x_r = \alpha_r + \beta_{r,r+1}x_{r+1} + \cdots + \beta_{r,n}x_n$$

where $\alpha_1, \ldots, \alpha_r$, and the β's are constants. This system does not say anything about the values of the variables $x_{r+1}, \ldots, x_n$; hence they can be given any arbitrary real values; and then the values of $x_1, \ldots, x_r$ can be computed by substituting those values for $x_{r+1}, \ldots, x_n$ in the above expressions to yield a solution to the system. Every solution of the system can be obtained this way. Hence a general solution for the original system is of the form:

$$\begin{pmatrix} x_1 \\ \vdots \\ x_r \\ x_{r+1} \\ \vdots \\ x_n \end{pmatrix} = \begin{pmatrix} \alpha_1 + \beta_{1,r+1}\bar{x}_{r+1} + \cdots + \beta_{1,n}\bar{x}_n \\ \vdots \\ \alpha_r + \beta_{r,r+1}\bar{x}_{r+1} + \cdots + \beta_{r,n}\bar{x}_n \\ \bar{x}_{r+1} \\ \vdots \\ \bar{x}_n \end{pmatrix}$$

where $\bar{x}_{r+1}, \ldots, \bar{x}_n$ are parameters that can take arbitrary real values. Such an expression is known as a **representation of the general solution to the system in parametric form with** $\bar{x}_{r+1}, \ldots, \bar{x}_n$ **as real valued parameters.**

In this representation, the variables $x_{r+1}, \ldots, x_n$ which can be given arbitrary real values are known as **independent** or **free** or **nonbasic variables.** The variables $x_1, \ldots, x_r$ whose values in the solution are computed from the values of independent variables are known as **dependent** or **basic variables** in this representation.

In the general solution given above, the expressions for the basic variables in terms of the nonbasic variables are listed in the order $x_1, \ldots, x_r$. Hence $x_1, x_2, \ldots, x_r$ are known as the **first basic variable, second basic variable,..., rth basic variable** respectively in this general solution.

When the basic variables are put in this order as a vector, $(x_1, \ldots, x_r)$, this vector is called the **basic vector** for this representation.

The particular solution obtained by giving all nonbasic variables the value zero, namely:

$$\begin{pmatrix} x_1 \\ \vdots \\ x_r \\ x_{r+1} \\ \vdots \\ x_n \end{pmatrix} = \begin{pmatrix} \alpha_1 \\ \vdots \\ \alpha_r \\ 0 \\ \vdots \\ 0 \end{pmatrix}$$

is known as the **basic solution** for the system of equations corresponding to the basic vector $(x_1, \ldots, x_r)$.

The choice of basic, nonbasic variables leading to such a representation of the general solution is usually not unique; i.e., the variables $x_1, \ldots, x_n$ can be partitioned into basic, nonbasic variables in many different ways leading to such a representation of the general solution. One important property is that in all such partitions, the number of basic variables, r will always be the same. This number r is called the **rank** of the original system of equations. The number of nonbasic variables in such a representation $= n - r$ (where n is the total number of variables in the system, and r is the number of basic variables in the representation) and is called the **dimension of the solution set for the system.**

As an example, consider the system:

$$x_1 + x_3 = 10$$
$$x_2 + x_3 = 20$$

For this system, one parametric representation of the general solution is:

$$\begin{pmatrix} x_1 \\ x_2 \\ x_3 \end{pmatrix} = \begin{pmatrix} 10 - \bar{x}_3 \\ 20 - \bar{x}_3 \\ \bar{x}_3 \end{pmatrix}$$

for arbitrary real values of the parameter $\bar{x}_3$. In this representation:

basic vector is: (x_1, x_2); nonbasic variable is: x_3; rank of system is 2; dimension of the solution set is 1. The basic solution of the system corresponding to this basic vector (x_1, x_2) is: $(x_1, x_2, x_3)^T = (10, 20, 0)^T$.

Another representation of the general solution of this system is:

$$\begin{pmatrix} x_1 \\ x_2 \\ x_3 \end{pmatrix} = \begin{pmatrix} \bar{x}_1 \\ 10 + \bar{x}_1 \\ 10 - \bar{x}_1 \end{pmatrix}$$

this one corresponds to the selection of (x_2, x_3) as the basic vector, and x_1 as the nonbasic variable; and this also exhibits the same rank of 2 for the system, and dimension of 1 for the solution set. Also, the basic solution of the system corresponding to the basic vector (x_2, x_3) is: $(x_1, x_2, x_3)^T = (0, 10, 10)^T$.

Efficient algorithms for determining the general solution of a given system of linear equations are discussed later on.

Exercises

1.8.1: Problem of the incorrectly cashed cheque: (Adopted from M. H. Greenblatt, *Mathematical Entertainments, A collection of Illuminating Puzzles New and Old*, T. Y. Crowell co., NY, 1965) Long time ago there was a man who cashed his pay cheque at a bank where the teller made the mistake of giving him as many dollars as there should have been cents, and as many cents as there should have been dollars. The man did not notice the error. He then made a 5 cent payphone call (phone rates were low in those days). Then he discovered that he had precisely twice as much money left as the pay cheque had been written for. Construct a linear equation model for this problem, find its general solution, and using the limits for the variables that come from the above facts on it deduce what the amount of his pay cheque is.

1.8.2: For each of the following problems construct a linear equation model, find its general solution, and using all the facts stated in the problem on it, deduce the required answer.

(i) A brother B and sister S are in their twenties now. When S was one year older than half of Bs present age, B was as old as S is now. Find their present ages.

(ii) A brother B is 6 years older than his sister S. The 1st digit in B's age is twice the 2nd digit in S's age; the 2nd digit in B's age is twice the 1st digit in S's age. Find their ages.

(iii)   A four digit number has the property that if you move the leftmost digit in it to the rightmost position, its value changes to 1+ three-fourths of the original number. Find the number.

(iv)   A two digit number is equal to 34+ one-third of the number obtained by writing the digits in it in reverse order. Find this number.

(v)   A kid's piggybank has a total of 70 coins consisting of pennies (cents), nickels (5-cent coins), and dimes (10-cent coins) for a total value of $1.17. Determine how many coins of each type it has.

(vi)   A brother B and sister S took the same examination in which there were several questions. Both B and S attempted all the questions. S got a third of the answers wrong. B had 5 of them wrong. Both of them put together got three-quarters of the questions right. Find out how many questions were there, and how many each answered correctly.

1.9   Linear Dependence Relation Among Constraints, Redundant Constraints, Linearly Independent Systems

Linear Combinations of Constraints

In a system with m linear equations, let R_1 to R_m denote the m equations. A **linear combination of these constraints** is of the form: $\sum_{i=1}^{m} \alpha_i R_i$, i.e., multiply R_i by α_i on both sides of the equality symbols and add over $i = 1$ to m. Here all α_i are arbitrary real numbers.

Consider the system of equations:

$$x_1 + x_3 = 10$$
$$x_2 + x_3 = 20$$

If $\bar{x} = (\bar{x}_1, \bar{x}_2, \bar{x}_3)^T$ is a solution to this system, it satisfies both the first and the second constraints, and hence also any linear combination of these constraints, namely:

$$\alpha_1(x_1 + x_3) + \alpha_2(x_2 + x_3) = 10\alpha_1 + 20\alpha_2$$

for every α_1, α_2 real. For example, taking $\alpha_1 = 5, \alpha_2 = 10$ leads to the linear combination of constraints in above system:

$$5x_1 + 10x_2 + 15x_3 = 250$$

which will be satisfied by every solution of the system.

For exactly the same reason, we have the following result:

Result 1.9.1: Every solution satisfies all linear combinations of equations in system: *Every solution of a system of linear equations satisfies every linear combination of constraints in the system.*

Redundant Constraints in a System of Linear Equations

Consider the following system of linear constraints in detached coefficient tableau form:

x_1	x_2	x_3	x_4	x_5	
1	1	1	1	1	10
1	-1	0	0	0	5
1	0	-1	0	0	15
1	0	0	-1	0	20
4	0	0	0	1	50

Let R_1 to R_5 denote the constraints in the system from top to bottom. The following relationship can be verified to hold among these constraints:

$$R_5 = R_1 + R_2 + R_3 + R_4.$$

In a system of linear equations, a particular constraint can be said to be a **redundant constraint** if it can be obtained as a linear combination of the other constraints. So, in the system above, R_5 can be treated as a redundant constraint. If we delete R_5 from that system, we get the following system:

x_1	x_2	x_3	x_4	x_5	
1	1	1	1	1	10
1	-1	0	0	0	5
1	0	-1	0	0	15
1	0	0	-1	0	20

Since the second system has one constraint less than the first, we have:

Solution set of 2nd system $\supset$ Solution set of 1st system

However, since R_5 is a linear combination of R_1 to R_4, by Result 1.9.1, every solution of the 2nd system satisfies R_5. Hence both systems above have the same solution set. Using exactly the same argument on a general system, we have the following result:

Result 1.9.2: Deletion of a redundant constraint leads to an equivalent system: *Consider a system of linear equations. If one of the equations in the system can be expressed as a linear combination of the other equations, that equation can be considered as a redundant constraint and deleted from the system without affecting the set of solutions; i.e., this deletion leads to an equivalent system with one less equation. The same process can be repeated on the remaining system.*

Linear Dependence Relations for a System of Linear Equations

In the system in the first tableau above, the constraints R_1 to R_5 satisfy $R_5 = R_1 + R_2 + R_3 + R_4$. So, if we take the linear combination of the constraints $R_1 + R_2 + R_3 + R_4 - R_5$, we get the equation:

$$0x_1 + 0x_2 + 0x_3 + 0x_4 + 0x_5 = 0$$

with all the coefficients of the variables on the LHS as 0, and the RHS constant 0. We will call this the **0 = 0 equation**.

A linear combination of these constraints R_1 to R_5 is of the form $\sum_{i=1}^{5} \alpha_i R_i$. In this linear combination if we take $\alpha_i = 0$ for all $i = 1$ to 5, we get the **trivial linear combination**, and it of course results in the 0 = 0 equation.

If at least one of the α_i is $\neq 0$, the linear combination is called a **nontrivial linear combination**.

In the same way, in a general system of linear equations with equations R_1 to R_m, a nontrivial linear combination of these equations is of the form: $\sum_{i=1}^{m} \alpha_i R_i$ with at least one $\alpha_i \neq 0$; and we say that the equation R_i **participates in this linear combination** iff $\alpha_i \neq 0$.

As an illustration, in the linear combination $R_1 + R_2 + R_3 + R_4 - R_5$ in the example above, all the equations R_1 to R_5 are participating because all of them have nonzero coefficients in it.

A nontrivial linear combination of a system of equations is said to be a **linear dependence relation** for the system if it leads to the 0 = 0 equation.

In the example considered above, the linear combination $R_1 + R_2 + R_3 + R_4 - R_5$ is a linear dependence relation because it is 0 = 0.

Among the equations participating in a linear dependence relation, any one can be expressed as a linear combination of the others. For instance, in the linear dependence relation $R_1 + R_2 + R_3 + R_4 - R_5$ in the example above, all the equations R_1 to R_5 are participating, and we can verify that:

$$R_1 = R_5 - R_2 - R_3 - R_4$$
$$R_2 = R_5 - R_1 - R_3 - R_4$$
$$R_3 = R_5 - R_2 - R_1 - R_4$$
$$R_4 = R_5 - R_2 - R_3 - R_1$$
$$R_5 = R_1 + R_2 + R_3 + R_4.$$

Therefore we have the following result:

Result 1.9.3: Any one constraint in a linear dependence relation can be considered redundant: *Among the constraints participating in a linear dependence relation for a system of linear equations. any one of them can be considered as a redundant equation and eliminated from the system without affecting the set of solutions.*

A system of linear equations is said to be **linearly dependent** if there is at least one linear dependence relation for it. In a linearly dependent system of linear equations, there is a redundant equation that can be eliminated without affecting the set of solutions.

A system of linear equations is said to be **linearly independent** if there is no linear dependence relation for it; i.e., if no nontrivial linear combination of constraints in it leads to the $0 = 0$ equation. For instance, the system obtained by deleting R_5 from the system considered above, which is reproduced below

x_1	x_2	x_3	x_4	x_5	
1	1	1	1	1	10
1	-1	0	0	0	5
1	0	-1	0	0	15
1	0	0	-1	0	20

is linearly independent. To verify this, the equations in this tableau from top to bottom are R_1 to R_4. The linear combination $\sum_{i=1}^{4} \alpha_i R_i$ is the equation:

$$(\alpha_1 + \alpha_2 + \alpha_3 + \alpha_4)x_1 + (\alpha_1 - \alpha_2)x_2 + (\alpha_1 - \alpha_3)x_3$$
$$+ (\alpha_1 - \alpha_4)x_4 + \alpha_1 x_5 = 10\alpha_1 + 5\alpha_2 + 15\alpha_3 + 20\alpha_4.$$

For this to be the $0 = 0$ equation, we must have the coefficient of x_5, $\alpha_1 = 0$. Given $\alpha_1 = 0$, the coefficients of x_2, x_3, x_4 are 0 only if $\alpha_2 = \alpha_3 =$

$\alpha_4 = 0$. Hence the only linear combination of the four equations in this system that is the $0 = 0$ equation is the trivial linear combination; i.e., the system is linearly independent.

A system of equations which is linearly dependent is said to be **minimally linearly dependent** if the system obtained by eliminating any one constraint from it is always linearly independent. Hence a system of linear equations is minimally linearly dependent iff it is linearly dependent, and there is no linear dependence relation for it in which at least one constraint does not participate. An example of a minimally linearly dependent system of linear equations is the system considered above consisting of equations R_1 to R_5, which is reproduced below:

x_1	x_2	x_3	x_4	x_5	
1	1	1	1	1	10
1	-1	0	0	0	5
1	0	-1	0	0	15
1	0	0	-1	0	20
4	0	0	0	1	50

Exercises

1.9.1: Show that the following systems of linear equations are linearly dependent. For each of them check whether it is minimally linearly dependent.

(i)
$$x_1 + x_2 + x_3 = 100$$
$$x_4 + x_5 + x_6 = 200$$
$$x_7 + x_8 + x_9 = 300$$
$$x_1 + x_4 + x_7 = 200$$
$$x_2 + x_5 + x_8 = 200$$
$$x_3 + x_6 + x_9 = 200$$

(ii)
$$x_1 = 6$$
$$x_2 = 7$$
$$x_1 + x_2 = 13$$
$$2x_1 + x_2 = 19$$

(iii)
$$x_1 + x_2 + x_3 + x_4 = 100$$
$$2x_1 = 20$$
$$x_1 + 2x_2 = 50$$
$$x_1 + x_2 + 2x_3 = 90$$
$$5x_1 + 4x_2 + 3x_3 + x_4 = 260$$

1.9.2: Show that the following systems of linear equations are linearly independent.

(i) $x_1 + x_2 + x_3 + x_4 + x_5 = -18$
        $x_2 + x_3 + x_4 + x_5 = -12$
        $x_3 + x_4 + x_5 = 9$
        $x_4 + x_5 = 6$

(ii) $x_1 + x_2 + x_3 = 800$
        $x_4 + x_5 + x_6 = 300$
        $x_1 + x_4 = 500$
        $x_2 + x_5 = 200$

1.10   The Fundamental Inconsistent Equation "0 = 1"

The equation $0 = 1$ is known as the **fundamental inconsistent equation**.

Consider the following system of linear equations. Let R_1 to R_4 denote the constraints in it from top to bottom.

x_1	x_2	x_3	x_4	x_5	
1	1	−1	1	1	4
1	2	4	3	2	6
2	3	3	4	3	9
−1	2	1	3	2	15

Taking the linear combination $R_1 + R_2 - R_3$ leads to the equation:

$$0x_1 + 0x_2 + 0x_3 + 0x_4 + 0x_5 = 1$$

which is the fundamental inconsistent equation $0 = 1$. If the above system of equations has any solution, it has to satisfy this inconsistent equation too, an impossibility. This implies that the above system has no solution.

We will show later (see item 3 in "Discussion of Some Features of the GJ Method", Section 1.16) that the converse must be true also. This leads to the following result:

Result 1.10.1: Fundamental condition for inconsistency: *A system of linear equations has no solution iff the fundamental inconsistent equation $0 = 1$ can be obtained as a linear combination of equations in it.*

An algorithm to determine whether this condition holds is given later.

Exercises

1.10.1: A problem from Bulgaria: A person A claims that he and his three brothers B, C, D live in a building, each occupying a separate apartment. Their statement contains the following details:

(i) Apartment of A has 3 windows and 2 doors.
(ii) B has as many windows as the doors of C.
(iii) C has as many windows as the doors of B.
(iv) D says that in total his brothers have the same number of windows as doors.

Represent these statements through a system of linear equations and solve it. What possible conclusion can you draw about the building from this?

1.11 Elementary Row Operations on a Tableau Leading to an Equivalent System

Row operations are the most heavily used computational tool today. It is truly amazing that such a simple tool plays such an important role in so many complex tasks. We now discuss two different **row operations** (also called **elementary row operations**). These are:

1. Scalar Multiplication: Multiply each entry in a row by the same nonzero scalar

2. Add a Scalar Multiple of a Row to Another: Multiply each element in a row by the same nonzero scalar and add the result to the corresponding element of the other row.

Here are the results for two rows in a general tableau:

Row						Con-dition
R_1	a_{11}	a_{12}	$\ldots$	a_{1n}	b_1	
R_2	a_{21}	a_{22}	$\ldots$	a_{2n}	b_2	
αR_1	αa_{11}	αa_{12}	$\ldots$	αa_{1n}	αb_1	$\alpha \neq 0$
$R_2 + \beta R_1$	$a_{21} + \beta a_{11}$	$a_{22} + \beta a_{12}$	$\ldots$	$a_{2n} + \beta a_{1n}$	$b_2 + \beta b_1$	$\beta \neq 0$

Here are numerical illustrations:

Row						
R_1	1	2	−1	2	1	−10
R_2	−1	1	−2	−1	1	15
$6R_1$	6	12	−6	12	6	−60
$R_2 + 2R_1$	1	5	−4	3	3	−5

There is another row operation called **row interchange** (interchanging two rows in the tableau) which we will discuss later in Section 1.20, as it is used by the Gaussian elimination method discussed there.

We state some important facts about row operations on the detached coefficient tableau for a system of linear equations in the following result:

Result 1.11.1: Row operations lead to an equivalent system, and nonzero columns remain nonzero

(a) *Since we are dealing with a system of equations, row operations lead to an equivalent system with the same set of solutions.*

(b) *A nonzero column vector in a tableau remains nonzero after some row operations are performed on the tableau.*

As discussed in Section 1.4, in the detached coefficient tableau for a system of linear equations, every column vector is always nonzero. From (b) of Result 1.11.1 we conclude that every column vector continues to remain nonzero after performing any sequence of row operations on the tableau.

Exercises

1.11.1: Perform the following row operations and record the results:

Row						
R_1	3	2	4	1	5	−7
R_2	1	−2	−1	1	0	−2
R_3	0	−4	−2	1	1	−7
R_4	−1	4	−3	4	−2	8
$R_2 + R_1$						
$R_3 + 2R_1$						
$R_4 - 2R_1$						
$(1/2)R_1$						

1.12 Memory Matrix to Keep Track of Row Operations

A tableau changes whenever row operations are performed on it. Clearly every row vector in the resulting tableau is a linear combination of row vectors in the original tableau, but we do not know what that linear combination is. Sometimes, when performing a series of row operations, it is a good idea to have the formula for each row vector in the current tableau as a linear combination of rows in the original tableau. A tool called **memory matrix** helps us to keep track of these linear combinations at the expense of slight extra work. Here we explain what this memory matrix is, how to set it up and how to use it.

Let $R_1, \ldots, R_m$ denote the row vectors in the original tableau.

If a row vector in the current tableau is the linear combination $\alpha_1 R_1 + \cdots + \alpha_m R_m$, the memory matrix will store this in detached coefficient form, i.e., it consists of column vectors labeled with $R_1, \ldots, R_m$ at the top, and in these column vectors it will have entries $\alpha_1, \ldots, \alpha_m$ in this row; and updates this whenever new row operations are carried out.

Algorithm: How to Set Up the Memory Matrix and Use It

BEGIN

To use this facility, before doing any row operations, on the left hand side of the original tableau open a memory matrix consisting of m column vectors labeled $R_1, \ldots, R_m$ at the top. For each $i = 1$ to m, the column vector under R_i should be the ith unit vector of dimension m. This will be the initial memory matrix set up on the original tableau itself. All it says is that the first row vector in the original tableau is $1R_1 + 0R_2 + \cdots + 0R_m = R_1$, the second row vector is $0R_1 + 1R_2 + 0R_3 + \cdots 0R_m = R_2$, etc.

Then, whenever you perform any row operations on the tableau, carry out exactly the same operations on the memory matrix also (this is the extra work in operating this memory matrix). This updates the memory matrix appropriately, and at any stage, the ith row of the memory matrix will contain the coefficients of $R_1, \ldots, R_m$ in the expression of this row in the current tableau as a linear combination of R_1 to R_m.

END

We will illustrate with an example of a detached coefficient tableau for a system consisting of 4 equations in 5 variables.

x_1	x_2	x_3	x_4	x_5	
1	−1	1	0	1	5
2	1	−2	1	−1	6
0	0	1	2	1	−2
1	2	0	0	0	3

Denoting the constraints from top to bottom as R_1 to R_4, we open the memory matrix as below.

Memory matrix				Original tableau					
R_1	R_2	R_3	R_4	x_1	x_2	x_3	x_4	x_5	
1	0	0	0	1	−1	1	0	1	5
0	1	0	0	2	1	−2	1	−1	6
0	0	1	0	0	0	1	2	1	−2
0	0	0	1	1	2	0	0	0	3

So, the information in the memory matrix on the original tableau indicates that the first row in the original tableau is $1R_1 + 0R_2 + 0R_3 + 0R_4 = R_1$, etc.

Now suppose we perform the following row operations:

Multiply Row 1 by 2
Subtract two times Row 1 from Row 2
Subtract Row 1 from Row 4.

We carry on the same operations on the memory matrix too. This leads to the current tableau which is the tableau after the above row operations are carried out.

Memory matrix				Current tableau					
R_1	R_2	R_3	R_4	x_1	x_2	x_3	x_4	x_5	
2	0	0	0	2	−2	2	0	2	10
−2	1	0	0	0	3	−4	1	−3	−4
0	0	1	0	0	0	1	2	1	−2
−1	0	0	1	0	3	−1	0	−1	−8

According to the memory matrix,

the first row in the current tableau is $2R_1$,

the second row is $-2R_1 + R_2$,

the third is R_3,

and the fourth is $-R_1 + R_4$

which are true. In the same way, we can continue performing additional row operations on the current tableau, and the memory matrix will update and always show the coeffients in the expressions of rows of the current tableau as linear combinations of rows in the original tableau.

Thus when performing a sequence of row operations on a tableau, we use the memory matrix if we need to have the expressions for rows in the current tableau as linear combinations of rows in the original tableau.

1.13 What to Do With "0 = 0", Or "0 = 1" Rows Encountered After Row Operations?

The "0 = 0" Equation

We start with an example first. Consider the detached coefficient tableau given below for a system of linear equations, in which we call the equations as R_1 to R_4.

Original Tableau-1

Eq. name	x_1	x_2	x_3	x_4	x_5	
R_1	1	1	1	1	1	−2
R_2	2	−1	2	−1	2	−4
R_3	5	−1	5	−1	5	−10
R_4	3	4	2	1	−2	6

We introduce the memory matrix, and then perform the following row operations: subtract Row 1 from Rows 3 and 4. This leads to the following tableau:

Memory matrix									
R_1	R_2	R_3	R_4	x_1	x_2	x_3	x_4	x_5	
1	0	0	0	1	1	1	1	1	−2
0	1	0	0	2	−1	2	−1	2	−4
−1	0	1	0	4	−2	4	−2	4	−8
−1	0	0	1	2	3	1	0	−3	8

In this tableau we subtract two times Row 2 from Rows 3 and 4. This leads to the following Current Tableau-1.

Memory matrix				Current Tableau-1					
R_1	R_2	R_3	R_4	x_1	x_2	x_3	x_4	x_5	
1	0	0	0	1	1	1	1	1	−2
0	1	0	0	2	−1	2	−1	2	−4
−1	−2	1	0	0	0	0	0	0	0
−1	−2	0	1	−2	5	−3	2	−7	16

In this tableau Row 3 has become the "0 = 0" equation. What should be done about it?

This implies that the set of 4 constraints in the original system is linearly dependent, and that a linear dependence relation for them has just been discovered by the row operations we carried out. Whenever a "0 = 0" row appears in the current tableau after carrying out some row operations on the detached coefficient tableau for a system of linear equations, it is an indication of the discovery of a linear dependence relation among the original constraints, and consequently the equation corresponding to this row in the original system is a redundant constraint.

That linear dependence relation can be read out from the entries in the memory matrix in the row of the "0 = 0" row in the current tableau. Here from Row 3 in the memory matrix in the current tableau we find that the linear dependence relation is: $-R_1 - 2R_2 + R_3$, because it leads to the "0 = 0" equation. The vector of coefficients of the equations in the original system (in the order in which the equations are listed in the original system) in this linear dependence relation is called the **evidence (or certificate) of redundancy** for treating the equation corresponding to Row 3 as a redundant constraint in the original system. The reason for the name is the fact that one can form the linear combination of the constraints in the original system with coefficients in this row and actually verify that it leads to the "0 = 0" equation. This evedence is provided by Row 3 of the memory matrix in the current tableau.

This evidence $(-1, -2, 1, 0)$ implies that among the constraints in the original tableau, we have the relation: $R_3 = R_1 + 2R_2$. So, R_3 can be eliminated from the original system as a redundant constraint, leading to an equivalent system. The effect of this is to eliminate the "0 = 0" row in the current row (i.e., third row) and the corresponding column, column 3 of R_3, from the memory matrix. After this elimination, the resulting equivalent original tableau, and current tableau are shown below.

Original Tableau-2

Eq. name	x_1	x_2	x_3	x_4	x_5	
R_1	1	1	1	1	1	−2
R_2	2	−1	2	−1	2	−4
R_4	3	4	2	1	−2	6

Memory matrix			Current Tableau-2					
R_1	R_2	R_4	x_1	x_2	x_3	x_4	x_5	
1	0	0	1	1	1	1	1	−2
0	1	0	2	−1	2	−1	2	−4
−1	−2	1	−2	5	−3	2	−7	16

The processing of these equivalent systems can now continue.

In the same way, whenever a "0 = 0" row is encountered after some operations on the detached coefficient tableau for a system of linear equations, do the following:

- Suppose the "0 = 0" equation is in Row r of the current tableau.
  It is an indication that the equation in Row r of the original system is a redundant constraint. If the memory matrix is being maintained, you can get an expression for this constraint as a linear combination of the other constraints from Row r of the memory matrix in the current tableau (this row of the memory matrix is the evidence for the redundancy of the equation in Row r), as illustrated in the above example.
- Eliminate Row r (i.e., the row of the "0 = 0" equation) from the current tableau, and also eliminate column r of the memory matrix if it is being maintained. Also, eliminate Row r from the original tableau. Continue processing the remaining tableaus, which represent an equivalent system with one less constraint.

  Equivalently, you can leave this row r in both the current and original tableaus (and row r and column r of the memory matrix too) as they are in these tableaus, but ignore them in any future computation to be carried out.

Equivalent Systems

Suppose we are given a system of linear equations, say System 1. Any system, say System 2, obtained by performing a series of row operations on System 1 and eliminating any redundant equations discovered during the process is said to be **equivalent** to System 1. All such systems have the same set of solutions as System 1.

Given a system of linear equations to solve, the strategy followed by algorithms discussed in the following sections is to transform this system through a series of row operations into an equivalent simpler one for which the general solution can be easily recognized.

The "$0 = 1$" Equation

Again we begin with an example. Consider the following system of equations in detached coefficient tableau form.

Original Tableau

Eq. name	x_1	x_2	x_3	x_4	x_5	
R_1	1	1	1	1	1	15
R_2	1	2	2	2	2	22
R_3	1	3	3	3	3	35
R_4	2	5	4	5	4	40

We introduce the memory matrix and perform the following row operations: subtract Row 1 from Rows 2 and 3; and subtract two times Row 1 from Row 4. This leads to the following tableau.

Memory matrix									
R_1	R_2	R_3	R_4	x_1	x_2	x_3	x_4	x_5	
1	0	0	0	1	1	1	1	1	15
-1	1	0	0	0	1	1	1	1	7
-1	0	1	0	0	2	2	2	2	20
-2	0	0	1	0	3	2	3	2	10

On this tableau subtract two times Row 2 from Row 3, and subtract three times Row 2 from Row 4. This leads to the following tableau.

Memory matrix									
R_1	R_2	R_3	R_4	x_1	x_2	x_3	x_4	x_5	
1	0	0	0	1	1	1	1	1	15
-1	1	0	0	0	1	1	1	1	7
1	-2	1	0	0	0	0	0	0	6
1	-3	0	1	0	0	-1	0	-1	-11

The third row in this tableau is the equation "0 = 6". So, from the memory matrix entries in this row we conclude that the linear combination $\frac{1}{6}(R_1 - 2R_2 + R_3)$ is the fundamental inconsistent equation "0 = 1", and hence that the original system has no solution.

The vector of coefficients of the equations in the original system (in the order in which the equations are listed in the original system) that produce the "0 = 1" equation or the "0 = a" equation for some $a \neq 0$ has been called by the obscure name **supervisor principle for infeasibility**. The idea behind this name is that if your supervisor is suspicious about your claim that the system is infeasible, he/she can personally verify the claim by obtaining the linear combination of constraints in the original system with this vector of coefficients and verify that it is "0 = 1" or "0 = a" for $a \neq 0$. But we will call this vector by the simpler name **evidence (or certificate) of infeasibility**. This evidence is provided by the row in the memory matrix in the current tableau that corresponds to the "0 = a" equation for $a \neq 0$.

In the same way for any system of equations, if we end up with a row corresponding to the equation "0 = a" where $a \neq 0$ after some row operations on it, we conclude that the system is inconsistent and has no solution, and terminate.

When row operations are carried out on the detached coefficient tableau of a system of linear equations without introducing the memory matrix, we will still be able to discover all the redundant constraints in the system, and determine if the system is infeasible; but then we will not have the evedences (or certificates) of redundancy or infeasibility. To have those evidences we need to introduce the memory matrix on the system before row operations are commenced, and carry out all the same row operations on the columns of the memory matrix as well.

1.14 The Elimination Approach to Solve Linear Equations

The classical **elimination approach** for solving systems of linear equations was figured out by Chinese and Indian mathematicians more than 2000 years ago. Without trying to list all possibilities that can occur individually, we want to present the main idea behind this approach, which is the following:

Algorithm: The Elimination Approach For Solving Linear Equations

BEGIN

Select any equation and take it out of the system. Select a variable which appears with a nonzero coefficient in it as the **basic variable** in this equation. Use this equation to derive the expression for this basic variable in terms of the other variables. Substitute this expression for the basic variable in all the other equations, thereby eliminating this variable from the remaining system.

Once the values of the other variables are determined from the remaining system, the corresponding value of this variable can be computed from the expression for it obtained above.

The remaining system is now smaller. It has one equation less, and one variable less than the previous system. If it has a "$0 = a$" equation in it for some $a \neq 0$, the original system is inconsistent (i.e., it has no solution), terminate. If it has some "$0 = 0$" equations, eliminate them. If the remaining system now has only one equation, obtain the general solution for it, and using the expressions for basic variables generated in previous steps obtain the general solution of the original system. If the remaining system consists of more than one equation, apply the elimination step on it and repeat the same way.

END

Example: Solving a System of Linear Equations by the Elimination Approach

We will apply this approach on the following system of 3 equations in 4 variables as an illustration:

Original System = System 1

$$
\begin{array}{rrrrcr}
2x_1 & +4x_2 & -6x_3 &   & = & 10 \\
 x_1 & -x_2 & +x_3 & -x_4 & = & 5 \\
-x_1 & +2x_2 & -2x_3 & +x_4 & = & 3 \\
\end{array}
$$

Selecting the first equation to take out from the system, and the variable x_1 with the coefficient of 2 as the basic variable in it, we get the expression:

$$ x_1 = \frac{1}{2}(10 - 4x_2 + 6x_3) = 5 - 2x_2 + 3x_3. $$

We now substitute this expression for x_1 in the remaining equations 2 and 3 leading to:

$$(5 - 2x_2 + 3x_3) - x_2 + x_3 - x_4 = 5$$
$$-(5 - 2x_2 + 3x_3) + 2x_2 - 2x_3 + x_4 = 3.$$

Rearranging terms, this leaves the reduced system:

System 2

$-3x_2$	$+4x_3$	$-x_4$	$=$	0
$4x_2$	$-5x_3$	$+x_4$	$=$	8

The reduced system, System 2 has two equations in three variables. Taking out the top equation from it, and the variable x_4 with the coefficient of -1 as the basic variable in it, we get the expression:

$$x_4 = -3x_2 + 4x_3.$$

Substituting this expression for x_4 in the remaining equation leads to the system with one equation only

$$4x_2 - 5x_3 + (-3x_2 + 4x_3) = 8$$

or

$$x_2 - x_3 = 8.$$

The remaining system is a single equation in two variables x_2, x_3. Selecting x_2 as the basic variable in it, we get the general solution for it as:

$$\begin{pmatrix} x_2 \\ x_3 \end{pmatrix} = \begin{pmatrix} 8 + \bar{x}_3 \\ \bar{x}_3 \end{pmatrix}$$

$\bar{x}_3$ is an arbitrary real valued parameter. Plugging this solution in the expression for x_4 above leads to the general solution

$$\begin{pmatrix} x_2 \\ x_3 \\ x_4 \end{pmatrix} = \begin{pmatrix} 8 + \bar{x}_3 \\ \bar{x}_3 \\ -24 + \bar{x}_3 \end{pmatrix}$$

for System 2. Plugging this in the expression for x_1 above leads to the general solution for the original system

$$\begin{pmatrix} x_1 \\ x_2 \\ x_3 \\ x_4 \end{pmatrix} = \begin{pmatrix} -11 + \bar{x}_3 \\ 8 + \bar{x}_3 \\ \bar{x}_3 \\ -24 + \bar{x}_3 \end{pmatrix}$$

where $\bar{x}_3$ is an arbitrary real parameter. In particular, taking $\bar{x}_3 = 0$ leads to the basic solution WRT the basic vector (x_1, x_4, x_2), which is:

$$\begin{pmatrix} x_1 \\ x_2 \\ x_3 \\ x_4 \end{pmatrix} = \begin{pmatrix} -11 \\ 8 \\ 0 \\ -24 \end{pmatrix}.$$

Exercises

1.14.1: Apply the elimination approach to get the general solution of the following systems. For each, check whether the solution obtained is unique, and explain your conclusions carefully.

(i)   An infeasible system
$$-x_1 + 7x_2 - 9x_3 = 10$$
$$x_1 - 8x_2 + 12x_3 = -15$$
$$2x_1 - 11x_2 + 7x_3 = 5$$
$$2x_1 - 12x_2 + 10x_3 = 0$$
$$3x_1 - 19x_2 + 19x_3 = -10$$

(ii) An infeasible system
$$3x_1 - x_2 + 2x_3 - 7x_4 = 10$$
$$2x_1 + 2x_2 + 15x_4 = 3$$
$$7x_1 + 3x_2 + 2x_3 + x_4 = 16$$
$$8x_1 + 4x_3 + x_4 = 30$$

(iii) A system with a unique solution
$$5x_1 - 7x_2 - x_3 = 5$$
$$-8x_1 + 15x_2 + 2x_3 = 15$$
$$-2x_1 + 9x_2 + x_3 = -13$$

(iv) A system with multiple solutions
$$4x_1 + 5x_2 - x_3 - 6x_4 + 3x_5 = 8$$
$$10x_1 - 6x_2 + x_3 + 13x_4 + 2x_5 = -4$$
$$24x_1 - 7x_2 + x_3 + 20x_4 + 7x_5 = 0$$
$$13x_1 + x_4 + 17x_5 = -5$$

1.14.2: Solve the following systems of equations through simple manipulations.

(i)

x_1	x_2	x_3	x_4	
1	7	3	5	16
8	4	6	2	−16
2	6	4	8	16
5	3	7	1	−16

(ii)

x_1	x_2	x_3	x_4	
1	1	1	0	4
0	1	1	1	−5
1	0	1	1	0
1	1	0	1	−8

1.15 The Pivot Step, a Convenient Tool For Carrying Out Elimination Steps

Here we discuss a step known as the **Gauss–Jordan pivot step** or **GJ pivot step** or sometimes just **pivot step** which offers a very convenient tool for carrying out the elimination steps discussed in the previous section. In a later section we will discuss a slightly different version of the pivot step called the Gaussian pivot step.

Consider the general system of m equations in n variables in detached coefficient tableau form. A GJ pivot step on this tableau is a computational tool specified by:

- A row of the tableau selected as the **pivot row** for the pivot step
- A column of the tableau selected as the **pivot column** for the pivot step
- The element in the pivot row and the pivot column as the **pivot element** for the pivot step, which should be nonzero.

The GJ pivot step itself consists of a series of row operations on the tableau to convert the pivot column into the unit column with a single nonzero entry of "1" in the pivot row. We give the operations for the general system first. We use the abbreviations "PR" for "pivot row", and "PC" for "pivot column", and indicate the pivot element by putting a box around it.

The general system in detached coefficient tableau form is:

Original Tableau

x_1	x_2	$\cdots$	x_s	$\cdots$	x_n	RHS constants	
a_{11}	a_{12}	$\cdots$	a_{1s}	$\cdots$	a_{1n}	b_1	
a_{21}	a_{22}	$\cdots$	a_{2s}	$\cdots$	a_{2n}	b_2	
$\vdots$	$\vdots$		$\vdots$		$\vdots$	$\vdots$	
a_{r1}	a_{r2}	$\cdots$	$\boxed{a_{rs}}$	$\cdots$	a_{rn}	b_r	PR
$\vdots$	$\vdots$		$\vdots$		$\vdots$	$\vdots$	
a_{m1}	a_{m2}	$\cdots$	a_{ms}	$\cdots$	a_{mn}	b_m	

$$\text{PC}$$

This GJ pivot step *is only possible if the pivot element* $a_{rs} \neq 0$. This pivot step involves the following row operations:

- for each $i \neq r$ such that $a_{is} = 0$, leave Row i unchanged
- for each $i \neq r$ such that $a_{is} \neq 0$, carry out Row $i -(\frac{a_{is}}{a_{rs}})$ Row r
- for $i = r$ carry out (Row r)$/a_{rs}$

The entries in the tableau will change as a result of the pivot step, we denote the new entries by $\bar{a}_{ij}, \bar{b}_i$ etc. with a bar on top

Tableau after pivot step

x_1	x_2	$\cdots$	x_s	$\cdots$	x_n	RHS constants
$\bar{a}_{11}$	$\bar{a}_{12}$	$\cdots$	0	$\cdots$	$\bar{a}_{1n}$	$\bar{b}_1$
$\bar{a}_{21}$	$\bar{a}_{22}$	$\cdots$	0	$\cdots$	$\bar{a}_{2n}$	$\bar{b}_2$
$\vdots$	$\vdots$		$\vdots$		$\vdots$	$\vdots$
$\bar{a}_{r1}$	$\bar{a}_{r2}$	$\cdots$	1	$\cdots$	$\bar{a}_{rn}$	$\bar{b}_r$
$\vdots$	$\vdots$		$\vdots$		$\vdots$	$\vdots$
$\bar{a}_{m1}$	$\bar{a}_{m2}$	$\cdots$	0	$\cdots$	$\bar{a}_{mn}$	$\bar{b}_m$

Hence the GJ pivot step eliminates the variable corresponding to the pivot column from all rows other than the pivot row.

We now provide some numerical examples. Consider the tableau given below. We will carry out the GJ pivot step on this tableau with Row 2 as the pivot row, Column 3 as the pivot column, and the boxed entry 2 in both of them as the pivot element.

Original Tableau

x_1	x_2	x_3	x_4	x_5	RHS	
-1	2	1	1	0	4	
4	-2	$\boxed{2}$	0	-4	-10	PR
3	7	0	-4	8	-2	
-2	4	-1	1	0	-7	
		PC				

This GJ pivot step involves the following row operations:

Row 1 $-(\frac{1}{2})$ Row 2
Row 4 $+(\frac{1}{2})$ Row 2
$(\frac{1}{2})$ Row 2

No change in Row 3 because the entry in pivot column in Row 3 is already zero.

Tableau after the pivot step

x_1	x_2	x_3	x_4	x_5	RHS
-3	3	0	1	2	9
2	-1	1	0	-2	-5
3	7	0	-4	8	-2
0	3	0	1	-2	-12

Notice that the pivot column, column of x_3, became the 2nd unit column (with a "1" entry in the pivot row, Row 2) after the pivot step.

Here is another example:

Original Tableau

x_1	x_2	x_3	x_4	x_5	RHS	
1	0	-4	2	1	-7	
-1	2	2	3	-1	8	
1	-2	-1	-5	0	-12	
$\boxed{-2}$	3	4	0	1	4	PR
PC						

Tableau after the pivot step

x_1	x_2	x_3	x_4	x_5	RHS
0	$\frac{3}{2}$	-2	2	$\frac{3}{2}$	-5
0	$\frac{1}{2}$	0	3	$-\frac{3}{2}$	6
0	$-\frac{1}{2}$	1	-5	$\frac{1}{2}$	-10
1	$-\frac{3}{2}$	-2	0	$-\frac{1}{2}$	-2

Exercises

1.15.1: Perform the GJ pivot steps indicated by the pivot row, pivot column, and the pivot element in a box, in the following tableaus.

(i)

x_1	x_2	x_3	x_4	x_5		
1	-2	-2	2	3	2	
0	1	2	$\boxed{-1}$	1	-7	PR
-12	-8	19	0	17	18	
3	-2	-1	-2	2	-3	
			PC			

(ii)

x_1	x_2	x_3	x_4	x_5		
0	21	48	37	0	73	
2	-4	5	6	1	8	
-3	-1	-2	-5	-1	-7	
4	-4	2	6	$\boxed{2}$	20	PR
				PC		

(iii)

x_1	x_2	x_3	x_4	x_5	x_6		
1	0	0	-1	-2	3	4	
0	1	0	1	-1	2	5	
0	0	1	$\boxed{2}$	1	-1	2	PR
			PC				

The GJ Pivot Step as a Tool for Eliminating a Variable From a System

In Section 1.14 we talked about eliminating a variable from a system of linear equations using one of the equations in which it appears with a nonzero coefficient. We will show that this is exactly what a GJ pivot step does. In fact, the GJ pivot step offers a very convenient tool for carrying out this elimination step when the system is expressed in detached coefficient form. In the following table we show the correspondence between various elements in the two steps.

Following this table, we reproduce the system solved in Section 1.14 by the elimination method. There, we first eliminated the variable x_1 from it using the first equation in the system.

Element in Elimination step	Correponding element in GJ pivot step
1. Eq. used to eliminate a variable	**Pivot row.** Row corresponding to that eq. in tableau is the PR in GJ pivot step.
2. Variable eliminated	**Pivot column.** Column corresponding to that variable in tableau is the PC in GJ pivot step.
3. Coeff. of variable eliminated, in eq. used to eliminate it	**Pivot element.**

Original System = System 1

$$\begin{array}{rrrrcr} 2x_1 & +4x_2 & -6x_3 & & = & 10 \\ x_1 & -x_2 & +x_3 & -x_4 & = & 5 \\ -x_1 & +2x_2 & -2x_3 & +x_4 & = & 3 \end{array}$$

After this elimination step, we got the following:

$x_1 = -2x_2 + 3x_3 + 5$ 　　　　Expression for x_1 in terms of other variables.

$$\begin{array}{r} -3x_2 + 4x_3 - x_4 = 0 \\ 4x_2 - 5x_3 + x_4 = 8 \end{array}$$ 　Remaining system after x_1 eliminated using 1st eq.

The PR, PC, and the boxed pivot element for the GJ pivot step on the detached coefficient tableau for the original system, corresponding to this elimination step are indicated below:

Original Tableau

x_1	x_2	x_3	x_4	RHS	
$\boxed{2}$	4	-6	0	10	PR
1	-1	1	-1	5	
-1	2	-2	1	3	
PC					

Carrying out this GJ pivot step leads to the following tableau

Tableau after pivot step

x_1	x_2	x_3	x_4	RHS
1	2	-3	0	5
0	-3	4	-1	0
0	4	-5	1	8

In this tableau, the equation corresponding to Row 1 is:

$$x_1 + 2x_2 - 3x_3 = 5$$

We get the expression for x_1 in terms of the other variables by transferring the terms of those variables to the right in this equation. Also, rows 2 and 3 in this tableau contain the reduced system after x_1 is eliminated from the original system.

Thus we see that the GJ pivot step eliminates the variable corresponding to the pivot column from all the equations other than that represented by the pivot row. Hence the GJ pivot step offers a very convenient computational tool for carrying out elimination steps.

In Section 1.16, we will formally discuss a method for solving systems of linear equations based on the elimination approach, using the GJ pivot step as the main computational tool.

Note: To carry out a sequence of pivot steps on a tableau, they are performed one after the other in sequence. Each pivot step leads to a new tableau, on which the next pivot step is performed. Usually the various tableaus obtained are listed one below the other in the order in which they are generated.

1.16 The Gauss–Jordan (GJ) Pivotal Method for Solving Linear Equations

We will now discuss the method for solving linear equations using the elimination approach based on GJ pivot steps. This is one of the most commonly used methods for solving linear equations. It is known as the **GJ (Gauss–Jordan) pivotal** (or **elimination**) **method** and also known as the **Row reduction algorithm** or **Row reduction method**.

Historical note on the GJ pivotal method: The essential ideas behind the method were figured out by Chinese and Indian mathematicians more than 2000 years ago, but they were unknown in Europe until the 19th century when the famous German mathematician Carl Friedrich Gauss discovered them. Gauss discovered the method while studying linear equations for approximating the orbits of celestial bodies, in particular the orbit of the asteroid Ceres (whose diameter is less than 1000 km). The positions of these bodies are known from observations which are subject to errors. Gauss was trying to approximate the values of the unknown coefficients in an algebraic formula for the orbit, that fit the observations as closely as possible, using the method of least squares described in Example 5 of Section 1.2. This method gives rise to a system of linear equations in the unknown coefficients. He had to solve a system of 17 linear equations in 17 unknowns, which he did using the elimination approach. His computations were so accurate that astronomers who lost track of the the asteroid Ceres, were able to locate it again easily.

Another German, Wilhelm Jordan, popularized the algorithm in a late 19th century book he wrote. From that time, the method has been popularly known as the Gauss–Jordan elimination method. ⋈

The method involves a series of GJ pivot steps on the detached coefficient tableau form. To solve a system of m equations in n variables, the method needs at most m (actually $\min\{m, n\}$, but as shown later, we can assume without loss of generality that $m \leq n$) GJ pivot steps.

At each pivot step, the variable corresponding to the pivot column is recorded as the basic variable selected in the pivot row, and is eliminated from all equations other than that represented by the pivot row (this is the property of the GJ pivot step). The equations corresponding to the expressions of each basic variable in terms of the other variables are kept in the tableau itself.

We first give the statement of the method, and then give illustrative examples for all the possibilities that can occur in the method. In the description of each step, instead of just giving a curt statement of the work to be done in that step, we also give some explanation of the reasons for it to help the reader understand it easily.

If you need evidence for each redundant constraint identified and eliminated during the method, and evidence for infeasibility when the method terminates with the infeasibility conclusion, you need to introduce the memory matrix into the computation as pointed out in Section 1.13. We omit those to keep the discussion simple. Also, the version of the method presented below is the simplest version from ancient times. A computationally more efficient version of the same method using the memory matrix is presented in Section 4.11 (it is based on computational improvements introduced by George Dantzig in his paper on the revised simplex method for linear programming). That version has the additional advantage of automatically producing the "evidences" without increasing the computational burden.

The GJ Pivot Method

BEGIN

Step 1: Put the system in detached coefficient tableau form and obtain the original tableau. Open a column on the left hand side to record the basic variables. Select the top row as the pivot row in the original tableau and go to Step 2.

Step 2: General Step: Let r denote the number of the present pivot row in the present tableau.

If the coefficients of all the variables in the pivot row are 0, go to Step 2.1 if the RHS constant in the pivot row, $\bar{b}_r = 0$, or to Step 2.2 if $\bar{b}_r \neq 0$. If at least one of the variables has a nonzero coefficient in the pivot row, go to Step 2.3.

> **Step 2.1: Redundant Constraint Elimination:** In the present tableau, the pivot row represents the equation "0 = 0", a redundant constraint. Eliminate the corresponding row (i.e., the one with the same number from the top) in the original tableau as a redundant constraint. If this pivot row is the last row in the present tableau, eliminate it from the present tableau and go to Step 3 with the resulting tableau. Otherwise, eliminate this row from the present tableau, in the resulting tableau select the next row as the pivot row and with these repeat Step 2. (Equivalently, you can leave these redundant rows in both the tableaus, but ignore them in future work.)

Step 2.2: Concluding Original System Inconsistent: In the present tableau, the pivot row represents the equation "$0 = \bar{b}_r$" where $\bar{b}_r \neq 0$, an inconsistent equation. So the original system is inconsistent and has no solution. Terminate.

Step 2.3: GJ Pivot Step: Select a variable with a nonzero coefficient in the pivot row in the present tableau as the basic variable in the pivot row (i.e., the rth basic variable) and record it. This variable is also called the **entering variable** for this pivot step. The column vector of this basic variable in the present tableau is the pivot column. Perform the GJ pivot step on the present tableau, and let the resulting tableau be the next tableau.

If row r is the last row in this next tableau, go to Step 3 with the next tableau. If row r is not the last row in the next tableau, with the next row in it as the pivot row repeat Step 2.

Step 3: Checking Uniqueness of Solution, and Writing the General Solution:

If every variable in the system has been selected as a basic variable in some row (i.e., there are no nonbasic variables at this stage), all the column vectors in the present tableau are unit vectors. So, the number of rows in the present tableau is n (n may be less than $m =$ number of equations in original system, if some redundant equations were eliminated during the method). In this case, the column vector of the basic variable in the ith row is the ith unit vector in R^n for $i = 1$ to n. So, each row vector in the present tableau represents the equation giving the value of a different variable in the solution to the system. Collect these values and put them in a column vector in the proper order of the variables to yield the solution. This solution is the unique solution of the system in this case.

If there are some nonbasic variables in the present tableau, the original system has an infinite number of solutions. In the equation corresponding to the ith row, if you transfer all the terms containing nonbasic variables to the right, it gives the expression for the ith basic variable in terms of the nonbasic variables. Collecting these expressions for all the basic variables together, we have a parametric representation for the general solution of the system as discussed in Section 1.8. The values of the nonbasic variables are the parameters in it, these can be given arbitrary real values to obtain a solution for the system, and all the solutions for the system can be obtained this way. In particular, making all the nonbasic variables equal to 0, leads to the basic solution of the system WRT present basic vector.

END

Numerical Examples for the GJ Method

We now provide numerical examples illustrating the three possible ways in which the method can terminate. In this method, on each tableau one GJ step is performed yielding the next tableau on which the next GJ step is performed, and so on. We provide the various tableaus one below the other indicating the pivot row (PR), pivot column (PC), and the boxed pivot element if a GJ pivot step is performed on it; or marking it as RC (redundant constraint to be eliminated), or IE (inconsistent equation) otherwise. We use the abbreviation "BV" for "basic variable".

We number all the rows (equations) serially from top to bottom, and record these numbers in a rightmost column, to make it easy to keep track of redundant constraints eliminated.

The original detached coefficient tableau for the original system is always the first one.

Example 1:

	x_1	x_2	x_3	x_4	x_5	RHS		Eq. no.
	-1	0	$\boxed{1}$	1	2	10	PR	1
	1	-1	2	-1	1	5		2
	1	-2	5	-1	4	20		3
	2	0	0	-2	1	25		4
	3	-2	5	-3	5	45		5
	5	-2	5	-5	6	70		6
BV			PC↑					
x_3	-1	0	1	1	2	10		1
	3	$\boxed{-1}$	0	-3	-3	-15	PR	2
	6	-2	0	-6	-6	-30		3
	2	0	0	-2	1	25		4
	8	-2	0	-8	-5	-5		5
	10	-2	0	-10	-4	20		6
		PC↑						
x_3	-1	0	1	1	2	10		1
x_2	-3	1	0	3	3	15		2
	0	0	0	0	0	0	RC	3
	2	0	0	-2	$\boxed{1}$	25	PR	4
	2	0	0	-2	1	25		5
	4	0	0	-4	2	50		6
					PC			

BV	x_1	x_2	x_3	x_4	x_5	RHS		Eq. no.
x_3	-5	0	1	5	0	-40		1
x_2	-9	1	0	9	0	-60		2
x_5	2	0	0	-2	1	25		4
	0	0	0	0	0	0	RC	5
	0	0	0	0	0	0	RC	6

Final Tableau

BV	x_1	x_2	x_3	x_4	x_5	RHS	Eq. no.
x_3	-5	0	1	5	0	-40	1
x_2	-9	1	0	9	0	-60	2
x_5	2	0	0	-2	1	25	4

So, the constraints eliminated as redundant constraints are equations 3, 5, 6 in the original tableau. When we eliminate the corresponding rows from the original tableau, we get the following detached coefficient tableau form with the redundant constraints deleted.

Original Tableau With Redundant Eqs. Deleted

x_1	x_2	x_3	x_4	x_5	RHS	Eq. no.
-1	0	1	1	2	10	1
1	-1	2	-1	1	5	2
2	0	0	-2	1	25	4

Since we have x_1 and x_4 as nonbasic variables in the final tableau, this system has an infinite number of solutions. The set of solutions has dimension 2. From the 1st, 2nd, 3rd rows in the final tableau, we get the expressions for the 1st, 2nd, 3rd basic variables x_3, x_2, x_5 respectively in terms of the nonbasic variables, leading to the parametric representation of the general solution of the system as:

$$\begin{pmatrix} x_1 \\ x_2 \\ x_3 \\ x_4 \\ x_5 \end{pmatrix} = \begin{pmatrix} \bar{x}_1 \\ -60 + 9\bar{x}_1 - 9\bar{x}_4 \\ -40 + 5\bar{x}_1 - 5\bar{x}_4 \\ \bar{x}_4 \\ 25 - 2\bar{x}_1 + 2\bar{x}_4 \end{pmatrix}$$

where $\bar{x}_1, \bar{x}_4$ are arbitrary real valued parameters. For example, making $\bar{x}_1, \bar{x}_4 = 0$ leads to the basic solution of the system WRT the basic vector (x_3, x_2, x_5) as $(x_1, x_2, x_3, x_4, x_5)^T = (0, -60, -40, 0, 25)^T$.

Notice also that the original system had six equations in five variables, but after eliminating three redundant equtions one at a time, the final tableau was left with only three equations. In the final tableau, the column vectors of the 1st, 2nd, 3rd basic variables x_3, x_2, x_5 are the 1st, 2nd, 3rd unit vectors in R^3 in that order.

Example 2:

	x_1	x_2	x_3	RHS		Eq. no.
	1	1	$\boxed{1}$	5	PR	1
	−1	3	1	6		2
	0	4	2	11		3
	2	3	1	10		4
BV			PC↑			
x_3	1	1	1	5		1
	−2	$\boxed{2}$	0	1	PR	2
	−2	2	0	1		3
	1	2	0	5		4
	PC↑					
x_3	2	0	1	$\frac{9}{2}$		1
x_2	−1	1	0	$\frac{1}{2}$		2
	0	0	0	0	RC	3
	$\boxed{3}$	0	0	4	PR	4
	PC↑					
Final Tableau						
x_3	0	0	1	$\frac{19}{6}$		1
x_2	0	1	0	$\frac{11}{6}$		2
x_1	1	0	0	$\frac{4}{3}$		4

So, the third constraint is eliminated as a redundant constraint. When we eliminate the corresponding row from the original tableau, we get the following detached coefficient tableau form with the redundant constraints deleted.

x_1	x_2	x_3	RHS	Eq. no.
1	1	1	5	1
−1	3	1	6	2
2	3	1	10	4

All the variables are basic variables, and there are no nonbasic variables. Hence this system has a unique solution which we read from the final tableau to be:

$$\begin{pmatrix} x_1 \\ x_2 \\ x_3 \end{pmatrix} = \begin{pmatrix} \frac{4}{3} \\ \frac{11}{6} \\ \frac{19}{6} \end{pmatrix}.$$

Example 3:

	x_1	x_2	x_3	x_4	x_5	RHS		Eq. no.
	2	1	−1	−2	0	4	PR	1
	4	1	5	6	4	9		2
	6	2	4	4	4	12		3
	10	3	9	10	8	22		4
BV		PC↑						
x_2	2	1	−1	−2	0	4		1
	2	0	6	8	4	5	PR	2
	2	0	6	8	4	4		3
	4	0	12	16	8	10		4
	PC							

Final Tableau — Inconsistency discovered

BV	x_1	x_2	x_3	x_4	x_5	RHS		Eq. no.
x_2	0	1	−7	−10	−4	−1		1
x_1	1	0	3	4	2	$\frac{5}{2}$		2
	0	0	0	0	0	−1	IE	3
	0	0	0	0	0	0		4

Since the 3rd row in the final tableau leads to the inconsistent equation "$0 = -1$", we terminate with the conclusion that the original system has no solution.

Optimal Selection of Pivot Columns and Rows in the GJ Method

Whenever Step 2.3 is carried out, we need to select one of the columns with a nonzero coefficient in the pivot row in the present tableau as the pivot column. How is the choice made?

In the GJ method we perform exactly one pivot step in each row of the tableau. In the description of the method given above, we have selected the rows of the tableaus as pivot rows in the order 1 to n; i.e., if Row r is the pivot row in the present tableau, we are taking Row $r + 1$ (or the first row different from the "$0 = 0$" equation below r) as the pivot row in the next tableau. Is this the best order or is there any advantage to be gained by using a different order?

We will provide brief answers to these questions now. But the answer depends on whether the situation under consideration is a small problem being solved by hand, or a large problem being solved on a digital computer. First, we review the main differences between these two situations.

When the pivot element is different from 1, division by the pivot element may generate fractions, and the calculations involved in the pivot step get very tedious to do manually. So, when all the data in the present tableau is integer, and the system is being solved by hand, we try to select the pivot column and row so that the pivot element is ± 1 as often as possible, or at least an integer with the smallest possible absolute value so that the fractions generated in the pivot step will not be too tedious. This is what you should do in solving the exercises given in this book (i.e., for the next pivot step select the pivot row among the rows in which a pivot step has not been carried out so far, and the pivot column among the columns in which a pivot step has not been carried out so far, so that the pivot element is ± 1 as often as possible, or an integer with the smallest absolute value).

Of course real world problems are always solved using digital computers. When writing a computer program for the GJ elimination method, the tediousness of the fractions generated is no longer an important consideration. Other highly critical factors become extremely important. Digital computation is not exact, as it involves finite precision arithmetic. For example, in computing the ratio $\frac{1}{3}$, digital computation would choose the answer to be 0.33 or 0.333 etc. depending on the number of digits of accuracy maintained. This introduces **round-off errors** which keep on accumulating as the method progresses, and it may cause the computed solution to be too far away from the unknown actual solution. In writing a computer program for the GJ method, the pivot row, pivot column choice is handled by very special rules to keep the accumulation of round-off errors in check as far as possible. Keeping round-off errors in check is the focus of the subject **Numerical Analysis**, which is outside the scope of this book; but we will give a summary of the most commonly used strategies called **full** or **complete pivoting** that work well.

Full Pivoting: For the next pivot step in the present tableau, select the pivot element to be an element of maximum absolute value among all the elements in the present tableau contained in rows in which a pivot step has not been carried out so far, and in columns in which a pivot step has not been carried out so far; breaking ties arbitrarily. The pivot row, pivot column for this pivot step are the row, column respectively of the pivot element.

In some applications, the GJ method is implemented in a way that in each pivot step, the pivot column is chosen first by some extraneous rule, and then the pivot row has to be selected among the rows in which a pivot step has not been carried out so far. In such implementations, the strategy that is commonly used is to find a maximum absolute value element in the present tableau among the entries in the pivot column contained in rows in which a pivot step has not been carried out so far, and select its row as the pivot row. This strategy is called **partial pivoting**.

Round-off error accumulation is a serious problem in the GJ method that limits its applicability to solve the very large scale systems that arise sometimes in real world applications. The Gaussian elimination method to be discussed in Section 1.20, implemented with other highly sophisticated numerical analysis techniques, is able to alleviate the round-off error problem somewhat better in such applications.

Obtaining the "0 = 1" and "0 = 0" Equations Encountered in GJ Method As Linear Combinations of Original Equations

In the description of the GJ method given above, whenever we discovered the "$0 = \bar{b}_r$" equations, we did not get the "evidence", i.e., the vector of coefficients in the expression of this equation as a linear combination of equations in the original system. In some applications it may be necessary to get these expressions. For this, the memory matrix should be introduced in the original tableau in Step 1 before beginning any work on the GJ method, as indicated in Section 1.12, 1.13.

Discussion on Some Features of the GJ Method

Here we summarize the main features of the GJ method. Some of these have already been mentioned above.

1. When applied on a system of m equations in n variables, the method performs at most one GJ pivot step in each row for a total of at most m pivot steps.

2. Whenever a "$0 = 0$" equation is encountered in the present tableau, a linear dependence relation among the equations in the original system has been identified. The row containing that equation corresponds to a redundant equation, and is eliminated from the present tableau and the original tableau.

   Thus the method identifies and eliminates redundant constraints one by one.

3. Whenever a "$0 = \bar{b}_r$" (for some $\bar{b}_r \neq 0$) equation is encountered in the present tableau, we conclude that the original system is inconsistent, i.e., it has no solution.

   If no inconsistency is encountered during the method (as evidenced by a "$0 = \bar{b}_r$" equation for some $\bar{b}_r \neq 0$) the method terminates with a solution for the system. This proves Result 1.10.1 discussed in Section 1.10.

4. Suppose the original system of equations is consistent, and there are p rows remaining in the final tableau. The method selects a basic variable for each of these rows, The basic variable in the ith row is called the ith **basic variable**, its column vector in the final tableau is the ith **unit vector** in R^p.

   Rearrange the variables and their column vectors in the final tableau so that the basic variables appear first in the tableau in their proper order, followed by nonbasic variables if any (in any order). Also rewrite the original tableau with the variables arranged in the same order.

   For instance, when we do this rearrangement for the systems in Examples 2, 1 above, we get the following:

Final tableau in Example 2 after
rearranging variables

Basic var.	x_3	x_2	x_1	RHS
x_3	1	0	0	$\frac{19}{6}$
x_2	0	1	0	$\frac{11}{6}$
x_1	0	0	1	$\frac{4}{3}$

Remaining Original
tableau for Example 2

x_3	x_2	x_1	RHS
1	1	1	5
1	3	-1	6
1	3	2	10

Final tableau in Example 1 after rearranging
variables

Basic var.	x_3	x_2	x_5	x_1	x_4	RHS
x_3	1	0	0	-5	5	-40
x_2	0	1	0	-9	9	-60
x_5	0	0	1	2	-2	25

Remaining original tableau for
Example 1

x_3	x_2	x_5	x_1	x_4	RHS
1	0	2	-1	1	10
2	-1	1	1	-1	5
0	0	1	2	-2	25

In the same way, for a general system, after rearrangement of variables, the final tableau will be of the following forms. We assume that the basic variables are $x_{i_1}, \ldots, x_{i_p}$ in that order and the remaining nonbasic variables are, if any, $x_{j_1}, \ldots, x_{j_{n-p}}$.

Then, the vector of variables $(x_{i_1}, \ldots, x_{i_p})$ is called the **basic vector of variables** selected by the method for the system. There may be several basic vectors of variables for the system, the one obtained in the method depends on the choice of pivot columns selected in Steps 2.3 during the method.

When no nonbasic variables left

Final tableau after rearrangement
of variables

Basic Vars.	x_{i_1}	x_{i_2}	$\cdots$	x_{i_p}	
x_{i_1}	1	0	$\cdots$	0	$\bar{b}_1$
x_{i_2}	0	1	$\cdots$	0	$\bar{b}_2$
$\vdots$	$\vdots$	$\vdots$	$\vdots$	$\vdots$	$\vdots$
x_{i_p}	0	0	$\cdots$	1	$\bar{b}_p$

When nonbasic variables exist

Final tableau after rearrangement of variables

Basic Vars.	x_{i_1}	x_{i_2}	...	x_{i_p}	x_{j_1}	...	$x_{j_{n-p}}$	
x_{i_1}	1	0	...	0	$\bar{a}_{1,j_1}$	...	$\bar{a}_{1,j_{n-p}}$	$\bar{b}_1$
x_{i_2}	0	1	...	0	$\bar{a}_{2,j_1}$	...	$\bar{a}_{2,j_{n-p}}$	$\bar{b}_2$
$\vdots$	$\vdots$	$\vdots$		$\vdots$	$\vdots$		$\vdots$	$\vdots$
x_{i_p}	0	0	...	1	$\bar{a}_{p,j_1}$	...	$\bar{a}_{p,j_{n-p}}$	$\bar{b}_p$

In both cases, since each row in the final tableau has a basic variable which only appears in that row and none of the other rows, the only linear combination of the constraints in the final tableau that leads to the "0 = 0" equation is the one in which the coefficients of all the constraints are zero. Hence the system of constraints in the final tableau in both cases is linearly independent. Since the system of constraints in the remaining original tableau (after the elimination of redundant constraints) is equivalent to the system in the final tableau, it is linearly independent as well.

In the first case, (which occurs when all the variables are basic, and there are no nonbasic variables) the final tableau consists of the unit columns arranged in their proper order, and the unique solution of the system is:

$$(x_{i_1}, \ldots, x_{i_p})^T = (\bar{b}_{i_1}, \ldots, \bar{b}_{i_p})^T$$

as can be read from the equations corresponding to the rows in the final tableau. The solution set consists of a single point and we say that its dimension is zero. In this case the final system and the remaining original system at the end (i.e., after the elimination of redundant constraints one by one until the final system has no more redundant constraints) have the same number of equations as variables, which are therefore called **square systems of linear equations**. Also, since they are linearly independent, they are called **nonsingular square systems of linear equations**.

In the second case, (which occurs when there is at least one nonbasic variable in the final tableau) there are an infinite number of solutions to the system. By writing the equations corresponding to the rows in the final tableau, and then transferring all the terms involving the nonbasic variables to the right in each equation, we get the general solution to our system as:

$$
\begin{pmatrix} x_{i_1} \\ x_{i_2} \\ \vdots \\ x_{i_p} \\ x_{j_1} \\ \vdots \\ x_{j_{n-p}} \end{pmatrix} = \begin{pmatrix} \bar{b}_1 - \bar{a}_{1,j_1}\bar{x}_{j_1} - \cdots - \bar{a}_{1,j_{n-p}}\bar{x}_{j_{n-p}} \\ \bar{b}_2 - \bar{a}_{2,j_1}\bar{x}_{j_1} - \cdots - \bar{a}_{2,j_{n-p}}\bar{x}_{j_{n-p}} \\ \vdots \\ \bar{b}_p - \bar{a}_{p,j_1}\bar{x}_{j_1} - \cdots - \bar{a}_{p,j_{n-p}}\bar{x}_{j_{n-p}} \\ \bar{x}_{j_1} \\ \vdots \\ \bar{x}_{j_{n-p}} \end{pmatrix}
$$

where $\bar{x}_{j_1}, \ldots, \bar{x}_{j_{n-p}}$, the values given to the nonbasic (independent) variables are arbitrary real parameters. In this representation for the general solution of the system, the variables may not be in the usual order of x_1 to x_n, but the variables can be reordered if desired. Except for the order of the variables, this representation is exactly the same parametric representation for the general solution as discussed in Section 1.8. By giving these parameters $\bar{x}_{j_1}, \ldots, \bar{x}_{j_{n-p}}$ appropriate real values in this representation, all possible solutions of the system can be generated. That's why in this case the **dimension of the set of solutions** is defined to be equal to the number of nonbasic (i.e., independent) variables in the final tableau. Also in this case, the number of equations in the final system is strictly less than the number of variables. Hence the final system and the remaining original system are **rectangular systems** with more variables than equations.

A Simple Data Structure

The phrase **data structure** refers to a strategy for storing information generated during the algorithm, and using it to improve its efficiency. The idea of recording, after a pivot step is completed, the variable corresponding to the pivot column as the basic variable in the pivot row, is a very simple data structure introduced by G. Dantzig in his pioneering 1947 paper on the simplex method for linear programming. It helps to keep track of the unit column vectors created by the pivot steps carried out in the algorithm, and makes it easier to directly read a solution of the system from the final tableau as described in Step 3.

When the system is consistent, the final tableau will have a basic variable associated with each remaining row in it; and the column associated with the rth basic variable will be the rth unit vector in the final tableau for all r. Dantzig used the term **canonical tableau** for the final tableau with a full set of unit vectors in it.

In describing the GJ elimination method, mathematics books on linear algebra usually state that in each step the pivot column is selected as the leftmost of the eligible columns. When the system is consistent and the method is carried out this way, the final tableau will have a full set of unit vectors occuring in their proper order all at the left end of the tableau, without the need for rearranging the variables and column vectors as mentioned above. The final tableau with this property is said to be in **RREF (reduced row echelon form)** in mathematical linear algebra literature. So RREF refers to a specific type of canonical tableau with all the unit column vectors appearing in their proper order at the left end of the tableau.

In fact it is not important that all the unit vectors in the final tableau appear at the left end, or in their proper order. Allowing the method to generate any canonical tableau (and not just the RREF) introduces the possibility of pivot column choice without having to choose it always as the leftmost of the eligible.

Basic Solution

In linear programming literature the final tableau obtained here (whether the column vectors and the variables are rearranged as discussed above or not) is commonly known as the **canonical tableau for the original system WRT the basic vector** $(x_{i_1}, \ldots, x_{i_p})$.

In the general solution for the system obtained above, if we fix all the nonbasic variables at zero, we get the solution

$$
\begin{pmatrix} x_{i_1} \\ x_{i_2} \\ \vdots \\ x_{i_p} \\ x_{j_1} \\ \vdots \\ x_{j_{n-p}} \end{pmatrix} = \begin{pmatrix} \bar{b}_1 \\ \bar{b}_2 \\ \vdots \\ \bar{b}_p \\ 0 \\ \vdots \\ 0 \end{pmatrix}
$$

known as the **basic solution for the system WRT the basic vector** $(x_{i_1}, \ldots, x_{i_p})$.

When we put the column vectors of the basic variables $x_{i_1}, \ldots, x_{i_p}$ in the remaining original tableau after all the redundant constraints are deleted, side by side in that order, we get a $p \times p$ square array of numbers which is known as the **basis matrix** or **basis** for the system corresponding to the basic vector $(x_{i_1}, \ldots, x_{i_p})$. The general concept of a matrix as a two-dimensional array of numbers is given later in Chapter 2.

There is another worked out example after this discussion.

5. One consequence of the results in 4 is that any consistent system of linear equations is either directly itself or is equivalent to another in which the number of equations is $\leq$ the number of variables. If a consistent system has more equations than variables, then some of them can be eliminated as redundant equations leading to an equivalent system in which the above condition is satisfied. Therefore, when discussing a consistent system of equations, we can always, without loss of generality, assume that the number of equations is $\leq$ the number of variables.

6. On a system of m equations in n variables, in the worst case all the $m(n + 1)$ entries may change in a pivot step, requiring up to $3m(n + 1)$ multiplication, division, or addition operations. Since the method may take up to m pivot steps on this system, it may require at most $3m^2(n+1)$ arithmetical operations.

The **computational complexity** of a method is a measure of the worst case computational effort needed by the method as a function of the size of the system as measured by the number of data elements, or the total number of binary digits of storage space needed to store all the data, etc. In giving the computational complexity, constant terms like 3 in $3m^2(n + 1)$ are usually omitted, and then we say that the computational complexity is of **Order** m^2n and write it as: $O(m^2n)$. It provides the order of an upper bound on the computational effort needed by the method in terms of the size of the problem. So, we can say that the computational complexity of the GJ method to solve a system of m linear equations in n variables is $O(m^2n)$.

Example 4:

We apply the GJ method to solve the following system of 5 linear equations in 5 variables. The detached coefficient tableau for the original system is the one at the top. PR (pivot row), PC (pivot column), RC (redundant constraint to be eliminated), BV (basic variable selected in this row) have the same meanings as before, and in each tableau the pivot element for the pivot step carried out in that tableau is boxed.

	x_1	x_2	x_3	x_4	x_5	RHS		Eq. no.
	0	1	[1]	−1	−2	3	PR	1
	1	0	1	1	−1	9		2
	1	1	0	−1	1	15		3
	1	1	2	0	−3	12		4
	2	2	2	−1	−2	27		5
			PC↑					
BV	x_1	x_2	x_3	x_4	x_5	RHS		Eq. no.
x_3	0	1	1	−1	−2	3		1
	[1]	−1	0	2	1	6	PR	2
	1	1	0	−1	1	15		3
	1	−1	0	2	2	6		4
	2	0	0	1	2	21		5
	PC↑							
x_3	0	1	1	−1	−2	3		1
x_1	1	−1	0	2	1	6		2
	0	[2]	0	−3	0	9	PR	3
	0	0	0	0	0	0		4
	0	2	0	−3	0	9		5
		PC ↑						
x_3	0	0	1	$\frac{1}{2}$	−2	$-\frac{3}{2}$		1
x_1	1	0	0	$\frac{1}{2}$	1	$\frac{21}{2}$		2
x_2	0	1	0	$-\frac{3}{2}$	0	$\frac{9}{2}$		3
	0	0	0	0	0	0	RC	4
	0	0	0	0	0	0	RC	5

Final tableau

	x_1	x_2	x_3	x_4	x_5	RHS	Eq. no.
BV	x_1	x_2	x_3	x_4	x_5	RHS	Eq. no.
x_3	0	0	1	$\frac{1}{2}$	−2	$-\frac{3}{2}$	1
x_1	1	0	0	$\frac{1}{2}$	1	$\frac{21}{2}$	2
x_2	0	1	0	$-\frac{3}{2}$	0	$\frac{9}{2}$	3

So, equations 4 and 5 are eliminated as redundant constraints. The remaining system of original constraints, after these redundant equations are deleted, is:

Remaining Original system after deletion of redundant eqs.

x_1	x_2	x_3	x_4	x_5	RHS	Eq. no.
0	1	1	−1	−2	3	1
1	0	1	1	−1	9	2
1	1	0	−1	1	15	3

The basic vector of variables selected is (x_3, x_1, x_2). The basis matrix corresponding to this basic vector is defined to be the square two dimensional array of numbers obtained by putting the column vectors of the basic variables x_3, x_1, x_2 in the remaining original system after the deletion of redundant constraints, in that order, which is:

$$B = \begin{pmatrix} 1 & 0 & 1 \\ 1 & 1 & 0 \\ 0 & 1 & 1 \end{pmatrix}.$$

The general concept of a matrix is discussed in Chapter 2. From the final tableau, we find that the general solution of the system is:

$$\begin{pmatrix} x_1 \\ x_2 \\ x_3 \\ x_4 \\ x_5 \end{pmatrix} = \begin{pmatrix} \frac{21}{2} - \frac{1}{2}\bar{x}_4 - \bar{x}_5 \\ \frac{9}{2} + \frac{3}{2}\bar{x}_4 \\ -\frac{3}{2} - \frac{1}{2}\bar{x}_4 + 2\bar{x}_5 \\ \bar{x}_4 \\ \bar{x}_5 \end{pmatrix}$$

where $\bar{x}_4, \bar{x}_5$ are real valued parameters representing the values of the free (nonbasic) variables x_4, x_5 in the solution. Making $\bar{x}_4 = 0, \bar{x}_5 = 0$ gives the basic solution of the system associated with the basic vector of variables

(x_3, x_1, x_2), which is:

$$
\begin{pmatrix} x_1 \\ x_2 \\ x_3 \\ x_4 \\ x_5 \end{pmatrix} = \begin{pmatrix} \frac{21}{2} \\ \frac{9}{2} \\ -\frac{3}{2} \\ 0 \\ 0 \end{pmatrix}.
$$

By selecting different basic variables in the various pivot steps, we can get different basic vectors of variables for the system.

Since there are two nonbasic variables in the final tableau in this example, the dimension of the set of solutions in this example is 2.

Exercises

In solving these exercises by hand you may want to use pivot elements of ± 1 as far as possible to avoid generating fractions that can make the work messy.

1.16.1: Systems with a unique solution: solve the following systems by the GJ method and verify that the solution obtained is the unique solution of the system.

(i)	x_1	x_2	x_3	x_4	
	-1	-2	0	0	1
	2	5	2	2	-3
	0	1	3	0	-4
	1	0	-1	1	3
	2	-2	1	-1	2

(ii)	x_1	x_2	x_3	
	3	1	4	5
	-4	-1	-6	-4
	-2	-1	0	-10

(iii)

x_1	x_2	x_3	x_4	
1	2	3	2	5
-1	-3	-5	-1	-4
1	0	4	1	11
1	2	-2	-2	-13

(iv)

x_1	x_2	x_3	
1	1	1	5
-1	3	1	6
0	4	2	11
2	3	1	10

1.16.2: Systems with multiple solutions, but no redundant constraints: Solve the following systems by the GJ method. For each obtain an expression for the general solution of the system, and find the dimension of the set of all solutions. From the general solution check whether you can obtain a nonnegative solution for the system by trying to generate nonnegative values to the nonbasic variables that will make all basic variables nonnegative using simple trial and error.

$$(i) \quad -x_2 + 3x_3 + 8x_4 - 7x_6 = -3$$
$$7x_1 + x_2 - 2x_3 - 10x_4 + 2x_5 - 3x_6 = -4$$
$$10x_1 + 2x_3 + 12x_4 + 5x_5 - 8x_6 = 2$$

(ii)

x_1	x_2	x_3	x_4	x_5	x_6	
1	1	0	-1	-1	1	-3
-1	0	1	2	-1	2	-7
0	3	2	3	-5	1	-10
1	1	0	2	-1	-2	8

(iii)

x_1	x_2	x_3	x_4	x_5	x_6	x_7	
2	-1	0	1	2	-3	-2	1
-3	1	-1	0	1	-2	-1	-4
2	0	3	-3	-1	0	2	7
0	2	1	-1	0	2	1	7

(iv)

x_1	x_2	x_3	x_4	x_5	x_6	x_7	
1	-2	0	1	0	-1	6	4
1	-1	2	-2	1	8	-17	-1
0	2	5	1	-5	3	-4	3

1.16.3: Systems with multiple solutions and redundant constraints: Solve the following systems by the GJ method, and obtain an expression for the general solution, and the dimension of the solution set.

(i)

x_1	x_2	x_3	x_4	x_5	x_6	
-1	0	1	-1	-1	2	-1
0	-1	1	0	1	1	2
-2	-3	5	-2	1	7	4
1	1	-3	2	0	-2	-2
-2	-3	4	-1	1	8	3
2	2	-4	1	2	-1	-4

(ii)

x_1	x_2	x_3	x_4	x_5	x_6	x_7	
2	0	3	0	4	5	-1	-10
0	3	-2	5	-3	2	1	10
4	6	2	10	2	14	0	0
3	-3	2	1	-1	-1	-1	-7
9	6	5	16	2	20	-1	-7
5	0	3	6	0	6	-1	-7

(iii)

x_1	x_2	x_3	x_4	x_5	
-2	7	5	1	4	38
-3	2	-2	2	9	84
-8	11	1	5	22	206
-7	16	8	4	17	160
-5	9	3	3	13	122

(iv)

x_{11}	x_{12}	x_{13}	x_{21}	x_{22}	x_{23}	
1	1	1	0	0	0	800
0	0	0	1	1	1	300
1	0	0	1	0	0	400
0	1	0	0	1	0	200
0	0	1	0	0	1	500

(v)

x_1	x_2	x_3	x_4	x_5	
-1	0	1	1	2	10
1	-1	2	-1	1	5
1	-2	5	-1	4	20
2	0	0	-2	1	25
3	-2	5	-3	5	45
5	-2	5	-5	6	70

1.16.4: Infeasible systems: solve the following systems by the GJ method.

(i)

x_1	x_2	x_3	x_4	x_5	x_6	
1	2	-1	0	3	1	3
-1	-3	2	2	-2	-3	4
0	-1	1	2	1	-2	7
-1	-4	3	4	-1	-5	9
1	1	0	2	4	-1	7

(ii)

x_{11}	x_{12}	x_{13}	x_{21}	x_{22}	x_{23}	
1	1	1	0	0	0	500
0	0	0	1	1	1	1000
1	0	0	1	0	0	600
0	1	0	0	1	0	500
0	0	1	0	0	1	500

(iii)

x_1	x_2	x_3	x_4	x_5	
1	1	0	0	0	3
0	1	1	0	0	1
0	0	1	1	0	1
0	0	0	1	1	4
1	0	0	0	1	5
1	1	1	1	1	7
1	2	1	1	1	9

(iv) $\quad -5x_1 - 2x_2 + 10x_3 + x_4 + 7x_5 + 2x_6 - 4x_7 = 2$

$\qquad -3x_1 + 3x_2 - 2x_3 - x_4 + 2x_5 + x_6 + 3x_7 = 3$

$\qquad -8x_1 + x_2 + 8x_3 + 9x_5 + 3x_6 - x_7 = 6$

$\qquad x_1 + 2x_3 - 3x_5 - 4x_6 + 2x_7 = 5$

$\qquad -7x_1 + x_2 + 10x_3 + 6x_5 - x_6 + x_7 = 8$

(v)

x_1	x_2	x_3	x_4	x_5	RHS
2	1	-1	-2	0	4
4	1	5	6	4	9
6	2	4	4	4	12
10	3	9	10	8	22

1.17 A Theorem of Alternatives for Systems of Linear Equations

Consider a system of linear equations. In the previous section, we have seen that the GJ method will always find a solution for this system if no inconsistent equation of the form "$0 = \alpha$" for some $\alpha \neq 0$ is encountered during the method.

If a "$0 = \alpha$" equation for some $\alpha \neq 0$ is found during the method, by dividing that equation by α we get the "$0 = 1$" equation. This provides a proof of Result 1.10.1 of Section 1.10 which states that a system of linear equations has no solution iff the "$0 = 1$" equation can be obtained as a linear combination of equations in the system.

Consider the general system of m equations in n variables $x = (x_1, \ldots, x_n)^T$ which is reproduced below in detached coefficient form.

General system

x_1	$\cdots$	x_j	$\cdots$	x_n	RHS constants
a_{11}	$\cdots$	a_{1j}	$\cdots$	a_{1n}	b_1
$\vdots$		$\vdots$		$\vdots$	$\vdots$
a_{i1}	$\cdots$	a_{ij}	$\cdots$	a_{in}	b_i
$\vdots$		$\vdots$		$\vdots$	$\vdots$
a_{m1}	$\cdots$	a_{mj}	$\cdots$	a_{mn}	b_m

Suppose we take a linear combination of these equations by multiplying the ith equation in this tableau by π_i and summing over $i = 1$ to m. The $\pi = (\pi_1, \ldots, \pi_m)$ is the vector of coefficients in this linear combination. This results in the following equation:

$$\sum_{j=1}^{n} (\pi_1 a_{ij} + \cdots + \pi_i a_{ij} + \cdots + \pi_m a_{mj}) x_j$$
$$= (\pi_1 b_1 + \cdots + \pi_i b_i + \cdots + \pi_m b_m).$$

This equation will be the "$0 = 1$" equation iff the coefficient of x_j in it, $\pi_1 a_{1j} + \cdots + \pi_m a_{mj}$ is 0 for all $j = 1$ to n, and $\pi_1 b_1 + \cdots + \pi_m b_m = 1$. That is, this linear combination is the "$0 = 1$" equation iff the (π_i) satisfy the following system of equations in detached coefficient form, which is known as the **alternate system** corresponding to the general system given above.

Alternate system

π_1	$\cdots$	π_i	$\cdots$	π_m	RHS
a_{11}	$\cdots$	a_{i1}	$\cdots$	a_{m1}	0
$\vdots$		$\vdots$		$\vdots$	$\vdots$
a_{1j}	$\cdots$	a_{ij}	$\cdots$	a_{mj}	0
$\vdots$		$\vdots$		$\vdots$	$\vdots$
a_{1n}	$\cdots$	a_{in}	$\cdots$	a_{mn}	0
b_1	$\cdots$	b_i	$\cdots$	b_m	1

Result 1.10.1 of Section 1.10 is often stated as a mathematical theorem about systems of linear equations in the following way:

Theorem of Alternatives for Systems of Linear Equations: *Consider the general system of m equations in n variables $x = (x_1, \ldots, x_n)^T$ given above. Using the same data in this system, we can construct the alternate system of $(n + 1)$ equations in m new variables $\pi = (\pi_1, \ldots, \pi_m)$ as shown above, such that the following property holds: the original system has no solution x iff the alternate system has a solution π.*

As explained in Section 1.13, any solution of the alternate system provides an evidence for the infeasibility of the original system.

As stated above, if the original system is solved by the GJ method after introducing the memory matrix in it, when the method terminates we will either have a solution of the original system, or of the alternate system.

As an example, consider the following system of 4 equations in 5 variables.

Original system

x_1	x_2	x_3	x_4	x_5	RHS
0	0	1	0	0	3
1	-2	2	-1	1	-8
-1	0	4	-7	7	16
0	-2	6	-8	8	6

The alternate system will have 4 variables $\pi_1, \pi_2, \pi_3, \pi_4$. It is:

Alternate system

π_1	π_2	π_3	π_4	RHS
0	1	-1	0	0
0	-2	0	-2	0
1	2	4	6	0
0	-1	-7	-8	0
0	1	7	8	0
3	-8	16	6	1

Let us denote the equations in the original system R_1 to R_4 from top to bottom. We introduce the memory matrix as discussed in Section 1.12, and apply the GJ method on it. In each step the PR (pivot row), PC (pivot column) are indicated and the pivot element is boxed; "BV" is "basic variable selected in row", and "IE" indicates an inconsistent equation discovered.

BV	Memory matrix									RHS	
	R_1	R_2	R_3	R_4	x_1	x_2	x_3	x_4	x_5		
	1	0	0	0	0	0	$\boxed{1}$	0	0	3	PR
	0	1	0	0	1	-2	2	-1	1	-8	
	0	0	1	0	-1	0	4	-7	7	16	
	0	0	0	1	0	-2	6	-8	8	6	
							PC ↑				
x_3	1	0	0	0	0	0	1	0	0	3	
	-2	1	0	0	$\boxed{1}$	-2	0	-1	1	-14	PR
	-4	0	1	0	-1	0	0	-7	7	4	
	-6	0	0	1	0	-2	0	-8	8	-12	
					PC ↑						

BV	Memory matrix									RHS	
	R_1	R_2	R_3	R_4	x_1	x_2	x_3	x_4	x_5		
x_3	1	0	0	0	0	0	1	0	0	3	
x_1	-2	1	0	0	1	-2	0	-1	1	-14	
	-6	1	1	0	0	$\boxed{-2}$	0	-8	8	-10	PR
	-6	0	0	1	0	-2	0	-8	8	-12	
						PC ↑					
x_3	1	0	0	0	0	0	1	0	0	3	
x_1	4	0	-1	0	1	0	0	7	-7	-4	
x_2	3	$-1/2$	$-1/2$	0	0	1	0	4	-4	5	
	0	-1	-1	1	0	0	0	0	0	-2	IE

The method terminates with the conclusion that the original system has no solution because of the inconsistent equation "$0 = -2$" in Row 4 of the final tableau. From the corresponding row of the memory matrix we conclude that the "$0 = -2$" equation is the linear combination $-R_2 - R_3 + R_4$ of equations in the original system. Dividing by -2, we conclude that the linear combination

$$\frac{1}{2}R_2 + \frac{1}{2}R_3 - \frac{1}{2}R_4$$

yields the inconsistent equation "$0 = 1$". Hence the coefficient vector of this linear combination, $\pi = (0, \frac{1}{2}, \frac{1}{2}, -\frac{1}{2})$ is a solution of the alternate system, which can be verified.

In the same way, for any system of linear equations, we can write the alternate system, and obtain a solution for it if the original system is inconsistent, by applying the GJ method on the original system with the memory matrix.

Exercises

1.17.1: For each of the following systems do the following: **(a)** Write the alternate system, **(b)** Solve the original system by the GJ method using the memory matrix, and obtain a solution for it if it is consistent, or a solution for the alternate system if the original system is inconsistent.

(i) $\begin{aligned} x_1 - 2x_2 + 2x_3 - x_4 + x_5 &= -8 \\ -x_1 + 4x_3 - 7x_4 + 7x_5 &= 16 \\ -2x_2 + 6x_3 - 8x_4 + 8x_5 &= 6 \end{aligned}$

(ii) $\begin{aligned} x_1 + x_2 + x_3 + x_4 &= 3 \\ -x_1 - 2x_2 + 2x_4 &= -4 \\ x_3 - x_4 &= 4 \\ 2x_1 + 3x_2 + x_4 &= 5 \end{aligned}$

(iii) $\begin{aligned} x_1 + x_2 - x_3 - x_4 &= -5 \\ -2x_2 + x_3 - x_4 &= -6 \\ x_1 - 2x_2 + 2x_4 &= 4 \\ 2x_1 - 3x_2 &= -4 \end{aligned}$

1.18 Special Triangular Systems of Linear Equations

In this section we consider square nonsingular systems of linear equations with the special **triangular property**, and a very efficient method for solving them.

Consider the general square system of n equations in n unknowns in detached coefficient form.

	x_1	x_2	$\ldots$	x_{n-1}	x_n	RHS	
	$\boxed{a_{11}}$	a_{12}	$\ldots$	$a_{1,n-1}$	a_{1n}	b_1	
	a_{21}	$\boxed{a_{22}}$	$\ldots$	$a_{2,n-1}$	a_{2n}	b_2	Above diagonal part
Below diagonal part	$\vdots$	$\vdots$		$\vdots$	$\vdots$	$\vdots$	
	a_{n1}	a_{n2}	$\ldots$	$a_{n,n-1}$	$\boxed{a_{nn}}$	b_n	

For $i = 1$ to n, the boxed coefficients a_{ii} in Row i and Column i of this tableau are known as the **diagonal entries in this tableau.** They constitute the **principal diagonal** or the **main diagonal** of the tableau.

Remember that the concept of diagonal entries and diagonal are only defined for square systems. There is no such concept for rectangular systems which are not square.

Ignoring the RHS constants vector, the coefficients of the variables in this square tableau are partitioned into three parts:

the diagonal part consisting of a_{ii} for $i = 1$ to n
below the diagonal part consisting of a_{ij} for $i > j$
above the diagonal part consisting of a_{ij} for $i < j$.

The square system is said to be an:

Upper triangular system	If all diagonal elements nonzero, and below diagonal part is all zero; i.e., $a_{ii} \neq 0$ for all i, and $a_{ij} = 0$ for all $i > j$.
Lower triangular system	If all diagonal elements nonzero, and above diagonal part is all zero; i.e., $a_{ii} \neq 0$ for all i, and $a_{ij} = 0$ for all $i < j$.
Triangular system	If the columns (or rows) of the tableau can be rearranged so that the system becomes an upper triangular system.

Here are some examples.

Upper triangular system

x_1	x_2	x_3	x_4	
1	1	−1	1	−10
0	−1	−2	1	−20
0	0	1	1	20
0	0	0	−1	10

Lower triangular system

x_1	x_2	x_3	x_4	x_5	
−1	0	0	0	0	5
1	−1	0	0	0	−5
−1	2	1	0	0	10
1	1	−1	1	0	10
−1	2	1	2	−1	−10

Triangular system

x_1	x_2	x_3	x_4	
0	0	0	−1	10
1	1	−1	1	−10
0	0	1	1	20
0	−1	−2	1	−20

The last system is triangular, because by arranging the rows in the order: Row 2, Row 4, Row 3, Row 1 it becomes the upper triangular first system.

The Back substitution Method for Solving an Upper Triangular System of Equations

Consider the general upper triangular system of equations given below. Here we know that $a_{ii} \neq 0$ for all i, but some of the other coefficients may be zero.

x_1	x_2	$\cdots$	x_{n-1}	x_n	RHS
a_{11}	a_{12}	$\cdots$	$a_{1,n-1}$	a_{1n}	b_1
0	a_{22}	$\cdots$	$a_{2,n-1}$	a_{2n}	b_2
$\vdots$	$\vdots$		$\vdots$	$\vdots$	$\vdots$
0	0	$\cdots$	$a_{n-1,n-1}$	$a_{n-1,n}$	b_{n-1}
0	0	$\cdots$	0	a_{nn}	b_n

From the equation represented by the last row we get

$$x_n = \frac{b_n}{a_{nn}}$$

in the solution. Then from the $(n-1)$th equation, we can get the value of x_{n-1} in the solution by substituting the value of x_n from above, in it.

In the same way, substituting the values of $x_{j+1}, \ldots, x_n$ in the solution in the jth equation, we can get the value of x_j in the solution, for $j = n-1, \ldots, 1$, in that order. Hence, using the equations in the tableau from the bottom to the top in that order, the variables are evaluated in the order x_n to x_1; this is the reason for the name **back substitution method**.

For an illustration, we will solve the upper triangular system given above, which is reproduced below.

Upper triangular system

x_1	x_2	x_3	x_4	
1	1	-1	1	-10
0	-1	-2	1	-20
0	0	1	1	20
0	0	0	-1	10

The last equation, "$-x_4 = 10$" yields $x_4 = -10$.

The 3rd equation "$x_3 + x_4 = 20$" yields $x_3 = 20 - x_4 = 20 - (-10) = 30$.

The 2nd equation "$-x_2 - 2x_3 + x_4 = -20$" yields $x_2 = 20 - 2x_3 + x_4 = 20 - 2(30) + (-10) = -50$.

The 1st equation "$x_1 + x_2 - x_3 + x_4 = -10$" yields $x_1 = -10 - x_2 + x_3 - x_4 = -10 - (-50) + 30 - (-10) = 80$.

So the solution of the system is $(x_1, x_2, x_3, x_4)^T = (80, -50, 30, -10)^T$.

In the same way, a lower triangular system of equations can be solved by using the equations in the tableau from top to bottom in that order, to evaluate the variables in the order x_1 to x_n. Commonly, this is also called back substitution method, but some textbooks refer to it as the forward substitution method.

A triangular system must satisfy the following property:

Property of a triangular system: There exists an equation in which only one variable has a nonzero coefficient.

To find the solution of a triangular system, we identify the equation in it satisfying the above property, and use it to get the value of the only variable,

x_s say, with a nonzero coefficient in it. Eliminate this equation from the system, and eliminate the variable x_s from the system by substituting its value in all the other equations.

The remaining system is also triangular, and hence it must also satisfy the above property, and hence the above step can be applied on it. The process continues in the same way until the values of all the variables in the solution are obtained.

Exercises

1.18.1: The tower of Brahma (or **The tower of Hanoi**) **puzzle:** This puzzle consists of 3 pegs, and n rings of different sizes stacked on one of them (in the traditional puzzle, $n = 64$). The rings are arranged in decreasing order of size with the largest one at the bottom and the smallest one at the top of the stack. The problem is to move the rings one at a time from one peg to another, never putting a larger one on a smaller one, and eventually transferring all 64 rings from one peg to another peg. Let

x_n = the minimum number of moves required to arrange the n rings in the same decreasing order of size on another peg, while never putting a larger ring on a smaller ring.

$(x_1, \ldots, x_8)^T$ can be shown to be a solution of the following system of linear equations.

$$x_1 = 1$$
$$x_2 - 2x_1 = 1$$
$$x_3 - 2x_2 = 1$$
$$x_4 - 2x_3 = 1$$
$$x_5 - 2x_4 = 1$$
$$x_6 - 2x_5 = 1$$
$$x_7 - 2x_6 = 1$$
$$x_8 - 2x_7 = 1$$

and in general $x_{n+1} - 2x_n = 1$ for all integers $n \geq 1$. Write the coefficient matrix of the system of equations for $n = 1$ to 8 given above, and verify that it is lower triangular. Obtain the solution of the system by back substitution.

Show that in general $x_n = 2^n - 1$.

1.18.2: Verify whether the following systems of equations are triangular, if so, obtain their solution by back substitution. If not explain why they are not triangular. (Here (i), (ii) are not triangular, while (iii), (iv) are.)

(i) $x_1 + x_2 + 3x_3 - x_4 + x_5 = 10$
$2x_3 + x_4 = 6$
$x_3 + 2x_4 - x_5 = 7$
$2x_5 = 10$
$x_5 = 5$

(ii) $2x_1 + 3x_2 + 2x_3 - 2x_4 = 5$
$2x_3 + 2x_4 = 6$
$x_1 + 2x_4 = 3$

(iii) $3x_3 - 3x_4 + 6x_5 = 30$
$x_5 = -2$
$2x_2 - 2x_3 - x_4 + x_5 = -83$
$2x_4 + 3x_5 = 10$
$4x_1 - 4x_4 - 8x_5 = 12$

(iv) $x_{11} + x_{12} = 3$
$x_{21} + x_{23} = 5$
$x_{32} + x_{34} = 9$
$x_{43} = 10$
$x_{11} + x_{21} = 4$
$x_{12} + x_{32} = 6$
$x_{23} + x_{43} = 12$

1.19   The Gaussian Pivot Step

Consider the general system of m equations in n variables in detached coefficient tableau form. Like the GJ pivot step discussed so far, the **G (Gaussian) pivot step** on this tableau is also specified by a row of the tableau selected as the pivot row, and a column of the tableau selected as the pivot column for it, with the pivot element (element in the pivot row and pivot column) nonzero. The G pivot step consists of a series of row operations on the tableau to convert the entries in the pivot column below the pivot row to zero. Notice that the pivot row and all the rows above it remain unchanged in a pivot step. Thus a G pivot step may involve a lot less work than a GJ pivot step with the same pivot row and pivot column, particularly if the pivot row is well below the top of the tableau. Using abbreviations PR (pivot row), PC (pivot column), we give the operations for a general system.

Original Tableau

x_1	x_2	$\cdots$	x_s	$\cdots$	x_n	RHS constants	
a_{11}	a_{12}	$\cdots$	a_{1s}	$\cdots$	a_{1n}	b_1	
a_{21}	a_{22}	$\cdots$	a_{2s}	$\cdots$	a_{2n}	b_2	
$\vdots$	$\vdots$		$\vdots$		$\vdots$	$\vdots$	
a_{r1}	a_{r2}	$\cdots$	$\boxed{a_{rs}}$	$\cdots$	a_{rn}	b_r	PR
$\vdots$	$\vdots$		$\vdots$		$\vdots$	$\vdots$	
a_{m1}	a_{m2}	$\cdots$	a_{ms}	$\cdots$	a_{mn}	b_m	

PC

The G pivot step involves the following row operations:

for each $i > r$, Row i - $\left(\frac{a_{is}}{a_{rs}}\right)$Row r.

Since the G pivot step involves division by the pivot element, it is possible only if the pivot element is nonzero. Denoting the entries in the tableau after the G pivot step by a_{ij}, b_i if they do not change, or by $\bar{a}_{ij}, \bar{b}_i$ if they do, we have:

Tableau after the G pivot step

x_1	x_2	$\cdots$	x_s	$\cdots$	x_n	RHS constants
a_{11}	a_{12}	$\cdots$	a_{1s}	$\cdots$	a_{1n}	b_1
a_{21}	a_{22}	$\cdots$	a_{2s}	$\cdots$	a_{2n}	b_2
$\vdots$	$\vdots$		$\vdots$		$\vdots$	$\vdots$
a_{r1}	a_{r2}	$\cdots$	a_{rs}	$\cdots$	a_{rn}	b_r
$\bar{a}_{r+1,1}$	$\bar{a}_{r+1,2}$	$\cdots$	0	$\cdots$	$\bar{a}_{r+1,n}$	$\bar{b}_{r+1}$
$\vdots$	$\vdots$		$\vdots$		$\vdots$	$\vdots$
$\bar{a}_{m1}$	$\bar{a}_{m2}$	$\cdots$	0	$\cdots$	$\bar{a}_{mn}$	$\bar{b}_m$

We provide a numerical example. The original tableau is at the top with PR, PC indicated and the pivot element boxed. The row operations performed on the original tableau in this G pivot step are: Row 4 $-(\frac{4}{2})$Row 3, Row 5 $-(\frac{-6}{2})$Row 3. The tableau after the G pivot step is given at the bottom.

Original tableau

x_1	x_2	x_3	x_4	x_5	x_6		
13	-12	11	0	-8	14	33	
-70	-80	33	-84	-17	8	-45	
-2	0	4	-2	$\boxed{2}$	6	-10	PR
2	1	-2	1	4	3	10	
-3	-1	1	3	-6	0	8	
				PC			

Tableau after the G pivot step

13	-12	11	0	-8	14	33	
-70	-80	33	-84	-17	8	-45	
-2	0	4	-2	2	6	-10	PR
6	1	-10	5	0	-9	30	
-9	-1	13	-3	0	18	-22	

We leave it to the reader to verify that this G pivot step may involve much less work than a GJ pivot step using the same pivot element.

Exercises

1.19.1: Perform the G pivot steps indicated by the pivot row (PR), pivot column (PC) indicated on the following tableaues.

(i)

			PC			
x_1	x_2	x_3	x_4	x_5		
-2	4	-4	6	2	-12	PR
2	1	3	-2	-1	10	
-6	3	6	0	2	0	
8	-2	1	-2	3	2	

(ii)

PC					
x_1	x_2	x_3	x_4		
1	-1	-2	3	-4	
0	3	-4	6	7	
3	9	-3	12	6	PR
-2	0	2	1	3	

(iii) In the following tableau, a G pivot step needs to be carried out in the pivot column (PC) marked. The pivot row (PR) is determined by the following procedure: identify the row with entry of maximum absolute value in the PC among rows with numbers ≥ 2, and interchange that row with row 2. After the row interchange, perform a G pivot step with row 2 as the PR.

x_1	x_2	x_3	x_4	x_5	
1	2	0	1	-2	-20
-1	3	2	0	-4	4
0	-4	-4	-2	6	-6
-18	12	-6	24	6	30
		PC			

(iv)   In the following tableau perform G pivot steps in all the rows in the order rows 1 to 4, with the diagonal element in that row as the pivot element; to convert the system into an upper triangular system. Then solve the resulting system by the back substitution method.

x_1	x_2	x_3	x_4	
1	-2	0	1	2
-1	1	-2	1	1
0	-1	-1	1	3
1	2	3	1	0

1.20   The Gaussian (G) Elimination Method for Solving Linear Equations

This method is based on G pivot steps, and the back substitution method of Section 1.18 to get the solution from the final tableau if the system turns out to be consistent.

The GJ method converted the columns of the basic variables into the various unit column vectors. The main point in the G method is to convert the column vectors of the basic variables in their proper order into upper triangular form so that the values of the basic variables can be computed by back substitution from the final tableau after giving the nonbasic variables arbitrary values at the user's choice, and transferring their terms to the right. For this reason, and the nature of the G pivot step, please note that the row interchanges mentioned in Steps 2 are necessary depending on the manner in which the next pivot element is choosen.

So, in the G elimination method, we also use the new row operation **row interchange** (interchanging two rows in the tableau) that we have not used so far in the GJ method.

The G elimination method needs less computational effort than the GJ method, and hence is the preferred among these two for solving large scale problems using computers. It is the most commonly used method for solving linear equations in scientific computation.

As in the discussion of the GJ method in Section 1.16, we do not introduce here the memory matrix to produce evidences of redundancy and infeasibility to keep the discussion simple. Also, the version of the method presented here is the simplest version. A computationally more efficient version based on the memory matrix is given in Section 4.11, it has the additional benefit of producing all evidences without additional computational burden.

The G Elimination Method

BEGIN

Step 1: Put the system in detached coefficient tableau form and obtain the original tableau. Open a column on the left hand side to record the basic variables. Give each row its original number (number in the original tableau) and store this number in a rightmost column. This number will not change as long as the row remains in the tableau, and moves with the row if the row is interchanged. This number may be different from the position of the row in the present tableau because of row interchanges taking place during the algorithm, or rows corresponding to redundant constraints being eliminated.

Set the row index r to 0.

Step 2: Pivot element selection: Among rows in positions $(r + 1)$ and below (counting from the top) in the present tableau, if there is a "$0 = \bar{b}_i$" row for some $\bar{b}_i \neq 0$, the original system is inconsistent and has no solution; terminate the method.

Among rows in positions $(r + 1)$ and below in the present tableau, if there is a "$0 = 0$" row (redundant constraint) eliminate it and all such rows from the present tableau. If all rows in positions $(r + 1)$ and below are eliminated in this process, go to Step 4.

If some rows in positions $(r+1)$ and below remain in the present tableau, each of them has at least one nonzero coefficient. Now select the pivot element for the next pivot step. One of the following strategies can be used for this.

1. **Full pivoting:** This strategy is used to help minimize accumulation of round-off errors when solving problems on a computer. Select a maximum absolute value coefficient in the present tableau in rows $(r+1)$ and below, as the pivot element. Suppose this element is $\bar{a}_{is}$ in row i and column of variable x_s in the present tableau. If $i \neq r+1$, interchange rows $r+1$ and i in the present tableau so that the pivot element is in row $r+1$ after the interchange. In the resulting tableau, select row $r+1$ as the pivot row and the column of x_s as the pivot column.

   When solving problems by hand, one can use a similar strategy, but select the pivot element to be an element of value ± 1, or one which is an integer of least absolute value in the same region in order to minimize the fractions generated.

2. **Partial pivoting:** In this strategy, the pivot column is first selected (could be by some other considerations) as a column with at least one nonzero entry in the present tableau among rows $r+1$ and below. Suppose it is column s of variable x_s. Then we select the pivot element to be an element of maximum absolute value among rows $r+1$ and below in the pivot column. If this element is $\bar{a}_{is}$ in row i and $i \neq r+1$, interchange rows $r+1$ and i in the present tableau so that the pivot element is in row $r+1$ after the interchange. In the resulting tableau, select row $r+1$ as the pivot row, column of x_s as the pivot column.

3. If solving problem by hand computation, select row $r+1$ as the pivot row in the present tableau, and a nozero coefficient in this row as the pivot element, and its column as the pivot column. This strategy is normally used when solving problems by hand, in this case we will look for pivot elements which are ± 1 if possible, or some integer of smallest absolute value to minimize fractions generated in the pivot step.

Step 3: Pivot step: Carry out the G pivot step in the present tableau with the pivot row, pivot column selected above, and record the variable corresponding to the pivot column as the basic variable in the pivot row in the resulting tableau. In this tableau, if row $r+1$ is the bottom most row, go with it to Step 4; otherwise increment r by 1 and go back to Step 2.

Step 4: Computing the solution: Let p be the number of rows in the final tableau, and $x_{i_1}, \ldots, x_{i_p}$ the basic variables in rows 1 to p in it. Then x_{i_t} is the tth basic variable and $(x_{i_1}, \ldots, x_{i_p})$ is the basic vector selected.

If there are no nonbasic variables at this stage, rearrange the variables and their columns in the final tableau in the order $x_{i_1}, \ldots, x_{i_p}$. After this rearrangement of variables, the final tableau has the following upper triangular form:

When no nonbasic variables left

Basic Vars.	x_{i_1}	x_{i_2}	$\cdots$	$x_{i_{p-1}}$	x_{i_p}	
x_{i_1}	$\bar{a}_{11}$	$\bar{a}_{12}$	$\cdots$	$\bar{a}_{1,p-1}$	$\bar{a}_{1p}$	$\bar{b}_1$
x_{i_2}	0	$\bar{a}_{22}$	$\cdots$	$\bar{a}_{2,p-1}$	$\bar{a}_{2p}$	$\bar{b}_2$
$\vdots$	$\vdots$	$\vdots$		$\vdots$	$\vdots$	$\vdots$
$x_{i_{p-1}}$	0	0	$\cdots$	$\bar{a}_{p-1,p-1}$	$\bar{a}_{p-1,p}$	$\bar{b}_{p-1}$
x_{i_p}	0	0	$\cdots$	0	$\bar{a}_{pp}$	$\bar{b}_p$

and it can be solved by the back substitution method discussed in Section 1.18 to obtain the unique solution of the original system in this case.

If there are some nonbasic variables in the final tableau, rearrange the variables and their column vectors in the final tableau so that the basic variables appear first in the tableau in their proper order, followed by non-basic variables if any (in any order). Then the final tableau assumes the following form:

When nonbasic variables exist

Final tableau after rearrangement of variables

Basic Vars.	x_{i_1}	x_{i_2}	$\cdots$	x_{i_p}	x_{j_1}	$\cdots$	$x_{j_{n-p}}$	
x_{i_1}	$\bar{a}_{11}$	$\bar{a}_{12}$	$\cdots$	$\bar{a}_{1p}$	$\bar{a}_{1,j_1}$	$\cdots$	$\bar{a}_{1,n-p}$	$\bar{b}_1$
x_{i_2}	0	$\bar{a}_{22}$	$\cdots$	$\bar{a}_{2p}$	$\bar{a}_{2,j_1}$	$\cdots$	$\bar{a}_{2,n-p}$	$\bar{b}_2$
$\vdots$	$\vdots$	$\vdots$		$\vdots$	$\vdots$		$\vdots$	$\vdots$
$x_{i_{p-1}}$	0	0	$\cdots$	$\bar{a}_{p-1,p}$	$\bar{a}_{p-1,j_1}$	$\cdots$	$\bar{a}_{p-1,n-p}$	$\bar{b}_{p-1}$
x_{i_p}	0	0	$\cdots$	$\bar{a}_{pp}$	$\bar{a}_{p,j_1}$	$\cdots$	$\bar{a}_{p,n-p}$	$\bar{b}_p$

In this case there are an infinite number of solutions to the system, and the dimension of the solution set is = number of nonbasic variables = $n - p$.

By giving the nonbasic variables $x_{j_1}, \ldots, x_{j_{n-p}}$ appropriate real values (0, if you want the basic solution WRT the present basic vector) and transferring all their terms to the right, we are left with an upper triangular system in the basic variables to compute their values in the corresponding solution by back substitution.

The final tableaus here, after rearrangement of variables as described above, are said to be in **echelon form** or **REF (row echelon form)** in mathematical literature.

In the original tableau, eliminate all the rows which do not appear among the original row numbers for rows in the final tableau (these rows are eliminated as those corresponding to redundant constraints during the algorithm). Now rearrange the remaining rows in the same order as the original row numbers in the final tableau. Then rearrange the variables and their column vectors in the same order as those in the final tableaus discussed above. This gives the remaining original tableau after the deletion of the redundant constraints, corresponding to the final tableau.

END

Numerical Examples of the G Method

We now provide numerical examples. We indicate the pivot row (PR), pivot column (PC), and the boxed pivot element when a G pivot step is performed on a tableau. RC (redundant constraint to be eliminated), IE (inconsistent equation) have the same meanings as before. We use the abbreviation "BV" for "basic variable". The indication "RI to Rt" on a row means that at that stage, that row will be interchanged with Row t in that tableau.

We number all the rows (equations) serially from top to bottom, and record these numbers in a rightmost column, to make it easy to keep track of redundant constraints eliminated.

The original detached coefficient tableau for the original system is always the first one.

Example 1:

In this example, we select the pivot column first, and then the pivot element in it using the "partial pivot strategy". Even though the present row may have the maximum absolute value element in the pivot column, we may select the pivot element as another one below it with the same absolute value to illustrate row interchanges.

BV	x_1	x_2	x_3	x_4	x_5	x_6	RHS		Eq. no.
	1	-2	1	1	1	-1	-4		1
	2	2	0	1	-1	-1	10		2
	2	-4	-2	0	-2	-2	-8	RI to R1	3
	0	-6	-2	-1	-1	-1	-18		4
	0	0	0	4	1	1	20		5
	1	4	-1	0	-2	0	14		6
	PC ↑								
	$\boxed{2}$	-4	-2	0	-2	-2	-8	PR	3
	2	2	0	1	-1	-1	10		2
	1	-2	1	1	1	-1	-4		1
	0	-6	-2	-1	-1	-1	-18		4
	0	0	0	4	1	1	20		5
	1	4	-1	0	-2	0	14		6
	PC ↑								

BV	x_1	x_2	x_3	x_4	x_5	x_6	RHS		Eq. no.
x_1	2	-4	-2	0	-2	-2	-8		3
	0	6	2	1	1	1	18		2
	0	0	2	1	2	0	0	RI to R2	1
	0	-6	-2	-1	-1	-1	-18		4
	0	0	0	4	1	1	20		5
	0	6	0	0	-1	1	18	6	
			PC ↑						
x_1	2	-4	-2	0	-2	-2	-8		3
	0	0	$\boxed{2}$	1	2	0	0	PR	1
	0	6	2	1	1	1	18		2
	0	-6	-2	-1	-1	-1	-18		4
	0	0	0	4	1	1	20		5
	0	6	0	0	-1	1	18		6
			PC ↑						
x_1	2	-4	-2	0	-2	-2	-8		3
x_3	0	0	2	1	2	0	0		1
	0	$\boxed{6}$	0	0	-1	1	18	PR	2
	0	-6	0	0	1	-1	-18		4
	0	0	0	4	1	1	20		5
	0	6	0	0	-1	1	18		6
	PC								

BV	x_1	x_2	x_3	x_4	x_5	x_6	RHS		Eq. no.
x_1	2	-4	-2	0	-2	-2	-8		3
x_3	0	0	2	1	2	0	0		1
x_2	0	6	0	0	-1	1	18		2
	0	0	0	0	0	0	0	RC	4
	0	0	0	4	1	1	20		5
	0	0	0	0	0	0	0	RC	6
x_1	2	-4	-2	0	-2	-2	-8		3
x_3	0	0	2	1	2	0	0		1
x_2	0	6	0	0	-1	1	18		2
	0	0	0	[4]	1	1	20	PR	5
				PC ↑					
x_1	2	-4	-2	0	-2	-2	-8		3
x_3	0	0	2	1	2	0	0		1
x_2	0	6	0	0	-1	1	18		2
x_4	0	0	0	4	1	1	20		5

Final tableau with variables rearranged

BV	x_1	x_3	x_2	x_4	x_5	x_6	RHS	Eq. no.
x_1	2	-2	-4	0	-2	-2	-8	3
x_3	0	2	0	1	2	0	0	1
x_2	0	0	6	0	-1	1	18	2
x_4	0	0	0	4	1	1	20	5

Since there are two nonbasic variables in the final tableau, this system has an infinite number of solutions, and the dimension of the solution set is two. The remaining original tableau after eliminating the redundant constraints 4 and 6, and rearranging the rows and the variables in the same order as those in the final tableau is given below:

Remaining original tableau with eqs. & variables rearranged

x_1	x_3	x_2	x_4	x_5	x_6	RHS	Eq. no.
2	-2	-4	0	-2	-2	-8	3
1	1	-2	1	1	-1	-4	1
2	0	2	1	-1	-1	10	2
0	0	0	4	1	1	20	5

After transferring the terms of the nonbasic (free) variables to the right, the final tableau becomes the following upper triangular system in the basic variables:

Upper triangular system to evaluate
basic variables

x_1	x_3	x_2	x_4	RHS
2	-2	-4	0	$-8 + 2x_5 + 2x_6$
0	2	0	1	$-2x_5$
0	0	6	0	$18 + x_5 - x_6$
0	0	0	4	$20 - x_5 - x_6$

Giving the nonbasic variables x_5, x_6 both zero values, leads to the basic solution WRT present basic vector (x_1, x_3, x_2, x_4) which is:

$$(x_1, x_2, x_3, x_4, x_5, x_6)^T = \left(\frac{-1}{2}, 3, \frac{-5}{2}, 5, 0, 0 \right)^T$$

Example 2:

Here also we use the pivot element choice strategy "partial pivot" as in Example 1 above and the same notation. "IE" indicates an inconsistent equation obtained in the method. The original tableau is the one at the top.

BV	x_1	x_2	x_3	x_4	x_5	RHS		Eq. no.
	0	-2	6	-8	8	6		1
	-1	0	4	-7	7	16		2
	1	-2	2	-1	1	-8	RI to R1	3
	0	-2	6	-8	8	8		4
	PC ↑							
	$\boxed{1}$	-2	2	-1	1	-8	PR	3
	-1	0	4	-7	7	16		2
	0	-2	6	-8	8	6		1
	0	-2	6	-8	8	8		4
	PC ↑							
x_1	1	-2	2	-1	1	-8		3
	0	$\boxed{-2}$	6	-8	8	8	PR	2
	0	-2	6	-8	8	6		1
	0	-2	6	-8	8	8		4
	PC ↑							

BV	x_1	x_2	x_3	x_4	x_5	RHS		Eq. no.
x_1	1	-2	2	-1	1	-8		3
x_2	0	-2	6	-8	8	8		2
	0	0	0	0	0	-2	IE	1
	0	0	0	0	0	0	RC	4

Since the third row in the final tableau is the inconsistent equation "$0 = -2$" we conclude that the original system is inconsistent, and has no solution.

Exercises

1.20.1: Solve the numerical exercises in Section 1.16 by the G elimination method.

1.21 Comparison of the GJ and the G Elimination Methods

A system of linear equations or the detached coefficient tableau for it is said to be a:

> **sparse system** if a large proportion of the coefficients in it are zero
> **dense system** otherwise.

For solving a dense system of m equations in n variables, if all the tableaus remain dense:

> GJ method needs at most $3m^2(n + 1)$ arithmetic operations
> G method needs at most $\frac{3m(m-1)(n+1)}{2} + \frac{m(m-1)}{2}$ arithmetic operations.

The G method cuts down the work needed by about half. Hence in practice, the G elimination method is the method of choice for solving linear equations.

However, for students going on to study linear programming, learning the details of the GJ method is important, as the original version of the simplex method for solving linear programs is based on the GJ elimination method and uses GJ pivot steps. But modern computer implementations of the simplex method are mostly based on highly sophisticated variants of G pivot steps for numerical stability.

1.22 Infeasibility Analysis

Consider a mathematical model which is a system of m linear equations in n variables, formulated for a real world application. In such a model, the coefficient vectors of variables usually come from things like properties of materials which are combined, etc., which are very hard to change. The RHS constants vector usually comes from requirements that are to be met, or targets to be achieved, etc., which are easier to modify if a need arises.

Suppose it turns out that the model is inconsistent, i.e., it has no solution. Mathematically there is nothing more that can be done on the current model. But the real world problem does not go away; it has to be solved somehow. In this situation, we have to investigate what practically feasible changes can be made in the model to modify it into a consistent or feasible system. The study of such changes is known as **infeasibility analysis**.

Since it is very hard to change the coefficient vectors of the variables, changes in them are rarely considered in applications. In most cases it is the RHS constants which are changed, and this is what we consider in this section. We will discuss what changes in the RHS constants vector will modify an inconsistent system into a consistent one.

Inconsistency of the original system is established when the GJ method is applied on it and it produces an inconsistent equation of the form "$0 = \alpha$" for some $\alpha \neq 0$ and reaches Step 2.2 in the method for the first time (a similar thing occurs if the G method is used instead of the GJ method). In the versions of the GJ (or the G) methods discussed earlier, when an inconsistent equation shows up for the first time, we terminated those methods once and for all with the conclusion that the original system has no solution.

In order to carry out infeasibility analysis, whenever an inconsistent equation shows up in the current tableau, we leave that inconsistent row in the tableau, but continue applying the method to perform pivot steps in the remaining rows. Subsequently, if other instances of inconsistency are encountered, do the same; i.e., leave those inconsistent rows in the tableau, but continue applying the method until pivot steps are carried out in all the rows in which they can be.

In the end we have the final tableau, and the remaining original tableau with the redundant constraints discovered during the method deleted. Since the original system is inconsistent, in the final tableau there are one or more inconsistent equations of the form "$0 = \alpha$" for some $\alpha \neq 0$. Suppose these are in rows $r_1, \ldots, r_t$ in the final tableau. Let the updated RHS constants in these rows in the final tableau be: $\bar{b}_{r_1}, \ldots, \bar{b}_{r_t}$, each of which is nonzero.

One possible way to modify the original RHS constants vector to make the model consistent is: for each $k = 1$ to t, replace the RHS constant b_{r_k} in the r_kth row of the remaining original tableau by $b_{r_k} - \bar{b}_{r_k}$. The effect of this change is to convert the updated RHS constants in all these rows in the final tableau to 0, i.e., the changes made convert all the inconsistent equations in the final tableau into redundant constraints.

To get a solution of the modified system, you can now strike off these rows $r_1, \ldots, r_t$ from the final tableau and the remaining original tableau as redundant constraints. With the remaining final tableau, go to the final step in the method to find a solution of the modified system.

Example:

Consider the following system of 7 equations in 7 variables in detached coefficient form in the first tableau below. We apply the GJ method to solve it. PR (pivot row), PC (pivot column), RC (redundant constraint to be eliminated), BV (basic variable selected in this row) have the same meanings as before, and in each tableau, the pivot element for the pivot step carried out in that tableau is boxed. "IE" indicates an inconsistent equation obtained in the method. We number all the rows (equations) serially from top to bottom, and record these numbers in a rightmost column, to make it easy to keep track of redundant constraints eliminated.

BV	x_1	x_2	x_3	x_4	x_5	x_6	x_7	RHS		Eq. no.
	$\boxed{1}$	0	1	−1	1	1	0	−7	PR	1
	0	−1	2	1	−1	0	1	8		2
	1	−2	5	1	−1	1	0	9		3
	1	1	0	2	1	0	0	10		4
	3	0	5	5	1	1	2	35		5
	0	0	1	3	1	0	0	15		6
	3	0	7	11	3	1	2	55		7
	PC ↑									
x_1	1	0	1	−1	1	1	0	−7		1
	0	$\boxed{-1}$	2	1	−1	0	1	8	PR	2
	0	−2	4	2	−2	0	0	16		3
	0	1	−1	3	0	−1	0	17		4
	0	0	2	8	−2	−2	2	56		5
	0	0	1	3	1	0	0	15		6
	0	0	4	14	0	−2	2	76		7
		PC ↑								

BV	x_1	x_2	x_3	x_4	x_5	x_6	x_7	RHS		Eq. no.
x_1	1	0	1	-1	1	1	0	-7		1
x_2	0	1	-2	-1	1	0	-1	-8		2
	0	0	0	0	0	0	0	0	RC	3
	0	0	$\boxed{1}$	4	-1	-1	1	25	PR	4
	0	0	2	8	-2	-2	2	56		5
	0	0	1	3	1	0	0	15		6
	0	0	4	14	0	-2	2	76		7
			PC ↑							
x_1	1	0	0	-5	2	2	-1	-32		1
x_2	0	1	0	7	-1	-2	1	42		2
x_3	0	0	1	4	-1	-1	1	25		4
	0	0	0	0	0	0	0	6	IE	5
	0	0	0	$\boxed{-1}$	2	1	-1	-10	PR	6
	0	0	0	-2	4	2	-2	-24		7
				PC ↑						

Final tableau

BV	x_1	x_2	x_3	x_4	x_5	x_6	x_7	RHS		Eq. no.
x_1	1	0	0	0	-8	-3	4	18		1
x_2	0	1	0	0	13	5	-6	-38		2
x_3	0	0	1	0	7	3	-3	-15		4
	0	0	0	0	0	0	0	6	IE	5
	0	0	0	1	-2	-1	1	10		6
	0	0	0	0	0	0	0	-4	IE	7

The original third constraint has been deleted as a redundant constraint. The remaining original tableau is:

x_1	x_2	x_3	x_4	x_5	x_6	x_7	RHS	Modified RHS	Eq. no.
1	0	1	-1	1	1	0	-7	-7	1
0	-1	2	1	-1	0	1	8	8	2
1	1	0	2	1	0	0	10	10	4
3	0	5	5	1	1	2	35	29	5
0	0	1	3	1	0	0	15	15	6
3	0	7	11	3	1	2	55	59	7

In the final tableau, the 4th and the 6th rows represent inconsistent equations with updated RHS constants of 6 and -4 respectively. To make the system feasible we need to subtract 6, and -4 from the corresponding entries in the remaining original tableau, leading to the "modified RHS constants vector" given on the right in the remaining original tableau.

The effect of this change is to modify the 4th and the 6th RHS constants in the final tableau to zero. This makes the corresponding constraints redundant, which we delete. So, the final tableau for the modified problem is:

BV	x_1	x_2	x_3	x_4	x_5	x_6	x_7	RHS
	Final tableau for modified problem							
x_1	1	0	0	0	-8	-3	4	18
x_2	0	1	0	0	13	5	-6	-38
x_3	0	0	1	0	7	3	-3	-15
x_4	0	0	0	1	-2	-1	1	10

from which we find that the general solution of the modified system is:

$$\begin{pmatrix} x_1 \\ x_2 \\ x_3 \\ x_4 \\ x_5 \\ x_6 \\ x_7 \end{pmatrix} = \begin{pmatrix} 18 + 8\bar{x}_5 + 3\bar{x}_6 - 4\bar{x}_7 \\ -38 - 13\bar{x}_5 - 5\bar{x}_6 + 6\bar{x}_7 \\ -15 - 7\bar{x}_5 - 3\bar{x}_6 + 3\bar{x}_7 \\ 10 + 2\bar{x}_5 + \bar{x}_6 - \bar{x}_7 \\ \bar{x}_5 \\ \bar{x}_6 \\ \bar{x}_7 \end{pmatrix}$$

where $\bar{x}_5, \bar{x}_6, \bar{x}_7$ are arbitrary real valued parameters.

Discussion on Modifying the RHS Vector to Convert an Inconsistent System Into a Feasible System

In the example of an inconsistent system given above, we modified the values of two RHS constants to make it consistent. Also we decreased the value of one RHS constant from 35 to 29, and increased the value of another from 55 to 59 for a total change of

$$|35 - 29| + |55 - 59| = 10 \text{ units.}$$

Usually, practitioners look for a modification in the RHS constants vector $b = (b_i)$ of an inconsistent system to $b' = (b'_i)$ to make it feasible, with

the smallest number of changes, i.e., one which minimizes the number of i for which $b_i \neq b'_i$, and
the smallest total change, which is $\sum_i |b_i - b'_i|$.

The number of changes made in the above process is equal to the number of inconsistent equations "$0 = \alpha$" for some $\alpha \neq 0$ encountered in the method. Surprisingly, the total number of inconsistent equations in the final tableau depends on the order in which rows of the tableau are chosen as pivot rows in the method. As an illustration of this, we consider the following example from Murty, Kabadi, Chandrasekaran [2001] (the equations are numbered for convenience in tracking them).

Eq. no.	
1	$x_1 = 1$
2	$x_2 = 1$
3	$x_3 = 1$
4	$x_1 + x_2 + x_3 = 2$
5	$x_1 + x_2 + x_3 = 3$
6	$x_1 + x_2 + x_3 = 3$

If we apply the GJ method on this system using the equations 1 to 6 in that order as pivot rows, we get the following final tableau with only one inconsistent equation:

Final tableau when eqs. used
as PR in order 1 to 6

x_1	x_2	x_3	RHS	Eq. no.
1	0	0	1	1
0	1	0	1	2
0	0	1	1	3
0	0	0	−1	4
0	0	0	0	5
0	0	0	0	6

On the other hand, if we apply the GJ method on this system using equations as pivot rows in the order 4, 1, 2, 3, 5, 6, we get the following final tableau with three inconsistent equations:

Final tableau when rows 4,1,2,3,5,6 are PRs in order

x_1	x_2	x_3	RHS	Eq. no.
1	0	0	1	4
0	1	0	1	1
0	0	1	0	2
0	0	0	1	3
0	0	0	1	5
0	0	0	1	6

Given an inconsistent system of equations, to find the order in which rows in the original tableau have to be choosen as pivot rows to get the smallest number of inconsistent equations in the final tableau is a hard combinatorial optimization problem which belongs to the realm of integer programming, and is outside the scope of this book.

Also, given an inconsistent system of linear equations with the original RHS constants vector $b = (b_i)$, finding the modification of b to $b' = (b'_i)$ that will make the system consistent while minimizing the sum of absolute changes $\sum_i |b_i - b'_i|$, can be posed as a linear programming problem and solved very efficiently using linear programming algorithms. This, again, is outside the scope of our book.

Exercises

1.22.1: Consider the following system of 6 equations in 7 variables being solved by the GJ method. We want to carry out infeasibility analysis on this system, if it is inconsistent.

Find the orders in which the rows of the tableau have to be choosen as pivot rows, in order to (i) minimize, (ii) maxiimize, the number of inconsistent equations encountered during the method.

$$x_1 + x_2 + x_3 + x_4 + x_5 + x_6 + x_7 = 3$$
$$x_1 + x_2 + x_3 + x_4 + x_5 + x_6 + x_7 = 3$$
$$x_1 + x_2 + x_3 + x_4 + x_5 + x_6 + x_7 = 3$$
$$x_1 + x_2 + x_3 + x_4 + x_5 + x_6 + x_7 = 2$$
$$x_1 + x_2 + x_3 + x_4 + x_5 + x_6 + x_7 = 3$$
$$x_1 + x_2 + x_3 + x_4 + x_5 + x_6 + x_7 = 3$$

1.22.2: Do infeasibility analysis for systems of linear equations given in Exercise 1.16.4. For each of those systems find the modified RHS constants vector that would make the system feasible as determined by the method discussed in this section, and a solution of the modified system.

1.23 Homogeneous Systems of Linear Equations

The general system of linear equations in detached coefficient form is;

x_1	$\cdots$	x_j	$\cdots$	x_n	RHS constants
a_{11}	$\cdots$	a_{1j}	$\cdots$	a_{1n}	b_1
$\vdots$		$\vdots$		$\vdots$	$\vdots$
a_{i1}	$\cdots$	a_{ij}	$\cdots$	a_{in}	b_i
$\vdots$		$\vdots$		$\vdots$	$\vdots$
a_{m1}	$\cdots$	a_{mj}	$\cdots$	a_{mn}	b_m

It is said to be a **homogeneous system** if the RHS constants $b_i = 0$ for all $i = 1$ to m; **nonhomogeneous system** if at least one $b_i \neq 0$.

For the homogeneous system, $x = 0$ is always a solution. The interesting question that one needs to find about a homogeneous system is whether $x = 0$ is its unique solution, or whether it has a nonzero solution (i.e., a solution $x \neq 0$).

Algorithm: How to Check if a Homogeneous System Has a Nonzero Solution

BEGIN

Apply the GJ method or the G elimination method on the system. Since $b_i = 0$ for all i in the original tableau, in all the tableaus obtained under these methods, all the RHS constants will always be zero. So inconsistent rows will not appear.

At the end, if all the variables have been selected as basic variables (i.e., there are no nonbasic variables), then $x = 0$ is the only solution for the homogeneous system; i.e., it has no nonzero solution.

On the other hand, if there is at least one nonbasic variable at the end, the original homogeneous system has nonzero solutions. A nonzero solution for the system can be obtained by giving one or more nonbasic variables specific nonzero values, and then finding the corresponding values for the basic variables from the final tableau. To be specific, suppose the final tableau has p rows with the basic variables $x_{i_1}, \ldots, x_{i_p}$ in that order, and let the nonbasic variables be $x_{j_1}, \ldots, x_{j_{n-p}}$. Then after rearranging the variables in the tableau so that the basic variables appear first in their proper order, the final tableau has the following form:

When nonbasic variables exist

Final tableau after rearrangement of variables							
Basic Vars.	x_{i_1}	x_{i_2}	$\cdots$	x_{i_p}	x_{j_1}	$\cdots$	$x_{j_{n-p}}$	
x_{i_1}	1	0	$\cdots$	0	$\bar{a}_{1,j_1}$	$\cdots$	$\bar{a}_{1,j_{n-p}}$	0
x_{i_2}	0	1	$\cdots$	0	$\bar{a}_{2,j_1}$	$\cdots$	$\bar{a}_{2,j_{n-p}}$	0
$\vdots$	$\vdots$	$\vdots$		$\vdots$	$\vdots$		$\vdots$	$\vdots$
x_{i_p}	0	0	$\cdots$	1	$\bar{a}_{p,j_1}$	$\cdots$	$\bar{a}_{p,j_{n-p}}$	0

This is the final tableau if the GJ method is used. If the G method is used, the portion of the tableau corresponding to basic variables will be an upper triangular tableau.

There are $n - p$ nonbasic variables here. By giving exactly one nonbasic variable the value 1, and making all other nonbasic variables zero, and computing the corresponding values of the basic variables by back substitution if the G method is used, or reading them directly from the final tableau if the GJ method is used, we get $n - p$ distinct nonzero solutions to the original homogeneous system. These solutions are of the following form:

$$
\begin{pmatrix} x_{i_1} \\ x_{i_2} \\ \vdots \\ x_{i_p} \\ x_{j_1} \\ x_{j_2} \\ \vdots \\ x_{j_{n-p}} \end{pmatrix}
=
\begin{pmatrix} -\bar{a}_{1,j_1} \\ -\bar{a}_{2,j_1} \\ \vdots \\ -\bar{a}_{p,j_1} \\ 1 \\ 0 \\ \vdots \\ 0 \end{pmatrix},
\begin{pmatrix} -\bar{a}_{1,j_2} \\ -\bar{a}_{2,j_2} \\ \vdots \\ -\bar{a}_{p,j_0} \\ 0 \\ 1 \\ \vdots \\ 0 \end{pmatrix}, \ldots,
\begin{pmatrix} -\bar{a}_{1,j_{n-p}} \\ -\bar{a}_{2,j_{n-p}} \\ \vdots \\ -\bar{a}_{p,j_{n-p}} \\ 0 \\ 0 \\ \vdots \\ 1 \end{pmatrix},
$$

In these solutions of the system, the variables may not be in the usual order of x_1 to x_n, but the variables can be reordered if desired. Every solution of the original homogeneous system is a linear combination of these $(n - p)$ solutions and the dimension of the set of all solutions is $(n - p)$.

END

Example 1:

We solve the following homogeneous system using the GJ pivotal method. The original detached coefficient tableau is the one at the top. We number the equations serially and record this number in the rightmost column. PR (pivot row), PC (pivot column), RC (redundant constraint to be eliminated), BV (basic variable selected in this row) have the same meaning as before, and the pivot elements are always boxed.

BV	x_1	x_2	x_3	x_4	RHS		Eq. no.
	$\boxed{1}$	2	0	1	0	PR	1
	-1	-3	1	0	0		2
	0	-1	1	1	0		3
	0	-2	3	1	0		4
	0	0	1	1	0		5
	PC $\uparrow$						

BV	x_1	x_2	x_3	x_4	RHS		Eq. no.
x_1	1	2	0	1	0		1
	0	$\boxed{-1}$	1	1	0	PR	2
	0	-1	1	1	0		3
	0	-2	3	1	0		4
	0	0	1	1	0		5
		PC ↑					
x_1	1	0	2	3	0		1
x_2	0	1	-1	-1	0		2
	0	0	0	0	0	RC	3
	0	0	$\boxed{1}$	-1	0	PR	4
	0	0	1	1	0		5
			PC ↑				
x_1	1	0	0	5	0		1
x_2	0	1	0	-2	0		2
x_3	0	0	1	-1	0		4
	0	0	0	$\boxed{2}$	0	PR	5
				PC ↑			
			Final Tableau				
x_1	1	0	0	0	0		1
x_2	0	1	0	0	0		2
x_3	0	0	1	0	0		4
x_4	0	0	0	1	0		5

Since all the variables are basic variables (i.e., there are no nonbasic variables at the end), this homogeneous system has $x = 0$ as the unique solution (i.e., it has no nonzero solution).

Example 2:

We will use the G elimination method to solve the homogeneous system for which the detached coefficient tableau is the one at the top. The notation is the same as in Example 1 above.

BV	x_1	x_2	x_3	x_4	x_5	RHS		Eq. no.
	$\boxed{1}$	0	0	1	1	0	PR	1
	0	1	0	1	0	0		2
	2	1	0	3	2	0		3
	1	1	1	1	0	0		4
	PC ↑							

BV	x_1	x_2	x_3	x_4	x_5	RHS		Eq. no.
x_1	1	0	0	1	1	0		1
	0	1	0	$\boxed{1}$	0	0	PR	2
	0	1	0	1	0	0		3
	0	1	1	0	-1	0		4
			PC $\uparrow$					
x_1	1	0	0	1	1	0		1
x_4	0	1	0	1	0	0		2
	0	0	0	0	0	0	RC	3
	0	$\boxed{1}$	1	0	-1	0	PR	4
	PC$\uparrow$							
			Final Tableau					
x_1	1	0	0	1	1	0		1
x_4	0	1	0	1	0	0		2
x_2	0	1	1	0	-1	0		4

Since there are two nonbasic variables x_3, x_5 in the final tableau, this homogeneous system has nonzero solutions. Rearranging the columns of the basic variables in their proper order, and transferring the columns of the nonbasic variables to the right, the final tableau becomes:

BV	x_1	x_4	x_2	RHS
x_1	1	1	0	$-x_5$
x_4	0	1	1	0
x_2	0	0	1	$-x_3 + x_5$

First fixing the nonbasic variables at values $x_3 = 1, x_5 = 0$, and then solving the above upper triangular system by back substitution for the values of the basic variables, we get the first solution given below. Then we fix the nonbasic variables at values $x_3 = 0, x_5 = 1$, and get the second solution given below. In writing these solutions, we rearranged the variables in their natural order x_1 to x_5.

$$\begin{pmatrix} x_1 \\ x_2 \\ x_3 \\ x_4 \\ x_5 \end{pmatrix} = \begin{pmatrix} -1 \\ -1 \\ 1 \\ 1 \\ 0 \end{pmatrix}, \begin{pmatrix} 0 \\ 1 \\ 0 \\ -1 \\ 1 \end{pmatrix}$$

The general solution of the system is a linear combination of these two solutions. Taking the coefficients of this linear combination to be $\bar{x}_3, \bar{x}_5$ respectively, the general solution is:

$$
\begin{pmatrix} x_1 \\ x_2 \\ x_3 \\ x_4 \\ x_5 \end{pmatrix} = \bar{x}_3 \begin{pmatrix} -1 \\ -1 \\ 1 \\ 1 \\ 0 \end{pmatrix} + \bar{x}_5 \begin{pmatrix} 0 \\ 1 \\ 0 \\ -1 \\ 1 \end{pmatrix}
$$

where $\bar{x}_3, \bar{x}_5$ are arbitrary real parameters representing the values of the nonbasic variables x_3, x_5 in the solution.

Exercises

1.23.1: In each of the following homogeneous system of linear equations, check whether it has a nonzero solution. If it does, find a nonzero solution, and an expression for the general nonzero solution of the system. Also, determine the dimension of the solution set in each case. ((i)–(iii) have no nonzero solution, the others do).

(i)

x_1	x_2	x_3	x_4	
1	-1	2	1	0
0	1	1	2	0
-1	0	-3	0	0
1	0	1	1	0
0	-1	3	5	0

(ii)

x_1	x_2	x_3	x_4	x_5	
1	2	0	-3	-1	0
-1	0	-1	2	0	0
0	-3	2	-2	1	0
-1	0	0	-1	2	0
0	1	-1	2	1	0
-1	0	0	-2	3	0

(iii)

x_1	x_2	x_3	
1	2	2	0
2	1	5	0
2	2	1	0

(iv)

x_1	x_2	x_3	x_4	x_5	x_6	
1	1	1	0	0	0	0
0	0	0	1	1	1	0
1	0	0	1	0	0	0
0	1	0	0	1	0	0
0	0	1	0	0	1	0

(v)

x_1	x_2	x_3	
1	2	2	0
2	1	5	0
7	8	16	0

(vi)

x_1	x_2	x_3	x_4	
-1	2	-2	3	0
2	-5	3	2	0
-1	3	0	-2	0
-4	10	-5	-1	0
-4	11	-4	-9	0

1.24 Solving Systems of Linear Equations And/Or Inequalities

Linear algebra techniques, the focus of this book, can directly handle only systems of linear equations. When there are also linear inequalities in the system, linear algebra techniques are not able to handle that system directly. To handle systems involving linear inequalities, we need linear programming techniques, which are outside the scope of this book.

1.25 Can We Obtain a Nonnegative Solution to a System of Linear Equations by the GJ or the G Methods Directly?

Nonnegativity restrictions on the variables are actually inequality constraints.

If a system of linear equations has a unique solution which happens to be nonnegative, the GJ or the G methods will of course find it. However, when a system of linear equations has multiple solutions, some of them may satisfy the nonnegativity constraints, others may not. In this case we cannot guarantee that the GJ or the G methods will find a nonnegative solution to the system of equations even if such a solution exists; or determine conclusively whether a nonnegative solution exists for the system.

As mentioned earlier, the problem of finding a nonnegative solution to a system of linear equations can be posed as a linear programming problem and solved using linear programming techniques. However, linear programming is outside the scope of this book.

1.26   Additional Exercises

1.26.1: Toy store problem: A toystore chain has several stores in the midwest. For the coming X-mas season, they need to put orders with their overseas suppliers before the end of May for delivery in time for the X-mas sales period.

Since unsold toys at the end of the X-mas season do not contribute much to the profit of the company, they base their order quantities quite close to the expected sales volume. From experience over the years they observed that the X-mas sales volume has a high positive correlation with the DJA = Dow Jones average (a measure of the economic status of the region prior to the sales period), and a high negative correlation with the % unemployment rate in the region. Following table gives data on the DJA during the months of Feb, Mar, Apr (these are independent variables x_1, x_2, x_3), the % unemployment in the region during this period (independent variable x_4), and the toy sales volume in the region in $ million during the X-mas sales season (dependent variable y) between 1990–2001.

From above discussion it is reasonable to assume that the expected value of y can be approximated by a function $a_0 + a_1x_1 + a_2x_2 + a_3x_3 + a_4x_4$, where the parameters a_1, a_2, a_3 are expected to be positive, and a_4 is expected to be negative. But ignore these sign restrictions on the parameters, and derive the normal equations for finding the best values for them that give the closest fit, by the least squares method.

Year	x_1	x_2	x_3	x_4	y
2001	10690	10185	10306	4.3	59
2000	10533	10525	10798	4.0	60
1999	9356	9550	10307	4.3	54
1998	8307	8664	8940	4.6	47
1997	6828	6727	6800	5.2	36
1996	5435	5527	5579	5.5	28
1995	3927	4080	4239	5.5	20
1994	3898	3723	3634	6.5	17
1993	3344	3405	3434	7.1	14
1992	3247	3253	3294	7.4	13
1991	2798	2903	2895	6.6	11
1990	2607	2665	2673	5.2	10

1.26.2: A brother B and his sister S stated the following facts about their family. S has the same number of brothers and sisters. B has no brothers, but has as many sisters as the sum of the number of brothers and sisters of S. Formulate the problem of determining how many sons and daughters their parents have using a system of linear equations.

1.26.3: The principle behind drain cleaners to unclog a sink is: they contain sodium hydroxide (NaOH) and aluminium (Al) which when mixed with water (H_2O) react to produce $NaAl(OH)_4$ and hydrogen gas that bubbles inside the plumbing to unclog the drain. So, NaOH, Al, H_2O react to produce $NaAl(OH)_4$ and H_2. Balance this reaction.

1.26.4: A test for arsenic (As) poisoning is based on combining hydrochloric acid (HCl), zinc (Zn), with a sample hair of the victim (which is likely to contain H_3AsO_4 if the victim suffers from arsenic poisoning). Zn, H_3AsO_4, and HCl combine to produce $ZnCl_2$, H_2O, and AsH_3 which is a gas called arsene gas whose presence can be easily detected. Balance this reaction.

1.26.5: A trucking company uses three types of trucks of different capacities. The company plans to use these trucks to haul various numbers of three different machines (machines A, B, C) from the shipyard where they are to be picked up, to the site of a new plant being set up.

Truck type 1 can haul one machine A, one machine B, and two copies of machine C in a full load.

Truck type 2 can haul one each of machines A, B, C in a full load.

Truck type 3 can haul one of each of machines A, C, and two copies of machine B in a full load.

12 copies of machine A, 17 copies of machine B, and 15 copies of machine C have to be picked up. Assuming that each truck sent can be fully loaded as mentioned above, find how many trucks of each type should be sent to pick up all the machines. Formulate the problem and obtain a solution.

1.26.6: There are three fretilizer types available (Fer 1, 2, 3) in bags. each bag contains the following amounts of the principal fertilizer components N, P, K.

Fertilizer type	Lbs. of following per bag		
	N	P	K
Fer 1	1	3	2
Fer 2	3	3	1
Fer 3	2	5	5

A farmer wants to apply exactly 25 lbs of N, 42 lbs of P, and 28 lbs of K on his field. Construct a model to determine how many bags of Fer 1, 2, 3 each to apply on the field to get exactly these amounts of N, P, K; and find the solution of that model.

1.26.7: (From C. Rorres and H. Anton, *Applications of Linear Algebra*, Wiley, 1984) Hooke's law of physics states that the length y of a uniform spring is a linear function of the force applied x; i.e., y can be approximated by $a + bx$, where the coefficient b is called the spring constant. A particular unstretched spring has length 6.1 (i.e., when $x = 0$, $y = 6.1$). Different forces were applied to this spring and it resulted in the following data.

$x =$ force applied in lbs.	$y =$ spring length under force
0	6.1
2	7.7
4	8.6
6	10.4

Derive the normal equations to find the best values for the parameters a, b that give the best fit for Hooke's law for this data using the method of least squares.

1.26.8: Find the value of b_3 for which the following system has at least one solution. Also, show that if the system has one solution, then it has an infinite number of solutions.

$$x_1 + x_2 = 2$$
$$2x_2 + x_3 = 3$$
$$x_1 - x_2 + (b_3 - 6)x_3 = -1$$
$$x_1 + 3x_2 + x_3 = b_3$$

1.26.9: Find the values of the parameter a for which the following system has no solutions, unique solution, infinite number of solutions.

$$x_1 + x_2 = 6$$
$$x_1 + x_3 = 7$$
$$ax_1 + x_2 + x_3 = 13$$

1.26.10: Develop a system of linear equations for which the general solution is $x = (x_1, x_2, x_3, x_4)^T = (-17 + 8\alpha - 7\beta, 28 - 13\alpha + 19\beta, \alpha, \beta)^T$, where α, β are arbitrary real valued parameters.

1.26.11: Consider the following system of linear equations. Find the conditions that the RHS constants have to satisfy for the system to have at least one solution. Also, show that if the system has a solution, then it has an infinite number of solutions.

$$x_1 + x_2 = b_1$$
$$x_2 + x_3 = b_2$$
$$x_1 + 2x_2 + x_3 = b_3$$
$$x_3 + x_4 = b_4$$
$$x_1 + 2x_2 + 2x_3 + x_4 = b_5$$

1.26.12: In the following homogeneous system α, β are real valued parameters. For which values of these parameters does the system have a nonzero solution? Explain. When those conditions are satisfied, find an expression for the general solution of the system.

$$x_1 + x_2 + x_3 = 0$$
$$-x_1 + 2x_2 - 2x_3 = 0$$
$$\alpha x_2 + \beta x_3 = 0$$

1.26.13: What conditions do the parameters a, b, c, d in the following system have to satisfy for this system to have at least one solution? If those conditions are satisfied, find an expression for the general solution of the system.

$$x_1 - x_2 + x_3 - x_4 = a$$
$$-x_1 + 2x_2 + 2x_3 + 2x_4 = b$$
$$x_2 + 3x_3 + x_4 = c$$
$$x_1 + 4x_3 = d$$

1.26.14: Find conditions that α has to satisfy so that the following system has at least one nonzero solution. When that condition holds, find the general solution of the system.

$$x_1 - x_2 + 2x_3 = 0$$
$$2x_1 - 3x_2 + x_3 = 0$$
$$\alpha x_2 - 3x_3 = 0$$

Reference

[1.1] K. G. Murty, S. N. Kabadi, and R. Chandrasekaran, "Infeasibility Analysis for Linear Systems, A Survey", *Arabian Journal of Science and Engineering*, 25, 1C (June 2000) 3–18.

Chapter 2

Matrices, Matrix Arithmetic, Determinants

2.1 A Matrix as a Rectangular Array of Numbers, Its Rows and Columns

A **matrix** is a rectangular array of real numbers. If it has m rows and n columns, it is said to be an $m \times n$ matrix, or a **matrix of order** or **size** $m \times n$. Here is a 3×5 matrix with three rows and five columns.

$$\begin{pmatrix} -1 & 7 & 8.1 & 3 & -2 \\ 0 & -2 & 0 & 5 & -7 \\ 1 & -1 & 3 & 0 & -8 \end{pmatrix}$$

The size or order of a matrix is always a pair of positive numbers, the first being the number of rows of the matrix, and the second the number of columns.

In mathematical literature, matrices are usually written within brackets of $(\ldots)$ or $[\ldots]$ type on the left and right.

History of the name "matrix": The word "matrix" is derived from the Latin word for "womb" or "a pregnant animal". In 1848 J. J. Sylvester introduced this name for a rectangular array of numbers, thinking of the large number of subarrays that each such array can generate (or "give birth to"). A few years later A. Cayley introduced matrix multiplication and the basic theory of matrix algebra quickly followed.

Initially, the concept of a matrix was introduced mainly as a tool for storing the coefficients of variables in a system of linear equations, in that case it is called the **coefficient matrix** or **matrix of coefficients** of the system. In 1858 A. Cayley wrote a memoir on the theory of linear transformations and matrices. With arithmetical operations on matrices defined, we can think of these operations as constituting an *algebra* of matrices, and the study of **matrix algebra** by Cayley and Sylvester in the 1850's attracted a lot of attention that led to important progress in matrix and determinant theory. ⋈

Consider the following system of linear equations.

$$3x_2 - 4x_4 + x_1 - x_3 = -10$$

$$2x_1 + 5x_3 - 2x_4 + 6x_2 = -5$$

$$4x_4 - x_1 + 8x_3 - 3x_2 = 7.$$

Since this is a system of 3 equations in 4 variables, its coefficient matrix will be of order 3×4. Each row of the coefficient matrix will correspond to an equation in the system, and each of its columns corresponds to a variable. So, in order to write the coefficient matrix of this system, it is necessary to arrange the variables in a particular order, say as in the column vector

$$x = \begin{pmatrix} x_1 \\ x_2 \\ x_3 \\ x_4 \end{pmatrix}$$

and write each equation with the terms of the variables arranged in this particular order x_1, x_2, x_3, x_4. This leads to

$$x_1 + 3x_2 - x_3 - 4x_4 = -10$$

$$2x_1 + 6x_2 + 5x_3 - 2x_4 = -5$$

$$-x_1 - 3x_2 + 8x_3 + 4x_4 = 7.$$

Now the coefficient matrix of the system $A = (a_{ij} : i = 1, \ldots, m; j = 1, \ldots, n)$, where a_{ij} is the coefficient of x_j in the ith equation, can be easily written. It is

$$\begin{pmatrix} 1 & 3 & -1 & -4 \\ 2 & 6 & 5 & -2 \\ -1 & -3 & 8 & 4 \end{pmatrix}$$

For $i = 1$ to 3, **row i of this matrix** A corresponds to the ith equation in the system, and we denote it by the symbol $A_{i\cdot}$. For $j = 1$ to 4, **column j of this matrix** A, denoted by the symbol $A_{\cdot j}$, corresponds to, or is associated with, the variable x_j in the system.

Thus the coefficient matrix of a system of linear equations is exactly the array of entries in the detached coefficient tableau representation of the system without the RHS constants vector. The rows of the coefficient matrix are the rows of the detached coefficient tableau without the RHS constant, its columns are the columns of the detached coefficient tableau.

In general, an $m \times n$ matrix D can be denoted specifically as $D_{m \times n}$ to indicate its order; it will have m row vectors denoted by $D_1., \ldots, D_m.$; and n column vectors denoted by $D_{.1}, \ldots, D_{.n}$. The matrix D itself can be viewed as the array obtained by putting its column vectors one after the other horizontally as in

$$D_{m \times n} = (D_{.1} \quad D_{.2} \ldots D_{.n})$$

or putting its row vectors one below the other vertically as

$$D_{m \times n} = \begin{pmatrix} D_{1.} \\ D_{2.} \\ \vdots \\ D_{m.} \end{pmatrix}.$$

A row vector by itself can be treated as a matrix with a single row. Thus the row vector

$$(3, \quad -4, \quad 7, \quad 8, \quad 9)$$

can be viewed as a matrix of order 1×5.

In the same way, a column vector can be viewed as a matrix with a single column. So, the column vector

$$\begin{pmatrix} -3 \\ 2 \\ 1 \end{pmatrix}$$

can be viewed as a matrix of order 3×1.

The entry in a matrix D in its ith row and jth column is called its (i, j)th entry, and denoted by a symbol like d_{ij}. Then the matrix D itself can be denoted by the symbol (d_{ij}). So, the $(2, 3)$th entry of the coefficient matrix A given above, $a_{23} = 5$.

Let b denote the column vector of RHS constants in a system of linear equations in which the coefficient matrix is A of order $m \times n$. Then the matrix $\mathcal{A}$ obtained by including b as another column at the right of A, written usually in a **partitioned** form as in

$$\mathcal{A} = (A|b)$$

is known as the **augmented matrix** of the system of equations.

A **partitioned matrix** is a matrix in which the columns and/or the rows are partitioned into various subsets. The augmented matrix $\mathcal{A}$ is written as a partitioned matrix with the columns associated with the variables in one subset of columns, and the RHS constants column vector by itself in another subset of columns.

As an example, consider the following system of 3 equations in 4 variables.

$$3x_2 - 4x_4 - x_3 = -10$$

$$2x_1 - 2x_4 + 6x_2 = -5$$

$$4x_4 - x_1 + 8x_3 = 7.$$

Arranging the variables in the order x_1, x_2, x_3, x_4, the system is

$$3x_2 - x_3 - 4x_4 = -10$$

$$2x_1 + 6x_2 - 2x_4 = -5$$

$$-x_1 + 8x_3 + 4x_4 = 7$$

So, for this system, the coefficient matrix A, and the augmented matrix are

$$A = \begin{pmatrix} 0 & 3 & -1 & -4 \\ 2 & 6 & 0 & -2 \\ -1 & 0 & 8 & 4 \end{pmatrix}, \quad \mathcal{A} = \begin{pmatrix} 0 & 3 & -1 & -4 & -10 \\ 2 & 6 & 0 & -2 & -5 \\ -1 & 0 & 8 & 4 & 7 \end{pmatrix}$$

2.2   Matrix-Vector Multiplication

The concept of the product of a matrix times a column vector in this specific order, is formulated so that we can write the system of equations with coefficient matrix A, column vector of variables x, and RHS constants vector b, in matrix notation as $Ax = b$. We define this concept next.

Definitions: Matrix-vector products: Consider a matrix D of order $m \times n$, and a vector p. The product Dp, written in this specific order, is only defined if p is a column vector in R^n (i.e., if the number of column vectors of D is equal to the number of rows in the column vector p), i.e., if $p = (p_1, \ldots, p_n)^T$. Then the product Dp is itself a column vector given by

$$Dp = \begin{pmatrix} D_{1.}p \\ \vdots \\ D_{m.}p \end{pmatrix}$$

$$= \sum_{j=1}^{n} p_j D_{.j}.$$

Also, if q is a vector, the product qD, written in this specific order, is only defined if q is a row vector in R^m (i.e., if the number of columns in q is equal to the number of rows in D), i.e., if $q = (q_1, \ldots, q_m)^T$, and the product qD is itself a row vector given by

$$qD = (qD_{.1}, \ldots, qD_{.n})$$

$$= \sum_{i=1}^{m} q_i D_{i.}.$$

We have given above two ways of defining each matrix-vector product; both ways yield the same result for the product in each case.

In this chapter we use superscripts to number various vectors in a set of vectors (for example, q^1, q^2, q^3 may denote three different vectors in R^n). We always use subscripts to indicate various entries in a vector, for example for a row vector $p \in R^3$, the entries in it will be denoted by p_1, p_2, p_3, so, $p = (p_1, p_2, p_3)$.

Example 1:

Let

$$q^1 = (1, 2, 1); \quad q^2 = (1, 2); \quad q^3 = (2, 2, 2, 1)$$

$$A = \begin{pmatrix} 1 & 3 & -1 & -1 & -4 \\ 2 & 6 & 5 & 5 & -2 \\ -1 & -3 & 8 & 8 & 4 \end{pmatrix}, \quad p^3 = \begin{pmatrix} 1 \\ 2 \end{pmatrix}, \quad p^4 = \begin{pmatrix} 0 \\ 1 \\ 1 \\ 1 \\ 1 \\ 1 \end{pmatrix}$$

$$p^1 = \begin{pmatrix} 1 \\ 0 \\ 1 \\ -1 \\ 2 \end{pmatrix}, \quad p^2 = \begin{pmatrix} -1 \\ 2 \\ 1 \\ 0 \\ 1 \end{pmatrix}.$$

The products $Ap^3, Ap^4, Aq^1, Aq^2, Aq^3, p^1A, p^2A, p^3A, p^4A, q^2A, q^3A$ are not defined. The reader is encouraged to explain the reasons for each. Here are the row vectors of matrix A.

$$A_{1.} = (1, 3, -1, -1, -4)$$
$$A_{2.} = (2, 6, 5, 5, -2)$$
$$A_{3.} = (-1, -3, 8, 8, 4).$$

The column vectors of matrix A are:

$$A_{.1} = \begin{pmatrix} 1 \\ 2 \\ -1 \end{pmatrix}, \quad A_{.2} = \begin{pmatrix} 3 \\ 6 \\ -3 \end{pmatrix}, \quad A_{.3} = \begin{pmatrix} -1 \\ 5 \\ 8 \end{pmatrix}$$

$$A_{.4} = \begin{pmatrix} -1 \\ 5 \\ 8 \end{pmatrix}, \quad A_{.5} = \begin{pmatrix} -4 \\ -2 \\ 4 \end{pmatrix}.$$

We have

$$Ap^1 = \begin{pmatrix} A_{1.}p^1 \\ A_{2.}p^1 \\ A_{3.}p^1 \end{pmatrix} = \begin{pmatrix} -7 \\ -2 \\ 7 \end{pmatrix}.$$

We also have

$$Ap^1 = 1A_{.1} + 0A_{.2} + 1A_{.3} - 1A_{.4} + 2A_{.5} = \begin{pmatrix} 1 \\ 2 \\ -1 \end{pmatrix} + \begin{pmatrix} -1 \\ 5 \\ 8 \end{pmatrix}$$

$$- \begin{pmatrix} -1 \\ 5 \\ 8 \end{pmatrix} + 2 \begin{pmatrix} -4 \\ -2 \\ 4 \end{pmatrix} = \begin{pmatrix} -7 \\ -2 \\ 7 \end{pmatrix}$$

giving the same result. In the same way verify that both definitions give

$$Ap^2 = \begin{pmatrix} 0 \\ 13 \\ 7 \end{pmatrix}.$$

Similarly

$$q^1 A = (q^1 A_{.1}, q^1 A_{.2}, q^1 A_{.3}, q^1 A_{.4}, q^1 A_{.5}) = (4, 12, 17, 17, -4)$$
$$= 1A_{1.} + 2A_{2.} + A_{3.}.$$

Some Properties of Matrix-Vector Products

Let A be an $m \times n$ matrix, $u^1, \ldots, u^r$ be all column vectors in R^n, and $\alpha_1, \ldots, \alpha_r$ real numbers. Then

$$A(u^1 + \cdots + u^r) = \sum_{t=1}^{r} Au^t$$

$$A(\alpha_1 u^1 + \cdots + \alpha_r u^r) = \sum_{t=1}^{r} \alpha_t A u^t.$$

We will provide a proof of the second relation above. (The first relation is a special case of the second obtained by taking $\alpha_t = 1$ for all t.) Let u denote a general column vector in R^n. From the definition

$$Au = \begin{pmatrix} A_{1.}u \\ \vdots \\ A_{m.}u \end{pmatrix}.$$

From Result 1.7.1 of Section 1.7, $A_{i.}u$ is a linear function of u, and hence from the definition of linear functions (see Section 1.7)

$$A_{i.}(\alpha_1 u^1 + \cdots + \alpha_r u^r) = \sum_{t=1}^{r} \alpha_t A_{i.}u^t.$$

Using this we see that $A\left(\sum_{t=1}^{r} \alpha_t u^t \right)$

$$= \begin{pmatrix} \sum_{t=1}^{r} \alpha_t A_{1.}u^t \\ \vdots \\ \sum_{t=1}^{r} \alpha_t A_{m.}u^t \end{pmatrix} = \sum_{t=1}^{r} \alpha_t \begin{pmatrix} A_{1.}u^t \\ \vdots \\ A_{m.}u^t \end{pmatrix} = \sum_{t=1}^{r} \alpha_t A u^t$$

proving the second relation above.

Also, if $v^1, \ldots, v^r$ are all row vectors in R^m, then we get the following results from similar arguments

$$(v^1 + \cdots + v^r)A = \sum_{t=1}^{r} v^t A$$

$$(\alpha_1 v^1 + \cdots + \alpha_r v^r)A = \sum_{t=1}^{r} \alpha_t v^t A$$

Exercises

2.2.1: Let

$$A = \begin{pmatrix} 1 & 0 & -1 \\ -2 & 1 & 2 \end{pmatrix}, \quad B = \begin{pmatrix} 1 & 1 & 2 \\ -1 & 0 & 3 \\ -2 & 2 & 1 \end{pmatrix}, \quad C = \begin{pmatrix} -2 & 1 \\ 1 & -2 \\ -2 & -3 \end{pmatrix}.$$

$$p^1 = \begin{pmatrix} -2 \\ 3 \end{pmatrix}, \quad p^2 = \begin{pmatrix} 0 \\ 3 \\ 2 \end{pmatrix}, \quad p^3 = \begin{pmatrix} 1 \\ 17 \\ -8 \\ 9 \end{pmatrix}.$$

$$q^1 = (-3, 2), \quad q^2 = (-2, -1, 2), \quad q^3 = (8, -7, 9, 15).$$

Mention which of Ax, Bx, Cx, yA, yB, yC are defined for $x, y \in \{p^1, p^2, p^3, q^1, q^2, q^3\}$ and which are not. Compute each one that is defined.

2.3 Three Ways of Representing the General System of Linear Equations through Matrix Notation

Consider the general system of m equations in n variables in detached coefficient tableau form.

x_1	$\cdots$	x_j	$\cdots$	x_n	RHS constants
a_{11}	$\cdots$	a_{1j}	$\cdots$	a_{1n}	b_1
$\vdots$		$\vdots$		$\vdots$	$\vdots$
a_{i1}	$\cdots$	a_{ij}	$\cdots$	a_{in}	b_i
$\vdots$		$\vdots$		$\vdots$	$\vdots$
a_{m1}	$\cdots$	a_{mj}	$\cdots$	a_{mn}	b_m

Let

$$A = (a_{ij}) \quad \text{the } m \times n \text{ coefficient matrix}$$
$$b = (b_1, \ldots, b_m)^T \quad \text{the column vector of RHS constants}$$
$$x = (x_1, \ldots, x_n)^T \quad \text{the column vector of variables}$$

in this system.

Representing the System as a Matrix Equation

Clearly, the above system of equations can be represented by the **matrix equation**

$$Ax = b.$$

The number of rows of the coefficient matrix A is the number of equations in the system, and the number of columns in A is the number of variables in the system.

From now on, we will use this notation to represent the general system of linear equations whenever it is convenient for us to do so.

Example 2:

Consider the following system in detached coefficient form.

x_1	x_2	x_3	
1	-2	-1	-5
-3	1	2	-7

Denoting the coefficient matrix, RHS constants vector, column vector of variables, respectively by A, b, x, we have

$$A = \begin{pmatrix} 1 & -2 & -1 \\ -3 & 1 & 2 \end{pmatrix}, \quad b = \begin{pmatrix} -5 \\ -7 \end{pmatrix}, \quad x = \begin{pmatrix} x_1 \\ x_2 \\ x_3 \end{pmatrix}$$

and we verify that this system of equations is: $Ax = b$ in matrix notation.

Representing the System by Equations

The ith row of the coefficient matrix A is $A_{i.} = (a_{i1}, \ldots, a_{in})$. Hence the equation by equation representation of the system is

$$A_{i.}x = b_i \quad i = 1 \text{ to } m.$$

Representing the System as a Vector Equation

Another way of representing the system of equations $Ax = b$ as a vector equation involving the column vectors of the coefficient matrix A and the RHS constants vector b is

$$x_1 A_{.1} + x_2 A_{.2} + \cdots + x_n A_{.n} = b$$

i.e., the values of the variables in a solution of the system x are the coefficients in an expression of b as a linear combination of the column vectors of the coefficient matrix A.

As an example, consider the system of equations $Ax = b$ in Example 2, where

$$A = \begin{pmatrix} 1 & -2 & -1 \\ -3 & 1 & 2 \end{pmatrix}, \quad b = \begin{pmatrix} -5 \\ -7 \end{pmatrix}, \quad x = \begin{pmatrix} x_1 \\ x_2 \\ x_3 \end{pmatrix}.$$

One representation of this system as a set of equations that should hold simultaneously is

$$A_{1.}x = b_1 \text{ which is } x_1 - 2x_2 - x_3 = -5$$

$$A_{2.}x = b_2 \text{ which is } -3x_1 + x_2 + 2x_3 = -7.$$

Another representation of this system as a vector equation is

$$x_1 A_{.1} + x_2 A_{.2} + x_3 A_{.3} = b$$

i.e. $\quad x_1 \begin{pmatrix} 1 \\ -3 \end{pmatrix} + x_2 \begin{pmatrix} -2 \\ 1 \end{pmatrix} + x_3 \begin{pmatrix} -1 \\ 2 \end{pmatrix} = \begin{pmatrix} -5 \\ -7 \end{pmatrix}.$

The basic solution of this system corresponding to the basic vector of variables (x_1, x_3) obtained by the GJ pivotal method is $\bar{x} = (17, 0, 22)^T$. It satisfies this vector equation because

$$17 \begin{pmatrix} 1 \\ -3 \end{pmatrix} + 0 \begin{pmatrix} -2 \\ 1 \end{pmatrix} + 22 \begin{pmatrix} -1 \\ 2 \end{pmatrix} = \begin{pmatrix} -5 \\ -7 \end{pmatrix}.$$

The general solution of this system is

$$\begin{pmatrix} x_1 \\ x_2 \\ x_3 \end{pmatrix} = \begin{pmatrix} 17 - 3\bar{x}_2 \\ \bar{x}_2 \\ 22 - 5\bar{x}_2 \end{pmatrix}$$

where $\bar{x}_2$ is an arbitrary real valued parameter, and it can be verified that this satisfies the above vector equation.

Cautionary Note: Care in writing matrix products: This is for those students who are still not totally familiar with matrix manipulations. In real number arithmetic, the equation

$$5 \times 4 = 20$$

can be written in an equivalent manner as

$$5 = \frac{20}{4}.$$

Carrying this practice to equations involving matrix products will most often result in drastic blunders. For example, writing

$$Ax = b \quad \text{as} \quad A = \frac{b}{x} \quad \text{or as} \quad x = \frac{b}{A} \quad \text{is a blunder}$$

because division by vectors is not defined, inverse of a rectangular matrix is not defined; and even if the matrix A has an inverse to be defined later, it is not denoted by $\frac{1}{A}$ but by a different symbol.

Therefore deducing that $x = \frac{b}{A}$ from $Ax = b$ is always a drastic blunder.

Exercises

2.3.1: For A, b given below, write the system of linear equations $Ax = b$ as a vector equation.

(i) $\quad A = \begin{pmatrix} 1 & 2 & 0 & 9 \\ -1 & 0 & 1 & 1 \\ 0 & 3 & 4 & -7 \end{pmatrix}, \quad b = \begin{pmatrix} 3 \\ 1 \\ -4 \end{pmatrix}.$

(ii) $\quad A = \begin{pmatrix} 1 & 1 & 1 & 1 \\ -1 & 2 & 1 & 2 \\ 1 & -2 & 1 & -2 \\ 0 & -1 & -2 & 0 \end{pmatrix}, \quad b = \begin{pmatrix} 9 \\ -7 \\ 0 \\ 4 \end{pmatrix}.$

2.4 Scalar Multiplication and Addition of Matrices, Transposition

Let $A = (a_{ij})$ be a matrix of order $m \times n$. Scalar multiplication of this matrix A by a scalar α involves multiplying each element in A by α. Thus

$$\alpha A = \alpha(a_{ij}) = (\alpha a_{ij}).$$

For instance

$$6 \begin{pmatrix} 1 & -2 & -1 \\ -3 & 1 & 2 \end{pmatrix} = \begin{pmatrix} 6 & -12 & -6 \\ -18 & 6 & 12 \end{pmatrix}.$$

Given two matrices A and B, their sum is only defined if both matrices are of the same order. So, if $A = (a_{ij})$ and $B = (b_{ij})$ are both of order $m \times n$, then their sum $A + B$ is obtained by summing up corresponding elements, and is equal to $(a_{ij} + b_{ij})$. For instance

$$\begin{pmatrix} 1 & -2 & -1 \\ -3 & 1 & 2 \end{pmatrix} + \begin{pmatrix} 1 & 0 \\ 0 & 1 \end{pmatrix} \qquad \text{is not defined.}$$

$$\begin{pmatrix} 1 & -2 & -1 \\ -3 & 1 & 2 \end{pmatrix} + \begin{pmatrix} 3 & -10 & 18 \\ 8 & -7 & 20 \end{pmatrix} = \begin{pmatrix} 4 & -12 & 17 \\ 5 & -6 & 22 \end{pmatrix}.$$

The operation of transposition of a matrix changes its row vectors into column vectors and vice versa. Thus if A is an $m \times n$ matrix, its transpose denoted by A^T is an $n \times m$ matrix obtained by writing the row vectors of A as column vectors in A^T in their proper order. For instance

$$\begin{pmatrix} 1 & -2 & -1 \\ -3 & 1 & 2 \end{pmatrix}^T = \begin{pmatrix} 1 & -3 \\ -2 & 1 \\ -1 & 2 \end{pmatrix}$$

$$\begin{pmatrix} 3 & 9 \\ 18 & -33 \end{pmatrix}^T = \begin{pmatrix} 3 & 18 \\ 9 & -33 \end{pmatrix}.$$

Earlier in Section 1.4 we defined the transpose of a row vector as a column vector and vice versa. That definition tallies with the definition of transpose for matrices given here when those vectors are looked at as matrices with a single row or column.

Some Properties of Matrix Addition and Transposition

Let $A^t = (a_{ij}^t)$ be a matrix of order $m \times n$ for $t = 1$ to r. The sum, and a linear combination with multipliers $\alpha_1, \ldots, \alpha_r$ respectively, of these r matrices are defined similarly. We have

$$A^1 + \cdots + A^r = \left(\sum_{t=1}^{r} a_{ij}^t \right)$$

$$\alpha_1 A^1 + \cdots + \alpha_r A^r = \left(\sum_{t=1}^{r} \alpha_t a_{ij}^t \right).$$

In finding the sum of two or more matrices, these matrices can be written in the sum in any arbitrary order, the result does not depend on this order. For instance

$$\begin{pmatrix} 1 & -2 & 3 \\ 0 & 8 & -7 \end{pmatrix} + \begin{pmatrix} 3 & 6 & -3 \\ -4 & -2 & -1 \end{pmatrix} + \begin{pmatrix} -1 & -8 & -3 \\ -2 & -4 & 13 \end{pmatrix}$$

$$= \begin{pmatrix} -1 & -8 & -3 \\ -2 & -4 & 13 \end{pmatrix} + \begin{pmatrix} 1 & -2 & 3 \\ 0 & 8 & -7 \end{pmatrix} + \begin{pmatrix} 3 & 6 & -3 \\ -4 & -2 & -1 \end{pmatrix}$$

$$= \begin{pmatrix} 3 & -4 & -3 \\ -6 & 2 & 5 \end{pmatrix}.$$

Let A be an $m \times n$ matrix, $u = (u_1, \ldots, u_n)^T$, $v = (v_1, \ldots v_m)$. Then it can be verified that

$$(Au)^T = u^T A^T \qquad (vA)^T = A^T v^T.$$

So, the transpose of a matrix vector product is the product of their transposes in reverse order. For example, let

$$A = \begin{pmatrix} 1 & -2 & 3 \\ 0 & 8 & -7 \end{pmatrix}, \quad u = \begin{pmatrix} 3 \\ 2 \\ -1 \end{pmatrix}, \quad v = (5, 10).$$

Then

$$Au = \begin{pmatrix} 1 & -2 & 3 \\ 0 & 8 & -7 \end{pmatrix} \begin{pmatrix} 3 \\ 2 \\ -1 \end{pmatrix} = \begin{pmatrix} -4 \\ 23 \end{pmatrix}$$

$$vA = (5, 10) \begin{pmatrix} 1 & -2 & 3 \\ 0 & 8 & -7 \end{pmatrix} = (5, 70, -55)$$

$$u^T A^T = (3, 2, -1) \begin{pmatrix} 1 & 0 \\ -2 & 8 \\ 3 & -7 \end{pmatrix} = (-4, 23)$$

$$A^T v^T = \begin{pmatrix} 1 & 0 \\ -2 & 8 \\ 3 & -7 \end{pmatrix} \begin{pmatrix} 5 \\ 10 \end{pmatrix} = \begin{pmatrix} 5 \\ 70 \\ -55 \end{pmatrix}.$$

So, we verify that $(Au)^T = u^T A^T$ and $(vA)^T = A^T v^T$.

Exercises

2.4.1: Write the transposes of the following matrices. Also, mention for which pair U, V of these matrices is $5U - 3V$ defined, and compute the result if it is.

$$A = \begin{pmatrix} 1 & -3 & 0 \\ -2 & 2 & -2 \end{pmatrix}, \quad B = \begin{pmatrix} 1 & -1 \\ -2 & 3 \end{pmatrix}, \quad C = \begin{pmatrix} 0 & 3 \\ -1 & -2 \\ 1 & -3 \end{pmatrix},$$

$$D = \begin{pmatrix} 2 & 3 \\ 1 & -2 \end{pmatrix}.$$

2.5   Matrix Multiplication and Its Various Interpretations

Rules for matrix multiplication evolved out of a study of transformations of variables in systems of linear equations. Let $A = (a_{ij})$ be the coefficient matrix of order $m \times n$ in a system of linear equations, b the RHS constants vector, and $x = (x_1, \ldots, x_n)^T$ the column vector of variables in the system. Then the system of equations is $Ax = b$ in matrix notation. Here are all the equations in it.

$$a_{11}x_1 + \cdots + a_{1n}x_n = b_1$$
$$\vdots \quad \vdots$$
$$a_{i1}x_1 + \cdots + a_{in}x_n = b_i$$
$$\vdots \quad \vdots$$
$$a_{m1}x_1 + \cdots + a_{mn}x_n = b_m.$$

Now suppose we transform the variables in this system using the transformation $x = By$ where $B = (b_{ij})$ is an $n \times n$ matrix, and $y = (y_1, \ldots y_n)$ are the new variables. So the transformation is:

$$x_1 = b_{11}y_1 + \cdots + b_{1n}y_n$$
$$\vdots \quad \vdots$$
$$x_j = b_{j1}y_1 + \cdots + b_{jn}y_n$$
$$\vdots \quad \vdots$$
$$x_n = b_{n1}y_1 + \cdots + b_{nn}y_n.$$

Substituting these expressions for $x_1, \ldots, x_n$ in the original system of equations and regrouping terms will modify it into the transformed system in terms of the new variables y. Suppose the transformed system is $Cy = b$ with C as its coefficient matrix. Alternately, substituting $x = By$ in $Ax = b$ gives $ABy = b$; hence we define the matrix product AB to be C.

We will illustrate this with a numerical example. Let

$$A = \begin{pmatrix} 1 & -2 & -1 \\ -3 & 1 & 2 \end{pmatrix}, \quad b = \begin{pmatrix} -5 \\ -7 \end{pmatrix}, \quad x = (x_1, x_2, x_3)^T$$

and consider the system of equations $Ax = b$, which is

$$x_1 - 2x_2 - x_3 = -5$$

$$-3x_1 + x_2 + 2x_3 = -7.$$

Suppose we wish to transform the variables using the transformation

$$x_1 = 10y_1 + 11y_2 + 12y_3$$

$$x_2 = 20y_1 + 21y_2 + 22y_3$$

$$x_3 = 30y_1 + 31y_2 + 32y_3$$

or $x = By$ where

$$B = \begin{pmatrix} 10 & 11 & 12 \\ 20 & 21 & 22 \\ 30 & 31 & 32 \end{pmatrix}, \quad y = \begin{pmatrix} y_1 \\ y_2 \\ y_3 \end{pmatrix}.$$

Substituting the expressions for x_1, x_2, x_3 in terms of y_1, y_2, y_3 in the two equations of the system above, we get

$$1(10y_1 + 11y_2 + 12y_3) - 2(20y_1 + 21y_2 + 22y_3)$$

$$-1(30y_1 + 31y_2 + 32y_3) = -5$$

$$-3(10y_1 + 11y_2 + 12y_3) + 1(20y_1 + 21y_2 + 22y_3)$$

$$+2(30y_1 + 31y_2 + 32y_3) = -7.$$

Regrouping the terms, this leads to the system

$$(1, -2, -1)(10, 20, 30)^T y_1 + (1, -2, -1)(11, 21, 31)^T y_2$$

$$+ (1, -2, -1)(12, 22, 32)^T y_3 = -5$$

$$(-3, 1, 2)(10, 20, 30)^T y_1 + (-3, 1, 2)(11, 21, 31)^T y_2$$

$$+ (-3, 1, 2)(12, 22, 32)^T y_3 = -7$$

or, $Cy = b$ where $y = (y_1, y_2, y_3)^T$ and

$$C = \begin{pmatrix} A_1.B_{.1} & A_1.B_{.2} & A_1.B_{.3} \\ A_2.B_{.1} & A_2.B_{.2} & A_2.B_{.3} \end{pmatrix}.$$

So, we should define the matrix product AB, in this order, to be the matrix C defined by the above formula. This gives the reason for the definition of matrix multiplication given below.

Matrix Multiplication

If A, B are two matrices, the product AB in that order is defined only if

the number of columns in A is equal to the number of rows in B

i.e., if A is of order $m \times n$, and B is of order $r \times s$, then for the product AB to be defined n must equal r; otherwise AB is not defined.

If A is of order $m \times n$ and B is of order $n \times s$, then $AB = C$ is of order $m \times s$, and the (i, j)th entry in C is the dot product $A_i.B_{.j}$.

Therefore matrix multiplication is **row by column multiplication**, and the (i, j)th entry in the matrix product AB is the dot product of the ith row vector of A and the jth column vector of B.

Examples of Matrix Multiplication

$$\begin{pmatrix} 1 & -2 & -1 \\ -3 & 1 & 2 \end{pmatrix} \begin{pmatrix} 1 & 0 \\ 0 & 1 \end{pmatrix} \quad \text{is not defined}$$

$$\begin{pmatrix} 1 & -2 & -1 \\ -3 & 1 & 2 \end{pmatrix} \begin{pmatrix} 10 & 11 & 12 \\ 20 & 21 & 22 \\ 30 & 31 & 32 \end{pmatrix} = \begin{pmatrix} -60 & -62 & -64 \\ 50 & 50 & 50 \end{pmatrix}$$

$$\begin{pmatrix} 1 & -1 & -5 \\ 2 & -3 & -6 \end{pmatrix} \begin{pmatrix} -2 & -4 \\ -6 & -4 \\ 3 & -2 \end{pmatrix} = \begin{pmatrix} -11 & 10 \\ -4 & 16 \end{pmatrix}.$$

The important thing to remember is that matrix multiplication is **not commutative**, i.e., the order in which matrices are multiplied is very important. When the product AB exists, the product BA may not even be defined, and even if it is defined, AB and BA may be of different orders; and even when they are of the same order they may not be equal.

Examples

Let

$$A = \begin{pmatrix} 1 & -2 & -1 \\ -3 & 1 & 2 \end{pmatrix}, \quad B = \begin{pmatrix} 10 & 11 & 12 \\ 20 & 21 & 22 \\ 30 & 31 & 32 \end{pmatrix},$$

$$D = \begin{pmatrix} -2 & -4 \\ -6 & -4 \\ 3 & -2 \end{pmatrix}, \quad E = \begin{pmatrix} 0 & 1 & 0 \\ 0 & 0 & 1 \\ 1 & 0 & 0 \end{pmatrix}.$$

AB is computed above, and BA is not defined.

$$AD = \begin{pmatrix} 7 & 6 \\ 6 & 4 \end{pmatrix}, \quad DA = \begin{pmatrix} 10 & 0 & -6 \\ 6 & 8 & -2 \\ 9 & -8 & -7 \end{pmatrix}$$

$$BE = \begin{pmatrix} 12 & 10 & 11 \\ 22 & 20 & 21 \\ 32 & 30 & 31 \end{pmatrix}, \quad EB = \begin{pmatrix} 20 & 21 & 22 \\ 30 & 31 & 32 \\ 10 & 11 & 12 \end{pmatrix}.$$

So, AD and DA are of different orders. BE and EB are both of the same order, but they are not equal.

Thus, when writing a matrix product, one should specify the order of the entries very carefully.

We defined matrix-vector multiplication in Section 2.2. That definition coincides with the definition of matrix multiplication given here when the vectors in the product are treated as matrices with a single row or column respectively.

In the matrix product AB we say that A **premultiplies** B; or B **post-multiplies** A.

Column-Row Products

Let $u = (u_1, \ldots, u_m)^T$ be a column vector in R^m, and $v = (v_1, \ldots, v_n)$ a row vector in R^n. Then the product uv in that order, known as the **outer product** of the vectors u, v is the matrix product uv when u, v are treated as matrices with one column and one row respectively, given by

$$uv = \begin{pmatrix} u_1v_1 & u_1v_2 & \cdots & u_1v_n \\ u_2v_1 & u_2v_2 & \cdots & u_2v_n \\ \vdots & & & \vdots \\ u_mv_1 & u_mv_2 & \cdots & u_mv_n \end{pmatrix}.$$

As an example

$$\begin{pmatrix} 1 \\ 0 \\ -1 \\ 2 \end{pmatrix} (3, 4) = \begin{pmatrix} 3 & 4 \\ 0 & 0 \\ -3 & -4 \\ 6 & 8 \end{pmatrix}.$$

Notice the difference between the inner (or dot) product of two vectors defined in Section 1.5, and the outer product of two vectors defined here. The inner product is always a real number, and it is only defined for a pair of vectors from the same dimensional space. The outer product is a matrix, and it is defined for any pair of vectors (they may be from spaces of different dimensions).

Matrix Multiplication Through a Series of Matrix-Vector Products

Let A, B be matrices of orders $m \times n$, $n \times k$ respectively. We can write the product AB as

$$AB = A(B_{.1} \vdots \ldots \vdots B_{.k})$$

$$= (AB_{.1} \vdots \ldots \vdots AB_{.k})$$

where for each $t = 1$ to k, $B_{.t}$ is the tth column vector of B, and $AB_{.t}$ is the column vector that is the result of a matrix-vector product as defined earlier in Section 2.2. Thus the product AB can be looked at as the matrix consisting of column vectors $AB_{.t}$ in the order $t = 1$ to k.

Another way of looking at the product AB is

$$AB = \begin{pmatrix} A_{1.} \\ \vdots \\ A_{m.} \end{pmatrix} B = \begin{pmatrix} A_{1.}B \\ \vdots \\ A_{m.}B \end{pmatrix}.$$

From these interpretations of matrix multiplication, we see that the

jth column of the product AB is $AB_{.j}$
ith row of the product AB is $A_{i.}B$.

Matrix Product as Sum of a Series of Column-Row Products

Let $A = (a_{ij})$, $B = (b_{ij})$ be two matrices such that the product AB is defined. So

the number of columns in $A =$ number of rows in $B = n$, say.

Let $A_{.1}, \ldots, A_{.n}$ be the column vectors of A; and $B_{1.}, \ldots, B_{n.}$ be the row vectors in B. For the sake of completeness suppose A is of order $m \times n$, and B is of order $n \times k$.

From the definition of the matrix product AB we know that its (i,j)th element is

$$\sum_{t=1}^{n} a_{it} b_{tj} = a_{i1} b_{1j} + \cdots + a_{in} b_{nj}.$$

So,

$$AB = \sum_{t=1}^{n} (\text{matrix of order } m \times k \text{ whose } (i,j)\text{th element is } a_{it} b_{tj}).$$

However, the matrix whose (i,j)th element is $a_{it} b_{tj}$ is the outer product $A_{.t} B_{t.}$. Hence

$$AB = \sum_{t=1}^{n} A_{.t} B_{t.} = A_{.1} B_{1.} + \cdots + A_{.n} B_{n.}.$$

This is the formula that expresses AB as a sum of n column-row products. As an example let

$$\begin{pmatrix} 1 & -2 & -1 \\ -3 & 1 & 2 \end{pmatrix}, \quad B = \begin{pmatrix} 10 & 11 & 12 \\ 20 & 21 & 22 \\ 30 & 31 & 32 \end{pmatrix}.$$

Then

$$A_{.1} B_{1.} = \begin{pmatrix} 1 \\ -3 \end{pmatrix} (10, 11, 12) = \begin{pmatrix} 10 & 11 & 12 \\ -30 & -33 & -36 \end{pmatrix}$$

$$A_{.2} B_{2.} = \begin{pmatrix} -2 \\ 1 \end{pmatrix} (20, 21, 22) = \begin{pmatrix} -40 & -42 & -44 \\ 20 & 21 & 22 \end{pmatrix}$$

$$A_{.3} B_{3.} = \begin{pmatrix} -1 \\ 2 \end{pmatrix} (30, 31, 32) = \begin{pmatrix} -30 & -31 & -32 \\ 60 & 62 & 64 \end{pmatrix}.$$

So

$$A_{.1} B_{1.} + A_{.2} B_{2.} + A_{.3} B_{3.} = \begin{pmatrix} -60 & -62 & -64 \\ 50 & 50 & 50 \end{pmatrix}$$

which is the same as the product AB computed earlier.

Exercises

2.5.1: Let A, B be $m \times n, n \times k$ matrices respectively. Show that

each column of AB is a linear combination of the columns of A
each row of AB is a linear combination of the rows of A.

The Product of a Sequence of 3 Or More Matrices

Let $A_1, A_2, \ldots, A_t$ be $t \geq 3$ matrices. Consider the matrix product

$$A_1 A_2 \ldots A_t$$

in this specific order. This product is defined iff

the product of every consecutive pair of matrices in this order is defined, i.e., for each $r = 1$ to $t - 1$, the number of rows in A_r is equal to the number of columns in A_{r+1}.

When these conditions are satisfied, this product of t matrices is computed recursively, i.e., take any consecutive pair of matrices in this order, say A_r an A_{r+1} for some $1 \leq r \leq t - 1$, and suppose $A_r A_{r+1} = D$. Then the above product is

$$A_1 \ldots A_{r-1} D A_{r+2} \ldots A_t$$

There are only $t - 1$ matrices in this product, it can be reduced to a product of $t - 2$ matrices the same way, and the same procedure is continued until the product is obtained as a single matrix.

When it is defined, the product $A_1 A_2 \ldots A_t$ is of order $m_1 \times n_t$ where

m_1 = number of rows in A_1, the leftmost matrix in the product
n_t = number of columns in A_t, the rightmost matrix in the product.

The product $A_1 A_2 \ldots A_t$ is normally computed by procedures known as **string computation**, of which there are two.

The **left to right string computation** for $A_1 A_2 \ldots A_t$ computes $D_r = A_1 \ldots A_r$ for $r = 2$ to t in that order using

$$D_2 = A_1 A_2$$

$$D_{r+1} = D_r A_{r+1}, \qquad \text{for } r = 2 \text{ to } t - 1$$

to obtain the final result D_t.

The **right to left string computation** for $A_1 A_2 \ldots A_t$ computes $E_r = A_r \ldots A_t$ for $r = t - 1$ to 1 in that order using

$$E_{t-1} = A_{t-1} A_t$$

$$E_{r-1} = A_{r-1} E_r, \qquad \text{for } r = t - 1 \text{ to } 2$$

to obtain the final result E_1, which will be the same as D_t obtained by the previous recursion.

These facts follow by repeated application of the following result.

Result 2.5.1: Product involving three matrices: *Let A, B, C be matrices of orders $m \times n, n \times p, p \times q$ respectively. Then $A(BC) = (AB)C$.*

To prove this result, suppose $BC = D$. Then from one of the interpretations of two-matrix products discussed earlier we know that

$$D = BC = (BC_{.1} \vdots \ldots \vdots BC_{.q}).$$

So,

$$A(BC) = AD = A(BC_{.1} \vdots \ldots \vdots BC_{.q}) = (A(BC_{.1}) \vdots \ldots \vdots A(BC_{.q}))$$

Also, let $C = (c_{ij})$. Then, for each $j = 1$ to q, $BC_{.j} = \sum_{i=1}^{p} c_{ij} B_{.i}$. So

$$A(BC_{.j}) = A \left(\sum_{i=1}^{p} c_{ij} B_{.i} \right) = \sum_{i=1}^{p} c_{ij} (AB_{.i}).$$

So, $A(BC)$ is an $m \times q$ matrix whose jth column is $\sum_{i=1}^{p} c_{ij}(AB_{.i})$ for $j = 1$ to q.

Also, from another interpretation of two-matrix products discussed earlier we know that

$$AB = (AB_{.1} \vdots \ldots \vdots AB_{.p}).$$

So,

$$(AB)C = (AB_{.1} \vdots \ldots \vdots AB_{.p}) \begin{pmatrix} C_{1.} \\ \vdots \\ C_{p.} \end{pmatrix} = (AB_{.1})C_{1.} + \cdots + (AB_{.p})C_{p.}$$

Notice that for each $i = 1$ to p, $(AB_{.i})C_{i.}$ is an outer product (a column-row product) that is an $m \times q$ matrix whose jth column is $c_{ij}(AB_{.i})$ for $j = 1$ to q. Hence, $(AB)C$ is an $m \times q$ matrix whose jth column is $\sum_{i=1}^{p} c_{ij}(AB_{.i})$

for $j = 1$ to q; i.e., it is the same as $A(BC)$ from the fact established above. Therefore

$$A(BC) = (AB)C$$

establishing the result.

Properties Satisfied By Matrix Products

Order important: The thing to remember is that the order in which the matrices are written in a matrix product is very important. If the order is changed, the product may not be defined, and even if it is defined, the result may be different. As examples, let

$$A = \begin{pmatrix} -3 & 0 & 1 \\ -2 & -6 & 4 \end{pmatrix}, \quad B = \begin{pmatrix} -5 & -10 \\ 20 & 30 \\ 1 & 2 \end{pmatrix}, \quad C = \begin{pmatrix} 3 \\ 4 \\ 5 \end{pmatrix}.$$

Then

$$AC = \begin{pmatrix} -4 \\ -10 \end{pmatrix}, \quad \text{and } CA \text{ is not defined}$$

$$AB = \begin{pmatrix} -14 & -28 \\ -106 & -152 \end{pmatrix}, \quad BA = \begin{pmatrix} 35 & 60 & -45 \\ -120 & -180 & 140 \\ -7 & -12 & 9 \end{pmatrix}$$

so, AB and BA are not even of the same order.

Associative law of multiplication: If A, B, C are matrices such that the product ABC is defined, then $A(BC) = (AB)C$.

Left and Right distributive laws: Let B, C be matrices of the same order. If A is a matrix such that the product $A(B + C)$ is defined, it is $= AB + AC$. If A is a matrix such that the product $(B + C)A$ is defined, it is $= BA + CA$. So, matrix multiplication distributes across matrix addition.

If α, β are scalars, and A, B two matrices such that the Product AB is defined, then, $(\alpha + \beta)A = \alpha A + \beta A$; and $\alpha AB = (\alpha A)B = A(\alpha B)$.

The product of two nonzero matrices may be zero (i.e., a matrix in which all the entries are zero). For example, for 2×2 matrices A, B both of which are nonzero, verify that $AB = 0$.

$$A = \begin{pmatrix} 1 & 1 \\ 1 & 1 \end{pmatrix}, \quad B = \begin{pmatrix} 1 & 1 \\ -1 & -1 \end{pmatrix}.$$

If A, B, C are matrices such that the products AB, AC are both defined; and $AB = AC$, we cannot, in general, conclude that $B = C$. Verify this for the following matrices A, B, C, we have $AB = AC$, even though $B \neq C$.

$$A = \begin{pmatrix} 1 & 1 \\ 1 & 1 \end{pmatrix}, \quad B = \begin{pmatrix} 1 & 1 \\ -1 & -1 \end{pmatrix}, \quad C = \begin{pmatrix} 2 & 2 \\ -2 & -2 \end{pmatrix}.$$

Matrix multiplication and transposes: If A, B are two matrices such that the product AB is defined, then the product $B^T A^T$ is defined and $(AB)^T = B^T A^T$. This can be verified directly from the definitions.

In the same way if the matrix product $A_1 A_2 \ldots A_r$ is defined, then $(A_1 A_2 \ldots A_r)^T = A_r^T A_{r-1}^T \ldots A_2^T A_1^T$.

Exercises

2.5.2: For every pair of matrices U, V among the following, determine whether UV is defined, and compute it if it is.

$$A = \begin{pmatrix} 1 & -3 & 2 & 0 \\ -1 & 2 & 3 & 4 \\ 2 & 1 & 1 & -1 \end{pmatrix}, \quad B = \begin{pmatrix} 0 & -2 & 1 \\ 1 & 0 & 2 \\ -1 & -3 & 0 \\ -2 & -7 & 6 \end{pmatrix},$$

$$C = \begin{pmatrix} 4 & -5 & 2 \\ -1 & 1 & -2 \\ 0 & 1 & -1 \end{pmatrix}, \quad D = \begin{pmatrix} 5 & -5 & 0 & 3 \\ -1 & -1 & -1 & -1 \\ 1 & 2 & 2 & 1 \\ 0 & -2 & -2 & -1 \end{pmatrix},$$

$$E = \begin{pmatrix} 2 & -3 \\ 4 & -2 \end{pmatrix}, \quad F = \begin{pmatrix} 1 & -2 & 3 & -4 \\ -1 & 3 & -2 & 2 \end{pmatrix}.$$

2.5.3: Sometimes when A, B are a pair of matrices, both products AB, BA may exist. These products may be of different orders, but even when they are of the same order they may or may not be equal. Verify these with the following pairs.

(i): $A = \begin{pmatrix} 1 & 7 & -7 & 2 \\ 0 & -5 & 3 & -1 \\ 1 & 1 & 1 & 1 \end{pmatrix}, \quad B = \begin{pmatrix} 1 & 1 & 1 \\ -1 & 1 & -1 \\ 1 & -1 & -1 \\ -1 & -1 & 1 \end{pmatrix}.$

(ii): $A = \begin{pmatrix} 1 & -1 \\ 1 & 1 \end{pmatrix}, \quad B = \begin{pmatrix} 2 & 3 \\ -3 & 2 \end{pmatrix}.$

(iii): $A = \begin{pmatrix} 8 & 3 \\ -4 & -2 \end{pmatrix}$, $B = \begin{pmatrix} 2 & 9 \\ 6 & 3 \end{pmatrix}$.

2.5.4: Compute the products $ADB, DBA, BAD, EFDB$ where the matrices A to F are given in Exercise 2.5.2.

2.6   Some Special Matrices

Square Matrices

A matrix of order $m \times n$ is said to be a

> **square matrix**   if $m = n$
> **rectangular matrix that is not square**   if $m \neq n$.

Hence a square matrix is a matrix of order $n \times n$ for some n. Since the number of rows and the number of columns in a square matrix are the same, their common value is called the **order of the square matrix**. Thus an $n \times n$ square matrix is said to be of order n. Here are examples of square matrices.

$$\begin{pmatrix} 2 & -3 \\ 1 & -10 \end{pmatrix}, \quad \begin{pmatrix} -3 & -4 & -5 \\ 6 & 7 & 8 \\ -9 & 10 & 11 \end{pmatrix}.$$

In a square matrix (a_{ij}) of order n, the entries $a_{11}, a_{22}, \ldots, a_{nn}$ constitute its **diagonal entries** or its **main diagonal**. All the other entries in this matrix are called **off-diagonal entries**. Here is a picture

$$\begin{pmatrix} \boxed{a_{11}} & a_{12} & \cdots & a_{1,n-1} & a_{1n} \\ a_{21} & \boxed{a_{22}} & \cdots & a_{2,n-1} & a_{2n} \\ \vdots & \vdots & & \vdots & \vdots \\ a_{n-1,1} & a_{n-1,2} & \cdots & \boxed{a_{n-1,n-1}} & a_{n-1,n} \\ a_{n1} & a_{n2} & \cdots & a_{n,n-1} & \boxed{a_{nn}} \end{pmatrix}.$$

A square matrix with its diagonal entries boxed.

Unit (Identity) Matrices

Unit matrices are square matrices in which all diagonal entries are "1" and all off-diagonal entries are "0". Here are the **unit matrices** (also called

identity matrices) of some orders.

$$\begin{pmatrix} 1 & 0 \\ 0 & 1 \end{pmatrix}, \quad \begin{pmatrix} 1 & 0 & 0 \\ 0 & 1 & 0 \\ 0 & 0 & 1 \end{pmatrix}, \quad \begin{pmatrix} 1 & 0 & 0 & 0 \\ 0 & 1 & 0 & 0 \\ 0 & 0 & 1 & 0 \\ 0 & 0 & 0 & 1 \end{pmatrix}.$$

The reason for the name is that when a matrix A is multiplied by the unit matrix I of appropriate order so that the product is defined, the resulting product IA or AI is A itself. Here are some examples.

$$\begin{pmatrix} 1 & 0 \\ 0 & 1 \end{pmatrix} \begin{pmatrix} 1 & -2 & -1 \\ -3 & 1 & 2 \end{pmatrix} = \begin{pmatrix} 1 & -2 & -1 \\ -3 & 1 & 2 \end{pmatrix}$$

$$\begin{pmatrix} 1 & -2 & -1 \\ -3 & 1 & 2 \end{pmatrix} \begin{pmatrix} 1 & 0 & 0 \\ 0 & 1 & 0 \\ 0 & 0 & 1 \end{pmatrix} = \begin{pmatrix} 1 & -2 & -1 \\ -3 & 1 & 2 \end{pmatrix}.$$

Hence, in the world of matrices, the unit matrices play the same role as the number "1" does in the world of real numbers. This is the reason for their name.

The unit matrix is usually denoted by the symbol I when its order can be understood from the context. If the order has to be indicated specifically, the unit matrix of order n is usually denoted by the symbol I_n.

In Chapter 1 we defined unit vectors as column vectors with a single nonzero entry of "1". From this it is clear that the unit vectors are column vectors of the unit matrix. Thus if I is the unit matrix of order n, then its column vectors $I_{.1}, I_{.2}, \ldots, I_{.n}$ are the 1st, 2nd, $\ldots$, nth unit vectors in R^n.

Permutation Matrices

A square matrix of order n in which all the entries are 0 or 1, and there is exactly a single "1" entry in each row and in each column is called a **permutation matrix** or an **assignment**. A permutation matrix can always be transformed into the unit matrix by permuting (i.e., rearranging in some order) its rows, or its columns. Here are some permutation matrices.

$$\begin{pmatrix} 0 & 1 \\ 1 & 0 \end{pmatrix}, \quad \begin{pmatrix} 0 & 1 & 0 \\ 1 & 0 & 0 \\ 0 & 0 & 1 \end{pmatrix}, \quad \begin{pmatrix} 0 & 1 & 0 & 0 \\ 1 & 0 & 0 & 0 \\ 0 & 0 & 0 & 1 \\ 0 & 0 & 1 & 0 \end{pmatrix}.$$

Premultiplying any matrix A by a permutation matrix of appropriate order so that the product is defined, permutes its rows. In the same way, postmultiplying any matrix A by a permutation matrix of appropriate order so that the product is defined, permutes its columns. Here are some examples.

$$\begin{pmatrix} 0 & 1 \\ 1 & 0 \end{pmatrix} \begin{pmatrix} 1 & -2 & -1 \\ -3 & 1 & 2 \end{pmatrix} = \begin{pmatrix} -3 & 1 & 2 \\ 1 & -2 & -1 \end{pmatrix}$$

$$\begin{pmatrix} 1 & -2 & -1 \\ -3 & 1 & 2 \end{pmatrix} \begin{pmatrix} 0 & 0 & 1 \\ 0 & 1 & 0 \\ 1 & 0 & 0 \end{pmatrix} = \begin{pmatrix} -1 & -2 & 1 \\ 2 & 1 & -3 \end{pmatrix}.$$

The unit matrix is also a permutation matrix. Also, it can be verified that if P is a permutation matrix, then

$$PP^T = P^T P = I, \quad \text{the unit matrix of the same order.}$$

Diagonal Matrices

A square matrix with all its off-diagonal entries zero, and all diagonal entries nonzero is called a **diagonal matrix**. Here are some examples.

$$\begin{pmatrix} 3 & 0 \\ 0 & -2 \end{pmatrix}, \quad \begin{pmatrix} -2 & 0 & 0 \\ 0 & -1 & 0 \\ 0 & 0 & -8 \end{pmatrix}, \quad \begin{pmatrix} 1 & 0 & 0 & 0 \\ 0 & 2 & 0 & 0 \\ 0 & 0 & 3 & 0 \\ 0 & 0 & 0 & 4 \end{pmatrix}.$$

Since all the off-diagonal entries in a diagonal matrix are known to be zero, if we are given the diagonal entries in it, we can construct the diagonal matrix. For this reason, a diagonal matrix of order n with diagonal entries $a_{11}, a_{22}, \ldots, a_{nn}$ is denoted by the symbol **diag**$\{a_{11}, a_{22}, \ldots, a_{nn}\}$. In this notation, the three diagonal matrices given above will be denoted by diag$\{3, -2\}$, diag$\{-2, -1, -8\}$, diag$\{1, 2, 3, 4\}$ respectively.

Premultiplying a matrix A by a diagonal matrix of appropriate order so that the product is defined, multiplies every entry in the ith row of A by the ith diagonal entry in the diagonal matrix; this operation is called **scaling the rows of A**. Similarly, postmultiplying a matrix A by a diagonal matrix scales the columns of A.

$$\begin{pmatrix} 3 & 0 \\ 0 & -2 \end{pmatrix} \begin{pmatrix} 1 & -2 & -1 \\ -3 & 1 & 2 \end{pmatrix} = \begin{pmatrix} 3 & -6 & -3 \\ 6 & -2 & -4 \end{pmatrix}$$

$$\begin{pmatrix} 1 & -2 & -1 \\ -3 & 1 & 2 \end{pmatrix} \begin{pmatrix} -2 & 0 & 0 \\ 0 & -1 & 0 \\ 0 & 0 & -8 \end{pmatrix} = \begin{pmatrix} -2 & 2 & 8 \\ 6 & -1 & -16 \end{pmatrix}.$$

Upper and Lower Triangular Matrices, Triangular Matrices

We already discussed upper, lower triangular tableaus; and triangular tableaus in Section 1.18. The corresponding matrices are called upper triangular, lower triangular, or triangular matrices. We give formal definitions below.

A square matrix is said to be **upper triangular** if all entries in it below the diagonal are zero, and all the diagonal entries are nonzero. It is said to be **lower triangular** if all the diagonal entries are nonzero, and all the entries above the diagonal are zero. It is said to be **triangular** if its rows can be rearranged to make it upper triangular. Here are examples.

$$
\begin{pmatrix} 1 & 1 & 1 & 1 \\ 0 & 1 & 1 & 1 \\ 0 & 0 & 1 & 1 \\ 0 & 0 & 0 & 1 \end{pmatrix}, \quad
\begin{pmatrix} 1 & 0 & 0 & 0 \\ 1 & 1 & 0 & 0 \\ 1 & 1 & 1 & 0 \\ 1 & 1 & 1 & 1 \end{pmatrix}
\begin{pmatrix} 0 & 0 & 0 & 1 \\ 0 & 0 & 1 & 1 \\ 1 & 1 & 1 & 1 \\ 0 & 1 & 1 & 1 \end{pmatrix}.
$$

The first matrix is upper triangular, the second is lower triangular, and the third is triangular.

Symmetric Matrices

A square matrix $D = (d_{ij})$ is said to be **symmetric** iff $D^T = D$, i.e., $d_{ij} = d_{ji}$ for all i, j.

A square matrix which is not symmetric is said to be an **asymmetric matrix**.

If D is an asymmetric matrix, $(1/2)(D + D^T)$ is called its **symmetrization**.

$$
\begin{pmatrix} 6 & 1 & -1 & -3 \\ 1 & 0 & -1 & 0 \\ -1 & -1 & -2 & 1 \\ -3 & 0 & 1 & 3 \end{pmatrix}, \quad
\begin{pmatrix} 0 & 2 & 0 \\ 0 & 0 & 2 \\ 2 & 0 & 0 \end{pmatrix}, \quad
\begin{pmatrix} 0 & 1 & 1 \\ 1 & 0 & 1 \\ 1 & 1 & 0 \end{pmatrix}.
$$

The first matrix is symmetric, the second is asymmetric, and the third is the symmetrization of the second matrix.

Submatrices of a Matrix

Let $A = (a_{ij})$ be a given matrix of order $m \times n$. Let $R \subset \{1, \ldots, m\}$, $C \subset \{1, \ldots, n\}$, in both of which the entries are arranged in increasing order. By deleting all the rows of A not in R, and all the columns of A not in C, we are left with a matrix which is known as the **submatrix of A determined by the subsets R, C of rows and columns** and denoted usually by A_{RC}. For example, let

$$A = \begin{pmatrix} 3 & -1 & 1 & 1 & 0 \\ 0 & 1 & 3 & 4 & 0 \\ 4 & -2 & 0 & 1 & 1 \\ 5 & -3 & 0 & 0 & 2 \end{pmatrix}.$$

Let $R = \{1, 3, 4\}$, $C = \{1, 2, 4, 5\}$. Then the submatrix corresponding to these subsets is

$$A_{RC} = \begin{pmatrix} 3 & -1 & 1 & 0 \\ 4 & -2 & 1 & 1 \\ 5 & -3 & 0 & 2 \end{pmatrix}.$$

Principal Submatrices of a Square Matrix

Let $D = (d_{ij})$ be a given square matrix of order n. Let $Q \subset \{1, \ldots, n\}$, in which the entries are arranged in increasing order. By deleting all the rows and columns of D not in Q, we are left with a matrix which is known as the **principal submatrix of D determined by the subset Q of rows and columns** and denoted usually by D_{QQ}. For example, let

$$D = \begin{pmatrix} 3 & -1 & 1 & 1 & 0 \\ 0 & 1 & 3 & 4 & 0 \\ 4 & -2 & 0 & 1 & 1 \\ 5 & -3 & 0 & 0 & 2 \\ 8 & -17 & -18 & 9 & 11 \end{pmatrix}.$$

Let $Q = \{2, 3, 5\}$. Then the principal submatrix of D determined by the subset Q is

$$D_{QQ} = \begin{pmatrix} 1 & 3 & 0 \\ -2 & 0 & 1 \\ -17 & -18 & 11 \end{pmatrix}.$$

Principal submatrices are only defined for square matrices, and the main diagonal of a principal submatrix is always a subset of the main diagonal of the original matrix.

Exercises

2.6.1: Let

$$
A = \begin{pmatrix}
-20 & 10 & 30 & 4 & -17 \\
18 & 0 & 3 & 2 & -19 \\
0 & 6 & -7 & 8 & 12 \\
13 & -6 & 19 & 33 & 14 \\
12 & -9 & 22 & 45 & 51
\end{pmatrix}.
$$

Write the submatrix of A corresponding to the subset of rows $\{2,4\}$ and the subset of columns $\{1,3,5\}$. Also write the principal submatrix of A determined by the subset of rows and columns $\{2,4,5\}$.

2.7 Row Operations and Pivot Steps on a Matrix

Since a matrix is a 2-dimensional array of numbers, it is like the tableau we discussed earlier. Rows and columns of the matrix are exactly like the rows and columns of a tableau.

Row operations on a matrix are exactly like row operations on a tableau. They involve the following:

1. Scalar Multiplication: Multiply each entry in a row by the same nonzero scalar

2. Add a Scalar Multiple of a Row to Another: Multiply each element in a row by the same nonzero scalar and add the result to the corresponding element of the other row.

3. Row interchange: Interchange two rows in the matrix.

Example:

Consider the matrix

$$
A = \begin{pmatrix}
-1 & 2 & 1 & 3 \\
0 & -3 & 9 & 7 \\
8 & 6 & -4 & -5
\end{pmatrix}.
$$

Multiplying $A_{3.} =$ Row 3 of A, by -2 leads to the matrix

$$
A' = \begin{pmatrix}
-1 & 2 & 1 & 3 \\
0 & -3 & 9 & 7 \\
-16 & -12 & 8 & 10
\end{pmatrix}.
$$

Performing the row operation $A_2. - 2A_1.$ leads to the matrix

$$A'' = \begin{pmatrix} -1 & 2 & 1 & 3 \\ 2 & -7 & 7 & 1 \\ 8 & 6 & -4 & -5 \end{pmatrix}.$$

A GJ (Gauss–Jordan) pivot step on a matrix is exactly like a GJ pivot step on a tableau. It is specified by choosing a row of the matrix as the pivot row, and a column of the matrix as the pivot column, with the element in the pivot row and pivot column called the pivot element which should be nonzero. The GJ pivot step then converts the pivot column into the unit column with a "1" entry in the pivot row and "0" entries in all other rows, by appropriate row operations.

A G (Gaussian) pivot step on a matrix is like the GJ pivot step, with the exception that it only converts all entries in the pivot column below the pivot row into zeros by appropriate row operations, but leaves all the rows above the pivot row and the pivot row itself unchanged.

2.8   Determinants of Square Matrices

History of determinants: Determinants are real valued functions of square matrices, i.e., associated with every square matrix is a unique real number called its **determinant**. Determinants are only defined for square matrices. There is no such concept for rectangular matrices that are not square. The definition of the determinant of a square array of numbers goes back to the end of the 17th century in the works of Seki Takakazu (also called Seki Kowa in some books) of Japan and Gottfried Leibnitz of Germany. Seki arrived at the notion of a determinant while trying to find common roots of algebraic equations. To find common roots of polynomials $f(x), g(x)$ of small degrees Seki got determinant expressions and published a treatize in 1674. Lebnitz did not publish the results of his studies related to determinants, but in a letter to l'Hospital in 1693 he wrote down the determinant condition of compatiability for a system of three linear equations in two unknowns. In Europe the first publication mentioning determinants was due to Cramèr in 1750 in which he gave a determinant expression for the problem of finding a conic through five given points (this leads to a system of linear equations).

Since then determinants have been studied extensively for their theoretical properties and their applications in linear algebra theory. Even though determinants play a very major role in theory, they have not been used that much in computational algorithms. Since the focus of this book is computational linear algebra, we will list all the important properties of determinants without detailed proofs. References for proofs of the results are provided for interested readers to pursue. ⊠

There are several equivalent ways of defining determinants, but all these satisfy the following fundamental result.

Result 2.8.1: One set of properties defining a determinant: *The determinant of a square matrix of order n is the unique real valued function of the matrix satisfying the following properties.*

(a) *If $A = (a_{ij})$ of order n is lower or upper triangular, then the determinant of A is the product of the diagonal entries of A.*
(b) *$det(A^T) = det(A)$ for all square matrices A.*
(c) *If A, B are two square matrices of order n, then $det(AB) = det(A)det(B)$.*

For $n = 1, 2$, determinants of square matrices of order n can be defined very easily.

The determinant of a 1×1 matrix (a_{11}) is defined to be a_{11}. So, $det(0) = 0$, $det(-4) = -4$, $det(6) = 6$, etc.

For a 2×2 matrix $A = \begin{pmatrix} a_{11} & a_{12} \\ a_{21} & a_{22} \end{pmatrix}$, $det(A)$ is defined to be $a_{11}a_{22} - a_{12}a_{21}$. For example

$$\det \begin{pmatrix} 3 & 2 \\ 1 & 4 \end{pmatrix} = 3 \times 4 - 2 \times 1 = 12 - 2 = 10$$

$$\det \begin{pmatrix} 1 & 4 \\ 3 & 2 \end{pmatrix} = 1 \times 2 - 4 \times 3 = 2 - 12 = -10$$

$$\det \begin{pmatrix} 1 & 4 \\ -3 & -12 \end{pmatrix} = 1 \times (-12) - (-3) \times 4 = -12 + 12 = 0.$$

The original concept of the determinant of a square matrix $A = (a_{ij})$ of order n for $n \geq 2$ consists of a sum of $n!$ terms, half with a coefficient of $+1$, and the other half with a coefficient of -1. We will now explain this concept.

Some preliminary definitions first. In linear programming literature, each permutation matrix of order n is referred to as an **assignment of**

order n. Here, for example, is an assignment of order 4.

$$\begin{pmatrix} 0 & 1 & 0 & 0 \\ 0 & 0 & 0 & 1 \\ 1 & 0 & 0 & 0 \\ 0 & 0 & 1 & 0 \end{pmatrix}.$$

Hence an assignment of order n is a square matrix of order n that contains exactly one nonzero entry of "1" in each row and column. Verify that there are $n!$ distinct assignments of order n.

We number the rows and columns of an assignment in natural order, and refer to the (i,j)th position in it as **cell** (i,j) (this is in row i and column j). We will represent each assignment by the subset of cells in it corresponding to the "1" entries written in natural order of rows of the matrix. For example, the assignment of order 4 given above corresponds to the set of cells $\{(1,2),(2,4),(3,1),(4,3)\}$ in this representation.

So, in this notation a general assignment or permutation matrix of order n can be represented by the set of cells $\{(1,j_1),(2,j_2),(3,j_3),\ldots,$ $(n,j_n)\}$ where $(j_1,j_2,\ldots,j_n)$ is a **permutation** of $(1,2,\ldots,n)$, i.e., an arrangement of these integers in some order. We will say that this permutation matrix and permutation correspond to each other. For example, the permutation matrix of order 4 given above corresponds to the permutation $(2,4,1,3)$.

There are $6 = 3!$ different permutations of $(1,2,3)$. These are: $(1,2,3)$, $(1,3,2),(2,1,3),(2,3,1),(3,1,2),(3,2,1)$. In the same way there are $24 = 4!$ permutations of $(1,2,3,4)$; and in general $n!$ permutations of $(1,2,\ldots,n)$.

Consider the permutation $(j_1,\ldots,j_n)$ corresponding to the permutation represented by the set of cells $\{(1,j_1),\ldots,(n,j_n)\}$. We will use the symbol p to represent either the permutation or the corresponding permutation matrix. The determinant of the square matrix $A = (a_{ij})$ of order n contains one term corresponding to the permutation p, it is

$$(-1)^{NI(p)} a_{1j_1} a_{2j_2} \ldots a_{nj_n}$$

where $NI(p)$ is a number called the **number of inversions** in the permutations p, which we will define below. The determinant of A is the sum of all the terms corresponding to all the $n!$ permutations.

In the permutation $(j_1,\ldots,j_n)$, consider a pair of distinct entries (j_r,j_s) where $s > r$. If $j_s < j_r$ $[j_s > j_r]$ we say that this pair contributes one $[0]$ inversions in this permutation. The total number of inversions in this permutation is obtained by counting the contributions of all pairs of the form (j_r, j_s) for $s > r$ in it. It is equal to $\sum_{r=1}^{n-1} q_r$ where

q_r = number of entries in $(j_{r+1}, \ldots, j_n)$ which are $< j_r$.

A permutation is called an **even (odd) permutation** if the number of inversions in it is an even (odd) integer. As an example consider the permutation $(6, 1, 3, 4, 5, 2)$. Here

$j_1 = 6$, all of j_2, j_3, j_4, j_5, j_6 are $< j_1$, so $q_1 = 5$.
$j_2 = 1$, no. entries among $(j_3, j_4, j_5, j_6) < j_2$ is 0, so $q_2 = 0$.
$j_3 = 3$, no. entries among $(j_4, j_5, j_6) < j_3$ is 1, so $q_3 = 1$.
$j_4 = 4$, no. entries among $(j_5, j_6) < j_4$ is 1, so $q_4 = 1$.
$j_5 = 5$, no. entries among $(j_6) < j_5$ is 1, so $q_5 = 1$.

So the number of inversions in this permutation is $5 + 0 + 1 + 1 + 1 = 8$, hence this permutation is an even permutation.

In the same way verify that the number of inversions in the permutation $(2, 4, 1, 3)$ is $1 + 2 + 0 = 3$, hence this is an odd permutation.

For example, for $n = 3$, there are $3! = 6$ permutations. These permutations, and the formula for the determinant of $A = (a_{ij})$ of order 3 are given below.

Permutation p	$NI(p)$	Term corresponding to p
$(1, 2, 3)$	0	$a_{11}a_{22}a_{33}$
$(1, 3, 2)$	1	$-a_{11}a_{23}a_{32}$
$(2, 3, 1)$	2	$a_{12}a_{23}a_{31}$
$(2, 1, 3)$	1	$-a_{12}a_{21}a_{33}$
$(3, 1, 2)$	2	$a_{13}a_{21}a_{32}$
$(3, 2, 1)$	3	$-a_{13}a_{22}a_{31}$
Determinant(A)		$a_{11}a_{22}a_{33} - a_{11}a_{23}a_{32} + a_{12}a_{23}a_{31}$ $-a_{12}a_{21}a_{33} + a_{13}a_{21}a_{32} - a_{13}a_{22}a_{31}$

In the same way, the original formula for the determinant of any square matrix of any order can be derived. However, since it is in the form of a sum of $n!$ terms, this formula is not used for computing the determinant when $n > 2$. There are much simpler equivalent formulas that can be used to compute determinants far more efficiently.

Recursive Definition of the Determinant of an $n \times n$ Matrix

We will now give a recursive definition of the determinant of an $n \times n$ matrix in terms of the determinants of its $(n-1) \times (n-1)$ submatrices. This definition follows from the one given above. We can use this definition to compute the determinant of a 3×3 matrix using the values of determinants of its 2×2 submatrices which themselves can be computed by the formula given above. The determinant of a matrix of order 4×4 can be computed using the determinants of its 3×3 submatrices, and so on.

Given a square matrix $A = (a_{ij})$, we will denote by A_{ij} the submatrix obtained by deleting the ith row and the jth column from A, this submatrix is called a **minor** of A.

Example:

Let

$$A = \begin{pmatrix} 1 & -2 & 0 & 0 \\ -1 & 7 & 8 & 5 \\ \hline 3 & 4 & -3 & -5 \\ 2 & 9 & -4 & 6 \end{pmatrix}$$

$$A_{32} = \begin{pmatrix} 1 & 0 & 0 \\ -1 & 8 & 5 \\ 2 & -4 & 6 \end{pmatrix}.$$

Given the square matrix $A = (a_{ij})$, the (i,j)th **cofactor** of A, denoted by C_{ij} is given by

$$C_{ij} = (-1)^{i+j} \text{determinant}(A_{ij})$$

where A_{ij} is the minor of A obtained by deleting the ith row and the jth column from A.

So, while the minor A_{ij} is a matrix, the cofactor C_{ij} is a real number.

Example:

Let $A = \begin{pmatrix} 3 & 1 & 2 \\ -2 & -1 & 1 \\ 8 & 1 & 0 \end{pmatrix}$. Then

$$A_{11} = \begin{pmatrix} -1 & 1 \\ 1 & 0 \end{pmatrix}, \quad C_{11} = (-1)^{1+1} \det \begin{pmatrix} -1 & 1 \\ 1 & 0 \end{pmatrix} = -1$$

$$A_{12} = \begin{pmatrix} -2 & 1 \\ 8 & 0 \end{pmatrix}, \quad C_{12} = (-1)^{1+2} \det \begin{pmatrix} -2 & 1 \\ 8 & 0 \end{pmatrix} = 8$$

$$A_{13} = \begin{pmatrix} -2 & -1 \\ 8 & 1 \end{pmatrix}, \quad C_{13} = (-1)^{1+3} \det \begin{pmatrix} -2 & -1 \\ 8 & 1 \end{pmatrix} = 6.$$

Definition: Cofactor expansion of a determinant: For $n \geq 2$, the determinant of the $n \times n$ matrix $A = (a_{ij})$ can be computed by a process called the **cofactor expansion** (or also **Laplace expansion**) **across any row** or **down any column**. Using the notation developed above, this expansion across the ith row is

$$\det(A) = a_{i1}C_{i1} + a_{i2}C_{i2} + \cdots + a_{in}C_{in}.$$

The cofactor expansion down the jth column is

$$\det(A) = a_{1j}C_{1j} + a_{2j}C_{2j} + \cdots + a_{nj}C_{nj}.$$

Example:

Let

$$A = \begin{pmatrix} 2 & 4 & 0 \\ 3 & 3 & -2 \\ 1 & -1 & 0 \end{pmatrix}.$$

Using cofactor expansion down the third column, we have

$$\det(A) = 0C_{13} - 2C_{23} + 0C_{33}$$

$$= -2(-1)^{2+3}\det\begin{pmatrix} 2 & 4 \\ 1 & -1 \end{pmatrix} = 2(2 \times -1 - 4 \times 1) = -12.$$

Using cofactor expansion across third row, we have

$$\det(A) = 1C_{31} - 1C_{32}$$

$$= (-1)^{3+1}\det\begin{pmatrix} 4 & 0 \\ 3 & -2 \end{pmatrix} - 1(-1)^{3+2}\det\begin{pmatrix} 2 & 0 \\ 3 & -2 \end{pmatrix}$$

$$= -8 - 4 = -12, \text{ same value as above.}$$

If a row or column has many zero entries, the cofactor expansion of the determinant using that row or column has many terms that are zero, and the cofactors in those terms need not be calculated. So, to compute the determinant of a square matrix, it is better to choose a row or column with the maximum number of zero entries, for cofactor expansion.

Exercises

2.8.1: Find the determinants of the following matrices with cofactor expansion using rows or columns that involve the least amount of computation.

$$(i) \quad \begin{pmatrix} 1 & 13 & 18 & -94 \\ 0 & -2 & -11 & 12 \\ 0 & 0 & 7 & 14 \\ 0 & 0 & 0 & -8 \end{pmatrix}, \quad (ii) \quad \begin{pmatrix} 0 & 0 & -4 & 3 \\ 3 & 0 & 0 & 8 \\ 2 & 1 & -2 & 0 \\ -1 & 3 & 4 & 0 \end{pmatrix}$$

$$(iii) \quad \begin{pmatrix} 0 & -8 & 0 & 0 \\ 4 & 0 & 0 & 0 \\ 0 & 0 & 0 & 3 \\ 0 & 0 & -1 & 0 \end{pmatrix}, \quad (iv) \quad \begin{pmatrix} 0 & 0 & -2 & -3 \\ 1 & 3 & 2 & -4 \\ 0 & 0 & 0 & -5 \\ 0 & -1 & -2 & 4 \end{pmatrix}.$$

Some Results On Determinants

1. Determinants of upper (lower) triangular matrices: *The determinant of an upper triangular, lower triangular, or diagonal matrix is the product of its diagonal entries.*

Let P be a permutation matrix. Its determinant is $(-1)^{NI(P)}$ *where* $NI(P)$ *is the number of inversions in the permutation corresponding to P.*

2. Adding a scalar multiple of a row (column) to another row (column) leaves determinant unchanged: *If a square matrix B is obtained by adding a multiple of a row to another row (or by adding a multiple of a column to another column) in a square matrix A, then det(B) = det(A).*

3. Determinant is 0 if matrix has a 0-row or 0-column: *If a square matrix has a zero row or a zero column, its determinant is 0.*

4. Determinant is 0 if a row (column) is a scalar multiple of another row (column): *If a row in a square matrix is a multiple of another row, or if a column in this matrix is a multiple of another column, then the determinant of that matrix is zero.*

5. Interchanging a pair of rows (columns) multiplies determinant by −1: *If a square matrix B is obtained by interchanging any two rows (or interchanging any two columns) of a square matrix A, then det(B) = −det(A).*

6. Multiplying a row (column) by a scalar also multiplies determinant by that scalar: *If a square matrix B is obtained by multiplying each element in a row or a column of a square matrix A by the same scalar α, then $det(B) = \alpha det(A)$.*

7. Multiplying all entries in a matrix of order n by scalar α, multiplies determinant by α^n: *If A is a square matrix of order n, and α is a scalar, then $det(\alpha A) = \alpha^n det(A)$.*

8. A matrix and its transpose have the same determinant: *For any square matrix A, we have $det(A^T) = det(A)$.*

9. Determinant of a product of two matrices is the product of their determinants: *If A, B are two square matrices of order n, then $det(AB) = (det(A))(det(B))$.*

An incomplete proof of this product rule for determinants was given by J. P. M. Binet in 1813. The proof was corrected by A. L. Cauchy in 1815. Hence this result is known as the **Cauchy–Binet theorem.**

10. The effect of a GJ pivot step on the determinant: *If the square matrix A' is obtained by performing a GJ pivot step on the square matrix A with a_{rs} as the pivot element, then $det(A') = (1/a_{rs})det(A)$.*

11. The effect of a G pivot step on the determinant: *If the square matrix A' is obtained by performing a G pivot step on the square matrix A, then*

> *$det(A') = det(A)$ if no row interchange was performed on A before this G pivot step;*
> *$det(A') = -det(A)$ if a row interchange was performed on A before this G pivot step.*

12. Linearity of the Determinant function in a single column or row of the matrix: *The determinant of a square matrix $A = (a_{ij})$ of order n is a linear function of any single row or any single column of A, when all the other entries in the matrix are fixed at specific numerical values.*

The columns of A are $A_{.j}$, $j = 1$ to n. Suppose, for some s, all the entries in $A_{.1}, \ldots, A_{.s-1}, A_{.s+1}, \ldots, A_{.n}$ are held fixed, and only the entries in $A_{.s}$ are allowed to vary over all possible real values. Then $det(A)$ is a linear function of the entries in $A_{.s} = (a_{1s}, \ldots, a_{ns})^T$. The reason for this is the following. Let C_{ij} denote the (i, j)th cofactor of A. Then by cofactor expansion down the sth column, we have

$$\det(A) = a_{1s}C_{1s} + a_{2s}C_{2s} + \cdots + a_{ns}C_{ns}.$$

The cofactors $C_{1s}, \ldots, C_{ns}$ depend only on entries in $A_{.1}, \ldots, A_{.s-1}$, $A_{.s+1}, \ldots, A_{.n}$ which are all fixed, and hence these cofactors are all constants here. Hence by the above equation and the definition of linear functions given in Section 1.7, $\det(A)$ is a linear function of $a_{1s}, \ldots, a_{ns}$, the entries in $A_{.s}$, which are the only variables here.

$A_{i.}, i = 1$ to n are the row vectors of A. A similar argument shows that if for some r, all the entries in $A_{1.}, \ldots, A_{r-1.}, A_{r+1.}, \ldots, A_{n.}$ are held fixed, and only the entries in $A_{r.}$ are allowed to vary over all possible real values, then $\det(A)$ is a linear function of the entries in $A_{r.} = (a_{r1}, \ldots, a_{rn})$.

This leads to the following result. Suppose in the square matrix A of order n, the column vectors $A_{.1}, \ldots, A_{.s-1}, A_{.s+1}, \ldots, A_{.n}$ are all fixed, and

$$A_{.s} = \beta b + \delta d$$

where b, d are two column vectors in R^n, and β, δ are two scalars. So, the column vectors of A are as given below

$$A = (A_{.1} \vdots \ldots \vdots A_{.s-1} \vdots \beta b + \delta d \vdots A_{.s+1} \vdots \ldots \vdots A_{.n}).$$

Define two matrices B, D with column vectors as given below.

$$B = (A_{.1} \vdots \ldots \vdots A_{.s-1} \vdots b \vdots A_{.s+1} \vdots \ldots \vdots A_{.n}),$$

$$D = (A_{.1} \vdots \ldots \vdots A_{.s-1} \vdots d \vdots A_{.s+1} \vdots \ldots \vdots A_{.n}).$$

So, the matrices B, D differ from A only in their sth column. Then by the linearity

$$\det(A) = \beta \det(B) + \delta \det(D).$$

Caution: On the linearity of a determinant: For a square matrix A, $\det(A)$ is a linear function when it is treated as a function of the entries in a single column, or a single row of A, while all the other entries in A are held fixed at specific numerical values. As a function of all the entries in A, $\det(A)$ is definitely not linear. That's why, if B, C are square matrices of order n, and $A = B + C$; these facts do not imply that $\det(A) = \det(B) + \det(C)$.

Example:

Let

$$E = \begin{pmatrix} 1 & -2 \\ 3 & 4 \end{pmatrix}, \quad F = \begin{pmatrix} -1 & -2 \\ -2 & 4 \end{pmatrix}, \quad A = \begin{pmatrix} 0 & -2 \\ 1 & 4 \end{pmatrix}.$$

$$B = \begin{pmatrix} 1 & -1 \\ 3 & 2 \end{pmatrix}, \quad C = \begin{pmatrix} -1 & -1 \\ -2 & 2 \end{pmatrix}.$$

Then, $A_{.2} = E_{.2} = F_{.2}$ and $A_{.1} = E_{.1} + F_{.1}$. We have $\det(E) = 10$, $\det(F) = -8$, and $\det(A) = 2 = \det(E) + \det(F)$, illustrating the linearity result stated above.

We also have $\det(B) = 5$, $\det(C) = -4$. Verify that even though $A = B + C$, we have $\det(A) \neq \det(B) + \det(C)$.

13. The determinant of a square matrix A is a multilinear function of A.

Definition of a multilinear function: Consider a real valued function $f(x^1, \ldots, x^r)$ of many variables which are partitioned into vectors $x^1, \ldots, x^r$; i.e., for each $k = 1$ to r, x^k is itself a vector of variables in R^{n_k} say. $f(x^1, \ldots, x^r)$ is said to be a **multilinear function** under this partition of variables if, for each $t = 1$ to r, the function

$$f(\bar{x}^1, \ldots, \bar{x}^{t-1}, x^t, \bar{x}^{t+1}, \ldots, \bar{x}^r)$$

obtained by fixing $x^k = \bar{x}^k$ where $\bar{x}^k$ is any arbitrary vector in R^{n_k} for each $k = 1, \ldots, t-1, t+1, \ldots, r$; is a linear function of x^t. Multilinear functions are generalizations of bilinear functions defined in Section 1.7.

Thus, from the above result we see that the determinant of a square matrix A is a multilinear function of A, when the entries in A are partitioned into either the columns of A or the rows of A.

The branch of mathematics dealing with the properties of multilinear functions is called **multilinear algebra**. The multilinearity property of the determinant plays an important role in many advanced theoretical studies of the properties of determinants.

14. Singular, nonsingular square matrices:

The square matrix A of order n is said to be a

> **singular square matrix**   if $\det(A) = 0$
> **nonsingular square matrix**   if $\det(A) \neq 0$.

The concepts of singularity, nonsingularity are only defined for square matrices, but not for rectangular matrices which are not square.

15. The inverse of a square matrix:

Definition of the inverse of a square matrix: Given a square matrix A of order n, a square matrix B of order n satisfying

$$BA = AB = I$$

where I is the unit matrix of order n, is called the **inverse** of A. If the inverse of A exists, A is said to be **invertible** and the inverse is denoted by the symbol A^{-1}. $\bowtie$

If A is invertible, its inverse is unique for the following reason. Suppose both B, D are inverses of A. Then

$$BA = AB = I; \quad DA = AD = I.$$

So, $B = BI = BAD = ID = D$.

Please note that the concept of the inverse of a matrix is defined only for square matrices, there is no such concept for rectangular matrices that are not square,

Also, if a square matrix A is singular (i.e., $\det(A) = 0$), then its inverse does not exist (i.e., A is not invertible) for the following reason. Suppose A is a singular square matrix and the inverse B of A exists. Then, since $BA =$ the unit matrix I, from Result 9 above we have

$$\det(B) \times \det(A) = 1$$

and since $\det(A) = 0$, the above equation is impossible. Hence a singular square matrix is not invertible.

Every nonsingular square matrix does have an inverse, in fact we will now derive a formula for its inverse in terms of its cofactors. This inverse is the matrix analogue of the reciprocal of a nonzero real number in real number arithmetic.

Let $A = (a_{ij})$ be a square matrix of order n, and for $i, j = 1$ to n let C_{ij} be the (i, j)th cofactor of A. Then by Laplace expansion we know that for all $i = 1$ to n

$$a_{i1}C_{i1} + \cdots + a_{in}C_{in} = \det(A)$$

and for all j

$$a_{1j}C_{1j} + \cdots + a_{nj}C_{nj} = \det(A).$$

Now consider the matrix A' obtained by replacing the first column of A by its second column, but leaving everything else unchanged. So, by its columns, we have

$$A' = (A_{.2} \vdots A_{.2} \vdots A_{.3} \vdots \ldots \vdots A_{.n}).$$

By Result 4 above we have $\det(A') = 0$. Since A' differs from A only in its first column, we know that C_{i1} is also the $(i, 1)$th cofactor of A' for all $i = 1$ to n. Therefore by cofactor expansion down its first column, we have

$$\det(A') = a_{12}C_{11} + a_{22}C_{21} + \cdots + a_{n2}C_{n1} = 0.$$

Using a similar argumant we conclude that for all $t \neq j$

$$a_{1t}C_{1j} + a_{2t}C_{2j} + \cdots + a_{nt}C_{nj} = 0$$

and for all $i \neq t$

$$a_{i1}C_{t1} + a_{i2}C_{t2} + \cdots + a_{in}C_{tn} = 0.$$

Now consider the square matrix of order n obtained by replacing each element in A by its cofactor in A, and then taking the transpose of the resulting matrix. Hence the (i, j)th element of this matrix is C_{ji}. This transposed matrix of cofactors was introduced by the French mathematician A. L. Cauchy in 1815 under the name *adjoint of A*. However, the term *adjoint* has acquired another meaning subsequently, hence this matrix is now called **adjugate** or **classical adjoint** of A, and denoted by $\mathbf{adj}(A)$. Therefore

$$\operatorname{adj}(A) = \begin{pmatrix} C_{11} & C_{21} & C_{31} & \ldots & C_{n1} \\ C_{12} & C_{22} & C_{32} & \ldots & C_{n2} \\ C_{13} & C_{23} & C_{33} & \ldots & C_{n3} \\ \vdots & \vdots & \vdots & & \vdots \\ C_{1n} & C_{2n} & C_{3n} & \ldots & C_{nn} \end{pmatrix}.$$

The equations derived above imply that $\operatorname{adj}(A)$ satisfies the following important property

$$\operatorname{adj}(A)A = A(\operatorname{adj}(A)) = \operatorname{diag}\{\det(A), \det(A), \ldots, \det(A)\}$$

where $\text{diag}\{\det(A), \det(A), \ldots, \det(A)\}$ is the diagonal matrix of order n with all its diagonal entries equal to $\det(A)$. So, if A is nonsingular, $\det(A) \neq 0$, and

$$\left(\frac{1}{\det(A)}\right) \text{adj}(A)A = A\left[\left(\frac{1}{\det(A)}\right) \text{adj}(A)\right] = I$$

i.e.,

$$A^{-1} = \left(\frac{1}{\det(A)}\right) \text{adj}(A) = \frac{1}{\det(A)} \begin{pmatrix} C_{11} & C_{21} & C_{31} & \cdots & C_{n1} \\ C_{12} & C_{22} & C_{32} & \cdots & C_{n2} \\ C_{13} & C_{23} & C_{33} & \cdots & C_{n3} \\ \vdots & \vdots & \vdots & & \vdots \\ C_{1n} & C_{2n} & C_{3n} & \cdots & C_{nn} \end{pmatrix}.$$

This provides a mathematical formula for the inverse of an invertible matrix. This formula is seldom used to compute the inverse of a matrix in practice, as it is very inefficient. An efficient method for computing the inverse based on GJ pivot steps is discussed in Chapter 4. This formula for the inverse is only used in theoretical research studies involving matrices.

Cramer's Rule for the solution of a square nonsingular system of linear equations:

Consider the system of linear equations

$$Ax = b.$$

It is said to be a square nonsingular system of linear equations if the coefficient matrix A in it is square nonsingular. In this case A^{-1} exists, and $x = A^{-1}b$ is the unique solution of the system for each $b \in R^n$. To show this, we see that $\bar{x} = A^{-1}b$ is in fact a solution because

$$A\bar{x} = AA^{-1}b = Ib = b.$$

If $\hat{x}$ is another solution to the system, then $A\hat{x} = b$, and multiplying this on both sides by A^{-1} we have

$$A^{-1}A\hat{x} = A^{-1}b = \bar{x}, \quad \text{i.e.,} \quad \bar{x} = A^{-1}A\hat{x} = I\hat{x} = \hat{x}$$

and hence $\hat{x} = \bar{x}$. So, $\bar{x} = A^{-1}b$ is the unique solution of the system.

Let $\bar{x} = A^{-1}b = (\bar{x}_1, \ldots, \bar{x}_n)^T$. Using

$$A^{-1} = \left(\frac{1}{\det(A)}\right) \mathrm{adj}(A) = \frac{1}{\det(A)} \begin{pmatrix} C_{11} & C_{21} & C_{31} & \cdots & C_{n1} \\ C_{12} & C_{22} & C_{32} & \cdots & C_{n2} \\ C_{13} & C_{23} & C_{33} & \cdots & C_{n3} \\ \vdots & \vdots & \vdots & & \vdots \\ C_{1n} & C_{2n} & C_{3n} & \cdots & C_{nn} \end{pmatrix}$$

where $C_{ij} =$ the (i,j)th cofactor of A, we get, for $j = 1$ to n

$$\bar{x}_j = \left(\frac{1}{\det(A)}\right)(b_1 C_{1j} + b_2 C_{2j} + \cdots + b_n C_{nj}).$$

For $j = 1$ to n, let $A_j(b)$ denote the matrix obtained by replacing the jth column in A by b, but leaving all other columns unchanged. So, by columns

$$A_j(b) = (A_{.1} \vdots A_{.2} \vdots \ldots \vdots A_{.j-1} \vdots b \vdots A_{.j+1} \vdots \ldots \vdots A_{.n}).$$

By cofactor expansion down the column b in $A_j(b)$ we see that

$$\det(A_j(b)) = b_1 C_{1j} + b_2 C_{2j} + \cdots + b_n C_{nj}$$

and hence from the above we have, for $j = 1$ to n

$$\bar{x}_j = \frac{\det(A_j(b))}{\det(A)}.$$

This leads to **Cramer's rule**, which states that the unique solution of the square nonsingular system of linear equations $Ax = b$ is $\bar{x} = (\bar{x}_1, \ldots, \bar{x}_n)$, where

$$\bar{x}_j = \frac{\det(A_j(b))}{\det(A)}$$

for $j = 1$ to n; with $A_j(b)$ being obtained by replacing the jth column in A by b, but leaving all other columns unchanged.

Historical note on Cramer's rule: Cramer's rule which first appeared in an appendix of the 1750 book *Introduction a L'analysedes Lignes Courbes*

Algebriques by the Swiss mathematician Gabriel Cramer, gives the value of each variable in the solution as the ratio of two determinants. Since it gives the value of each variable in the solution by an explicit determinantal ratio formula, Cramer's rule is used extensively in theoretical research studies involving square nonsingular systems of linear equations. It is not used to actually compute the solutions of systems in practice, as the methods discussed in Chapter 1 obtain the solution much more efficiently.

Example:

Consider the following system of equations in detached coefficient form.

x_1	x_2	x_3	b
1	-1	1	-5
0	2	-1	10
-2	1	3	-13

Let A denote the coefficient matrix, and for $j = 1$ to 3, let $A_j(b)$ denote the matrix obtained by replacing $A_{.j}$ in A by b. Then we have

$$\det(A) = \det \begin{pmatrix} 1 & -1 & 1 \\ 0 & 2 & -1 \\ -2 & 1 & 3 \end{pmatrix} = 9;$$

$$\det(A_1(b)) = \det \begin{pmatrix} -5 & -1 & 1 \\ 10 & 2 & -1 \\ -13 & 1 & 3 \end{pmatrix} = 18;$$

$$\det(A_2(b)) = \det \begin{pmatrix} 1 & -5 & 1 \\ 0 & 10 & -1 \\ -2 & -13 & 3 \end{pmatrix} = 27;$$

$$\det(A_3(b)) = \det \begin{pmatrix} 1 & -1 & -5 \\ 0 & 2 & 10 \\ -2 & 1 & -13 \end{pmatrix} = -36.$$

So, by Cramer's rule, the unique solution of the system is

$$\bar{x} = \left(\frac{18}{9}, \frac{27}{9}, \frac{-36}{9} \right)^T = (2, 3, -4)^T.$$

Example:

Consider the following system of equations in detached coefficient form.

x_1	x_2	x_3	b
1	−1	0	30
0	2	2	0
−2	1	−1	−60

Here the determinant of the coefficient matrix is 0, hence it is singular. So, Cramer's rule cannot be applied to find a solution for it.

Example:

Consider the following system of equations in detached coefficient form.

x_1	x_2	x_3	b
1	−1	0	20
10	12	−2	−40

This system has three variables but only two constraints, it is not a square system. Hence Cramer's rule cannot be applied to find a solution for it.

An Algorithm for Computing the Determinant of a General Square Matrix of Order n Efficiently

   BEGIN

   Step 1: Perforing G pivot steps on the matrix to make it triangular: Let A be a given matrix of order n. To compute its determinant, perform G pivot steps in each row of A beginning at the top. Select pivot elements using any of the strategies discussed in Step 2 of the G elimination method in Section 1.20. Make row interchanges as indicated there, and a column interchange to bring the pivot element to the diagonal position (i.e., if the pivot element is in position (p, q), interchange columns p and q before carrying out the pivot step so that the pivot element comes to the diagonal (p, p) position). After the pivot step, enclose the pivot element in a box. At any stage, if the present matrix contains a row with all its entries zero, then $\det(A) = 0$, terminate.

   Step 2: Computing the determinant: If a pivot step is performed in every row, in the final matrix the diagonal elements are the boxed pivot elements used in the various pivot steps. For $t = 1$ to n, let $\hat{a}_{tt}$ be the boxed pivot element in the final matrix in row t. Let r denote the total number of row and column interchanges performed. Then

$$\det(A) = (-1)^r \prod_{t=1}^{n} [\hat{a}_{tt}]$$

END

Example:

Consider the matrix

$$A = \begin{pmatrix} 0 & 1 & 3 & 0 \\ 4 & -1 & -1 & 2 \\ 6 & 2 & 10 & 2 \\ -8 & 0 & -2 & -4 \end{pmatrix}.$$

To compute det (A) using G pivot steps, we will use the partial pivoting strategy discussed in Section 1.20 for selecting the pivot elements, performing pivot steps in Columns 2, 3, 4, 1 in that order. We indicate the pivot row (PR), pivot column (PC), and put the pivot element in a box after any row and column interchanges needed are performed. The indication "RI to Rt" on a row means that at that stage that row will be interchanged with Row t in that tableau. Similarly "CI of Cp & Cq" indicates that at that stage columns p and q will be interchanged in the present tableau.

Carrying out Step 1

0	1	3	0		
4	−1	−1	2		
6	2	10	2	RI to R1	
−8	0	−2	−4		
	PC↑				
6	2	10	2	PR	CI of C1 & C2
4	−1	−1	2		
0	1	3	0		
−8	0	−2	−4		
	PC↑				
2	6	10	2	PR	
−1	4	−1	2		
1	0	3	0		
0	−8	−2	−4		
PC↑					

Step 1 continued

2	6	10	2	
0	7	4	3	CI of C2 & C3
0	−3	−2	−1	
0	−8	−2	−4	

| | | PC↑ | | |

2	10	6	2	
0	4	7	3	PR
0	−2	−3	−1	
0	−2	−8	−4	

| | PC↑ | | | |

2	10	6	2	
0	4	7	3	
0	0	$\frac{1}{2}$	$\frac{1}{2}$	
0	0	$\frac{-9}{2}$	$\frac{-5}{2}$	RI to R3

| | | | PC↑ | |

2	10	6	2	
0	4	7	3	
0	0	$\frac{-9}{2}$	$\frac{-5}{2}$	PR CI of C3 & C4
0	0	$\frac{1}{2}$	$\frac{1}{2}$	

| | | | PC↑ | |

2	10	2	6	
0	4	3	7	
0	0	$\boxed{\frac{-5}{2}}$	$\frac{-9}{2}$	PR
0	0	$\frac{1}{2}$	$\frac{1}{2}$	

| | | PC↑ | | |

Step 1 continued

$\boxed{2}$	10	2	6
0	$\boxed{4}$	3	7
0	0	$\boxed{\frac{-5}{2}}$	$\frac{-9}{2}$
0	0	0	$\boxed{\frac{-4}{10}}$

Carrying out Step 2: We made a total of 5 row and column interchanges. So, we have

$$\det(A) = (-1)^5 \left[2 \times 4 \times \frac{-5}{2} \times \frac{-4}{10} \right] = -8.$$

Example:

Consider the matrix

$$A = \begin{pmatrix} 1 & 1 & 3 & 1 \\ 1 & -1 & 3 & 3 \\ 4 & -2 & 12 & 10 \\ 4 & -1 & 13 & 18 \end{pmatrix}$$

To compute $\det(A)$ using G pivot steps, we use the same notation as in the example above. We use the same partial pivoting strategy for selecting the pivot element from the pivot column indicated.

Carrying out Step 1

1	1	3	1		
1	-1	3	3		
4	-2	12	10		RI to R1
4	-1	13	18		
	PC↑				
4	-2	12	10	PR	CI of C1 & C2
1	-1	3	3		
1	1	3	1		
4	-1	13	18		
	PC↑				

Step 1 continued

-2	4	12	10	PR
-1	1	3	3	
1	1	3	1	
-1	4	13	18	
PC↑				
-2	4	12	10	
0	-1	-3	-2	
0	3	9	6	RI to R2
0	2	7	13	
	PC↑			
-2	4	12	10	
0	3	9	6	PR
0	-1	-3	-2	
0	2	7	13	
	PC↑			
-2	4	12	10	
0	3	9	6	
0	0	0	0	
0	0	1	9	

Since all the entries in the third row in the last tableau are all zero, $\det(A) = 0$ in this example.

Importance of Determinants

For anyone planning to go into research in mathematical sciences, a thorough knowledge of determinants and their properties is essential, as determinants play a prominent role in theoretical research. However, determinants do not appear often nowadays in actual numerical computation. So, the study of determinants is very important for understanding the theory and the results, but not so much for actual computational purposes.

Exercises:

2.8.2: Let $A_{n \times n} = (a_{ij})$ where $a_{ij} = i + j$. What is $\det(A)$?

2.8.3: What is the determinant of the following $n \times n$ matrix?

$$\begin{pmatrix} 1 & 2 & \cdots & n \\ n+1 & n+2 & \cdots & 2n \\ \vdots & \vdots & & \vdots \\ n^2 - n + 1 & \cdots & \cdots & n^2 \end{pmatrix}$$

2.8.4: For any real p, q, r prove that the determinant of the following matrix is 0.

$$\begin{pmatrix} p^2 & pq & pr \\ qp & q^2 & qr \\ rp & rq & r^2 \end{pmatrix}$$

2.8.5: The following square matrix of order n called the **Vandermonde matrix** is encoutered often in research studies.

$$V(x_1, \ldots, x_n) = \begin{pmatrix} 1 & x_1 & x_1^2 & \cdots & x_1^{n-1} \\ \vdots & \vdots & \vdots & & \vdots \\ 1 & x_n & x_n^2 & \cdots & x_n^{n-1} \end{pmatrix}$$

Prove that the determinant of the Vandermonde matrix, called the Vandermonde determinant, is $= \prod_{i>j}(x_i - x_j)$. (Use the results on determinants stated above to simplify the determinant and then apply Laplace's expansion across the 1st row).

2.8.6: The following square matrix of order n is called the **Frobinious matrix** or the **companion matrix** of the polynomial $p(\lambda) = \lambda^n - a_{n-1}\lambda^{n-1} - a_{n-2}\lambda^{n-2} - \ldots - a_1\lambda - a_0$.

$$A = \begin{pmatrix} 0 & 1 & 0 & \cdots & 0 & 0 \\ 0 & 0 & 1 & \cdots & 0 & 0 \\ \vdots & \vdots & \ddots & \ddots & \ddots & \vdots \\ 0 & 0 & 0 & \ddots & 1 & 0 \\ 0 & 0 & 0 & \cdots & 0 & 1 \\ a_0 & a_1 & a_2 & \cdots & a_{n-2} & a_{n-1} \end{pmatrix}$$

Prove that $\det(\lambda I - A) = p(\lambda)$, where I is the identity matrix of order n.

2.8.7: A square matrix $A = (a_{ij})$ of order n is said to be a **skew-symmetric matrix** if $a_{ij} = -a_{ji}$ for all i, j (i.e., $a_{ij} + a_{ji} = 0$ for all i, j).

Prove that the determinant of a skew-symmetric matrix of odd order is 0.

Prove that the determinant of a skew-symmetric matrix of even order does not change if a constant α is added to all its elements.

In a skew-symmetric matrix $A = (a_{ij})$ of even order, if $a_{ij} = 1$ for all $j > i$, find $\det(A)$.

2.8.8: Let $A = (a_{ij}), B = (b_{ij})$ be two square matrices of the same order n, where $b_{ij} = (-1)^{i+j} a_{ij}$ for all i, j. Prove $\det(A) = \det(B)$.

2.8.9: Let $A = (a_{ij})$ be a square matrix of order n, and $s_i = \sum_{j=1}^{n} a_{ij}$ for all i. So, s_i is the sum of all the elements in row i of A. Prove that the determinant of the following matrix is $(-1)^{n-1}(n-1)\det(A)$.

$$\begin{pmatrix} s_1 - a_{11} & \cdots & s_1 - a_{1n} \\ \vdots & & \vdots \\ s_n - a_{n1} & \cdots & s_n - a_{nn} \end{pmatrix}.$$

2.8.10: Basic minors of a matrix: There are many instances where we have to consider the determinants of square submatrices of a matrix, such determinants are called **minors** of the matrix. A minor of order p of a matrix A is called a **basic minor** of A if it is nonzero, and all minors of A of orders $\geq p + 1$ are zero; i.e., a basic minor is a nonzero minor of maximal order, and its order is called the *rank of A*.

If the minor of A defined by rows $i_1, \ldots, i_k$ and columns $j_1, \ldots, j_k$ is a basic minor, then show that the set of row vectors $\{A_{i_1.}, \ldots, A_{i_k.}\}$ is linearly independent, and all other rows of A are linear combinations of rows in this set.

2.8.11: The **trace** of a square matrix $A = (a_{ij})$ of order n is $\sum_{i=1}^{n} a_{ii}$, the sum of its diagonal elements. If A, B are square matrices of order n, prove that $\mathrm{trace}(AB) = \mathrm{trace}(BA)$.

2.8.12: If the sum of all the elements of an invertible square matrix A in every row is s, prove that the sum of all the elements in every row of A^{-1} is $1/s$.

2.9   Additional Exercises

2.9.1: A, B are two matrices of order $m \times n$. Mention conditions under which $(A + B)^2$ is defined. Under these conditions, expand $(A + B)^2$ and write it as a sum of individual terms, remembering that matrix multiplication is not commutative.

2.9.2: Consider the following matrices. Show that FE and EF both exist but they are not equal.

$$F = \begin{pmatrix} 1 & 0 & 0 \\ 0 & 1 & 0 \\ 0 & 3 & 1 \end{pmatrix}, \quad E = \begin{pmatrix} 1 & 0 & 0 \\ -2 & 1 & 0 \\ 0 & 0 & 1 \end{pmatrix}.$$

2.9.3: Is it possible to have a square matrix $A \neq 0$ satisfying the property that $A \times A = A^2 = 0$? If so find such matrices of orders 2 and 3.

2.9.4: Show that the product of two lower triangular matrices is lower triangular, and that the product of upper triangular matrices is upper triangular.

Show that the inverse of a lower triangular matrix is lower triangular, and that the inverse of an upper triangular matrix is upper triangular.

Show that the sum of symmetric matrices is symmetric. Show that any linear combination of symmetric matrices of the same order is also symmetric.

Construct numerical examples of symmetric matrices A, B of the same order such that the product AB is not symmetric. Construct numerical examples of symmetric matrices whose product is symmetric.

If A, B are symmetric matrices of the same order, prove that the product AB is symmetric iff $AB = BA$.

Show that the inverse of a symmetric matrix is also symmetric.

For any matrix A show that both the products $AA^T, A^T A$ always exist and are symmetric.

Prove that the product of two diagonal matrices is also diagonal.

2.9.5: Let $C = AB$ where A, B are of orders $m \times n, n \times p$ respectively. Show that each column vector of C is a linear combination of the column vectors of A, and that each row vector of C is a linear combination of row vectors of B.

2.9.6: If both the products AB, BA of two matrices A, B are defined then show that each of these products is a square matrix. Also, if the sum of these products is also defined show that A, B must themselves be square matrices of the same order.

Construct numerical examples of matrices A, B where both products AB, BA are defined but are unequal square matrices of the same order.

Construct numerical examples of matrices A, B both nonzero such that $AB = 0$.

Construct numerical examples of matrices A, B, C where $B \neq C$ and yet $AB = AC$.

2.9.7: Find the matrix X which satisfies the following matrix equation.

$$\begin{pmatrix} 1 & 1 & -1 \\ -1 & 1 & 1 \\ 1 & -1 & 1 \end{pmatrix} X = \begin{pmatrix} 11 & -5 & -7 \\ -5 & -7 & 11 \\ -7 & 11 & -5 \end{pmatrix}.$$

2.9.8: Consider the following square symmetric matrix $A = (a_{ij})$ of order n in which $a_{11} \neq 0$. Perform a G pivot step on A with a_{11} as the pivot element leading to $\bar{A}$. Show that the matrix $\mathcal{A}$ of order $n-1$ obtained from $\bar{A}$ by striking off its first row and first column is also symmetric.

$$A = \begin{pmatrix} \boxed{a_{11}} & a_{12} & \cdots & a_{1n} \\ a_{21} & a_{22} & \cdots & a_{2n} \\ \vdots & \vdots & \cdots & \vdots \\ a_{n1} & a_{n2} & \cdots & a_{nn} \end{pmatrix}, \quad \bar{A} = \begin{pmatrix} a_{11} & a_{12} & \cdots & a_{1n} \\ 0 & \bar{a}_{22} & \cdots & \bar{a}_{2n} \\ \vdots & \vdots & \cdots & \vdots \\ 0 & \bar{a}_{n2} & \cdots & \bar{a}_{nn} \end{pmatrix}.$$

2.9.9: A square matrix A is said to be **skew symmetric** if $A^T = -A$. Prove that if A is skew symmetric and A^{-1} exists then A^{-1} is also skew symmetric.

If A, B are skew symmetric matrices of the same order show that $A+B$, $A - B$ are also skew symmetric. Show that any linear combination of skew matrices of the same order is also skew symmetric.

Can the product of skew symmetric matrices be skew symmetric?

Using the fact that $A = (1/2)((A + A^T) + (A - A^T))$ show that every square matrix is the sum of a symmetric matrix and a skew symmetric matrix.

If A is a skew symmetric matrix, show that $I + A$ must be invertible.

2.9.10: List all the 24 permutations of $\{1, 2, 3, 4\}$ and classify them into even, odd classes.

Find the number of inversions in the permutation $(7, 4, 2, 1, 6, 3, 5)$.

2.9.11: Show that the determinant of the elementary matrix corresponding to the operation of multiplying a row of a matrix by α is α.

Show that the determinant of the elementary matrix corresponding to the operation of interchanging two rows in a matrix is -1.

Show that the determinant of the elementary matrix corresponding to the operation of adding a scalar multiple of a row of a matrix to another is 1.

Show that the determinant of the pivot matrix corresponding to a GJ pivot step with a_{rs} as the pivot element is $1/a_{rs}$.

2.9.12: Construct a numerical example to show that $\det(A + B)$ is not necessarily equal to $\det(A) + \det(B)$.

By performing a few row operations show that the following determinants are 0

$$
\begin{vmatrix} 5 & 4 & 3 \\ 1 & 2 & 3 \\ 1 & 1 & 1 \end{vmatrix}, \qquad
\begin{vmatrix} 2 & 3 & 0 & 0 \\ 3 & 2 & 2 & 0 \\ -5 & -5 & 0 & 2 \\ 0 & 0 & 1 & 1 \end{vmatrix}.
$$

If A is an integer square matrix and $\det(A) = \pm 1$, then show that A^{-1} is also an integer matrix.

Show that the following determinant is 0 irrespective of what values the a_{ij}s have.

$$
\begin{vmatrix}
a_{11} & a_{12} & a_{13} & 0 & 0 & 0 \\
a_{21} & a_{22} & a_{23} & 0 & 0 & 0 \\
a_{31} & a_{32} & a_{33} & 0 & 0 & 0 \\
a_{41} & a_{42} & a_{43} & 0 & 0 & 0 \\
a_{51} & a_{52} & a_{53} & 0 & a_{55} & a_{56} \\
a_{61} & a_{62} & a_{63} & a_{64} & a_{65} & a_{66}
\end{vmatrix}.
$$

2.9.13: In a small American town there are three newspapers that people buy, NY, LP, CP. These people make highly predictable transitions from one newspaper to another from one quarter to the next as described below.

Among subscribers to NY: 80% continue with NY, 7% switch to CP, and 13% switch to LP.

Among subscribers to LP: 78% continue with LP, 12% switch to NY, 10% switch to CP.

Among subscribers to CP: 83% continue with CP, 8% switch to NY, and 9% switch to LP.

Let $x(n) = (x_1(n), x_2(n), x_3(n))^T$ denote the vector of the number of subscribers to NY, LP, CP in the nth quarter. Treating these as continuous variables, find the matrix A satisfying: $x(n+1) = Ax(n)$.

2.9.14: There are two boxes each containing different numbers of bags of three different colors as represented in the following matrix A (rows of A corespond to boxes, columns of A correspond to colors of bags) given in tableau form.

	White	Red	Black
Box 1	3	4	2
Box 2	2	2	5

Each bag contains different numbers of three different fruit as in the following matrix B given in tabular form.

	Mangoes	Sapotas	Bananas
White	3	6	1
Red	2	4	4
Black	4	3	3

Show that rows, columns of the matrix product $C = AB$ correspond to boxes, fruit respectively, and that the entries in C are the total number of various fruit contained in each box.

2.9.15: Express the row vector $(a_{11}b_{11} + a_{12}b_{21}, a_{11}b_{21} + a_{12}b_{22})$ as a product of two matrices.

2.9.16: A, B are two square matrices of order n. Is $(AB)^2 = A^2 B^2$?

2.9.17: Plant 1, 2, each produce both fertilizers Hi-ph, Lo-ph simultaneously each day with daily production rates (in tons) as in the following matrix A represented in tableau form.

	Plant 1	Plant 2
Hi-ph	100	200
Lo-ph	200	300

In a certain week, each plant worked for different numbers of days, and had total output in tons of $(\text{Hi-ph}, \text{Lo-ph})^T = b = (1100, 1800)^T$.

What is the physical interpretation of the vector x which is the solution of the system $Ax = b$?

2.9.18: The % of N, P, K in four different fertilizers is given below.

	N	P	K
Fertilizer 1	10	10	10
2	25	10	5
3	30	5	10
4	10	20	20

Two different mixtures are prepared by combining various quantities in the following way.

| | Lbs in mixture, of | | | |
	Fertilizer 1	2	3	4
Mixture 1	100	200	300	400
2	300	500	100	100

Derive a matrix product that expresses the % of N, P, K in the two mixtures.

2.9.19: Bacterial growth model: A species of bacteria has this property: after the day an individual is born, it requires a maturation period of two complete days, then on 3rd days it divides into 4 new individuals (after division the original individual no longer exists).

Let $x^k = (x_1^k, x_2^k, x_3^k)^T$ denote number of individuals for whom day k is 1st, 2nd, 3rd complete day after the day of their birth.

Find a square matrix A of order 3 which expresses x^{k+1} as Ax^k. For any k express x^k in terms of A and x^1.

2.9.20: Bacterial growth model with births and deaths: Consider bacterial population described in Exercise 2.9.19 above with two exceptions.

1. Let x_i^k be continuous variables here.

2. In 1st full day after their birth day each bacteria either dies with probability 0.25 or lives to 2nd day with probability 0.75.

Each bacteria that survives 1st full day of existence after birth, will die on 2nd full day with probability 0.2, or lives to 3rd day with probability 0.8.

The scenario on 3rd day is the same as described in Excercise 19, i.e., no death on 3rd day but only division into new individuals.

Again find matrix A that expresses x^{k+1} as Ax^k here, and give an expression for x^k in terms of A and x^1.

2.9.21: A 4-year community college has 1st, 2nd, 3rd, 4th year students. From past data they found that every year, among the students in each year's class, 80% move on to the next year's class (or graduate for 4th year students); 10% drop out of school; 5% are forced to repeat that year; and the remaining 5% are asked to take the next year off from school to work and get strong motivation to continue studies, and return to college into the same year's class after that work period. Let $x(n) = (x_1(n), x_2(n), x_3(n), x_4(n))^T$ denote the vector of the number of 1st year, 2nd year, 3rd year, 4th year students in the college in the nth year, treat each of these as continuous variables. Assume that the college admits 100 first year students, and 20 transfer students from other colleges into the 2nd year class each year. Find matrices A, b that expresses $x(n+1)$ as $Ax(n) + b$.

2.9.22: Olive harvest: Harvested olives are classified into sizes 1 to 5. Harvest season usually lasts 15 days, with harvesting carried out only on days 1, 8, and 15 of the season. Of olives of sizes 1 to 4 on trees not harvested on days 1, 8 of the season, 10% grow to the next size, and 5% get rotten and drop off, in a week.

At the beginning of the harvest season an olive farm estimated that they have 400, 400, 300, 300, 40 kg of sizes 1 to 5 respectively on their trees at that time. On days 1, 8 of the harvest season they plan to harvest all the size 5 olives on the trees, and 100, 100, 50, 40 kg of sizes 1 to 4 respectively; and on the final 15th day of the season, all the remaining olives will be harvested.

Let $x^r = (x_1^r, x_2^r, x_3^r, x_4^r, x_5^r)^T$ denote the vector of kgs of olives of sizes 1 to 5 on the trees, before harvest commences, on the rth harvest day, $r = 1, 2, 3$.

Find matrices A^1, A^2, b^1, b^2 such that $x^2 = A^1(x^1 - b^1), x^3 = A^2(x^2 - b^2)$. Also determine the total quantity of olives of various sizes that the farm will harvest that year.

2.9.23: A population model: (From A. R. Majid, *Applied Matrix Models*, Wiley, 1985) (i): An animal herd consists of three cohorts: newborn (age 0 to 1 year), juveniles (age 1 to 2 years), and adults (age 2 or more years).

Only adults reproduce to produce 0.4 newborn per adult per year (a herd with 50 male and 50 female adults will produce 40 newborn calves annually, half of each sex). 65% of newborn survive to become juveniles, 78% of juveniles survive to become adults, and 92% of adults survive to live another year.

Let $x^k = (x_1^k, x_2^k, x_3^k)^T$ denote the vector of the number of newborn, juvenile, and adults in the kth year. Find a matrix A that expresses x^{k+1} as Ax^k for all k.

A system like this is called a *discrete time linear system* with A as the transition matrix from k to $k + 1$. Eigenvalues and eigenvectors of square matrices discussed in Chapter 6 are very useful to determine the solutions of such systems, and to study their limiting behavior.

(ii): Suppose human predators enter and colonize this animal habitat. Let x_4^k denote the new variable representing the number of humans in the colony in year k. Human population grows at the rate of 2.8% per year, and each human eats 2.7 adult animals in the herd per year. Let $X^k = (x_1^k, x_2^k, x_3^k, x_4^k)^T$. Find the matrix B such that $X^{k+1} = BX^k$.

2.9.24: Fibonacci numbers: Consider the following system of equations: $x_1 = 1$, $x_2 = 1$, and $x_{n+1} = x_n + x_{n-1}$ for $n \geq 2$.

Write the 3rd equation for $n = 2, 3, 4$; and express this system of 5 equations in $(x_1, \ldots, x_5)^T$ in matrix form.

For general n, find a matrix B such that $(x_{n+1}, x_{n+2})^T = B(x_n, x_{n+1})^T$. Then show that $(x_{n+1}, x_{n+2})^T = B^n(x_1, x_2)^T$.

Also for general n find a matrix A such that $(x_{n+1}, x_{n+2}, x_{n+3})^T = A(x_n, x_{n+1}, x_{n+2})^T$ Hence show that $(x_{n+1}, x_{n+2}, x_{n+3})^T = A^n(x_1, x_2, x_3)^T$.

x_n is known as the nth Fibonacci number, all these are positive integers. Find x_n for $n \leq 12$, and verify that x_n grows very rapidly. The growth of x_n is similar to the growth of populations that seem to have no limiting factors.

2.9.25: For any u, v between 1 to n define E_{uv} to be the square matrix of order n with the (u, v)th entry $= 1$, and every other entry $= 0$.

Let $(j_1, \ldots, j_n)$ be a permutation of $(1, \ldots, n)$, and let $P = (p_{r,s})$ be the permutation matrix of order n corresponding to it (i.e., $p_{rj_r} = 1$ for all r, and all other entries in P are 0).

Then show that $PE_{uv}P^{-1} = E_{j_u j_v}$. (Remember that for a permutation matrix P, $P^{-1} = P^T$.)

2.9.26: Find conditions on b_1, b_2, b_3 so that the following system has at least one solution. Does $b_1 = 5$, $b_2 = -4$, $b_3 = -2$ satisfy this condition? If it does find the general solution of the system in this case in parametric form.

x_1	x_2	x_3	x_4	
1	−1	−1	1	b_1
2	−3	0	1	b_2
8	−11	−2	5	b_3

2.9.27: Fill in the missing entries in the following matrix product.

$$\begin{pmatrix} 1 & 0 & 1 \\ & 2 & -1 \\ 1 & & 3 \end{pmatrix} \begin{pmatrix} & & 1 \\ & 0 & \\ 1 & 0 & 2 \end{pmatrix} = \begin{pmatrix} 0 & -1 & 3 \\ 1 & 0 & 0 \\ 2 & -2 & 9 \end{pmatrix}$$

2.9.28: If A, B are square matrices of order n, is it true that $(A-B)(A+B) = A^2 - B^2$? If not under what conditions will it be true?
Expand $(A+B)^3$.

2.9.29: Determine the range of values of a, b, c for which the following matrix is triangular.

$$\begin{pmatrix} 5 & 6 & b & 0 \\ 0 & a & 2 & -1 \\ c & 0 & -1 & 2 \\ 0 & 0 & 0 & 3 \end{pmatrix}$$

2.9.30: In each case below, you are given a matrix A and its product Ax with an unknown vector x. In each case determine whether you can identify the vector x unambiguously.

$$\text{(i)} \quad A = \begin{pmatrix} 1 & -1 \\ 1 & 2 \end{pmatrix}, \quad Ax = \begin{pmatrix} -1 \\ 11 \end{pmatrix}$$

$$\text{(ii)} \quad A = \begin{pmatrix} 1 & 1 & -1 \\ -1 & 1 & 1 \\ -1 & 5 & 1 \end{pmatrix}, \quad Ax = \begin{pmatrix} 4 \\ 2 \\ 14 \end{pmatrix}$$

$$\text{(iii)} \quad A = \begin{pmatrix} 1 & 1 & -1 \\ -1 & 1 & 1 \\ 0 & 3 & 1 \end{pmatrix}, \quad Ax = \begin{pmatrix} 0 \\ 0 \\ 0 \end{pmatrix}$$

$$\text{(iv)} \quad A = \begin{pmatrix} 1 & 1 & 2 \\ -1 & 0 & 1 \\ 1 & 3 & 4 \end{pmatrix}, \quad Ax = \begin{pmatrix} 0 \\ 0 \\ 0 \end{pmatrix}$$

2.9.31: Inverses of partitioned matrices: Given two square matrices of order n, to show that B is the inverse of A, all you need to do is show that $AB = BA = $ the unit matrix of order n.

Consider the partitioned square matrix M of order n, where A, D are square submatrices of orders r, $n-r$ respectively. In the following questions, you are asked to show that M^{-1} is of the following form, with the entries as specified in the question.

$$M = \begin{pmatrix} A & B \\ C & D \end{pmatrix}$$

$$M^{-1} = \begin{pmatrix} E & F \\ G & H \end{pmatrix}$$

(i) If M, A are nonsingular, then $E = A^{-1} - FCA^{-1}$, $F = -A^{-1}BH$, $G = -HCA^{-1}$, and $H = (D - CA^{-1}B)^{-1}$.

(ii) If M, D are nonsingular, then $E = (A - BD^{-1}C)^{-1}$, $F = -EBD^{-1}$, $G = -D^{-1}CE$, $H = D^{-1} - GBD^{-1}$.

(iii) If $B = 0$, $r = n-1$, and $D = \alpha$ where $\alpha = \pm 1$, and A is invertible, then $E = A^{-1}$, $F = 0$, $G = -\alpha CA^{-1}$, $H = \alpha$.

2.9.32: Find the determinant of the following matrices.

$$\begin{pmatrix} 0 & 1 & 0 & \cdots & 0 \\ 0 & 0 & 1 & \cdots & 0 \\ \vdots & \vdots & \vdots & \ddots & \vdots \\ 0 & 0 & 0 & \cdots & 1 \\ p_1 & p_2 & p_3 & \cdots & p_n \end{pmatrix}, \quad \begin{pmatrix} a+b & c & 1 \\ b+c & a & 1 \\ c+a & b & 1 \end{pmatrix}$$

$$\begin{pmatrix} 1+\alpha_1 & \alpha_2 & \cdots & \alpha_n \\ \alpha_1 & 1+\alpha_2 & \cdots & \alpha_n \\ \vdots & \vdots & \ddots & \vdots \\ \alpha_1 & \alpha_2 & \cdots & 1+\alpha_n \end{pmatrix}$$

2.9.33: Consider following matrices:

$A_n = (a_{ij})$ of order n with all diagonal entries $a_{ii} = n$, and all off-diagonal entries $a_{ij} = 1$ for all $i \neq j$.

$B_n = (b_{ij})$ of order n with all diagonal entries $b_{ii} = 2$; $b_{i+1,i} = b_{i,i+1} = -1$ for all $i = 1$ to $n - 1$; and all other entries 0.

$C_n = (c_{ij})$ of order n with $c_{i,i+1} = 1$ for all $i = 1$ to $n-1$; and all other entries 0.

Show that $C_n + C_n^T$ is nonsingular iff n is even.

Develop a formula for determinant(A_n) in terms of n.

Show that determinant$(B_n) = 2$ determinant$(B_{n-1}) -$ determinant (B_{n-2}).

2.9.34: If A, B, C are all square matrices of the same order, C is nonsingular, and $B = C^{-1}AC$, show that $B^k = C^{-1}A^kC$ for all positive integers k.

If A is a square matrix and AA^T is singular, then show that A must be singular too.

If $u = (u_1, \ldots, u_n)^T$ is a column vector satisfying $u^T u = 1$, show that the matrix $(I - 2uu^T)$ is symmetric and that its square is I.

If A, B are square matrices of the same order, and A is nonsingular, show that determinant$(ABA^{-1}) =$ determinant(B).

2.9.35: Consider the following system of equations. Without computing the whole solution for this system, find the value of the variable x_2 only in that solution.

$$x_1 + 2x_2 + x_3 = 13$$

$$x_1 + x_2 + 2x_3 = -9$$

$$2x_1 + x_2 + x_3 = -12.$$

Reference

[2.1] M. Golubitsky and M. Dellnitz, *Linear Algebra and Differential Equations Using MATLAB* (Brooks/Cole, Pacific Grove, CA, 1999).

Chapter 3

n-Dimensional Geometry

3.1 Viewing n-tuple Vectors as Points of n-Dimensional Space R^n by Setting Up Cartesian Coordinate Axes

In 1637 the French philosopher and mathematician Rene' Descartes outlined a very simple idea for viewing n-tuple vectors as points in n-dimensional space using a frame of reference which is now called the **Cartesian coordinate system**. It is a very revolutionary idea that helped unify the two main branches of mathematics, algebra and geometry. We will now explain this idea for $n = 1, 2, 3$, and higher values.

The Case $n = 1$

The case $n = 1$ is very direct. Each vector of dimension 1 is a real number, x, say. Draw a straight line (the **one dimensional real line**) and select any point on it as the **origin** which will be represented by (or corresponds to) 0. Select a unit of length for measurement, say 1 inch = 1 unit. The origin divides the real line into two half lines (on the right and left hand sides of the origin), select one of these half lines to represent positive numbers, and the other to represent negative numbers. By the universally accepted convention, if $x > 0$, then it is represented by the point exactly x inches to the right of the origin on the real line; and if $x < 0$, then it is represented by the point exactly $|x|$ inches to the left of the origin. This way, there is a one-to-one correspondence between points on the real line and real numbers, i.e., one dimensional vectors. See Figure 3.1.

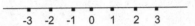

Figure 3.1   One dimensional vectors represented as points on the real line.

The Case $n = 2$

We will now explain the concept for $n = 2$. In this case also, it is possible to draw the axes of coordinates on a two dimensional sheet of paper and verify the one-to-one correspondence between the vectors and points in the space.

Select a point on the two dimensional sheet of paper as the origin corresponding to the 0-vector, $(0, 0)$. Draw two perpendicular straight lines through the origin, one horizontal stretching from left to right, and the other vertical stretching from top to bottom. These two perpendicular straight lines constitute the **Cartesian coordinate axes system** (also called **rectangular coordinate system**) for the 2-dimensional space represented by the sheet of paper; the lines are called **axes** or **coordinate axes**, the horizontal and the vertical axes. Now assign each component of the vector (x_1, x_2) to an axis. Suppose we assign x_1 to the horizontal axis, and x_2 to the vertical axis. Then the horizontal axis is called the x_1-**axis**, and the vertical axis called the x_2-**axis** in the coordinate system.

Select a unit of length for measurement, say 1 inch = 1 unit, on each of the axes (the units for the two axes may be different). By convention, on the horizontal axis, points to the right of the origin have $x_1 > 0$; and those to the left have $x_1 < 0$. Similarly, on the vertical axis, points above the origin have $x_2 > 0$, and those below the origin correspond to $x_2 < 0$. On each axis mark out the points which are an integer number of inches from the origin and write down the number corresponding to them as in Figure 3.2.

To get the unique point corresponding to the vector $x = (x_1, x_2)$, do the following:

if $x_1 = 0$ from the origin, go to the point corresponding to x_2 on the x_2-axis itself (i.e., if $x_2 > 0$ it is x_2 inches above the origin, if $x_2 < 0$ it is $|x_2|$ inches below the origin), this is the point corresponding to the vector x

if $x_1 \neq 0$,   go to the point corresponding to x_1 on the x_1-axis, P say. From P move x_2 inches above P if $x_2 > 0$, or $|x_2|$ inches below P if $x_2 < 0$, along the straight line parallel to the x_2-axis through P. The resulting point is the one corresponding to the vector x.

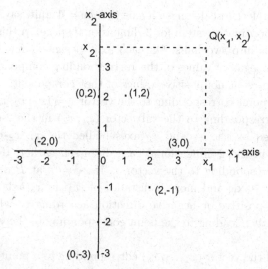

Figure 3.2   The Cartesian (rectangular) coordinate system for the 2-dimensional plane. The angles are all $90°$ ($\pi/2$ in radians). 2-dimensional vectors $x = (x_1, x_2)$ are represented as points in the 2-dimensional plane.

Now to find the 2-dimensional vector corresponding to a point Q in the 2-dimensional plane under this frame of reference, do the following: drop perpendiculars from Q to each of the coordinate axes, and measure the length of each perpendicular in the unit choosen for the axis to which it is parallel (inches here). Then, this point Q corresponds to the vector

$x = (x_1, x_2)$ where $|x_1|$, $|x_2|$ are the lengths of the perpendiculars parallel to the x_1, x_2 axes respectively, and the signs of x_1, x_2 are chosen appropriately depending on the quadrant in which q lies.

this vector x is called the **vector of Cartesian coordinates** or the **coordinate vector** of the point Q.

Clearly each 2-dimensional vector corresponds to a unique point in the 2-dimensional space and vice versa, once the rectangular coordinate system is choosen.

The Case $n = 3$

For $n = 3$, to represent 3-dimensional vectors $x = (x_1, x_2, x_3)$ as points in 3-dimensional space, the procedure is very similar. We select a point in 3-dimensional space as the origin representing the vector $0 = (0, 0, 0)$, and draw three mutually perpendicular axes, the x_1-axis, x_2-axis, x_3-axis through it, and select a scale on each axis (1 inch = 1 unit, say). This gives the Cartesian coordinate system for 3-dimensional space. In this, the origin divides each axis into two halfs. On each axis, designate one of the halfs to represent nonnegative values of the corrresponding component, and the other half to represent nonpositive values of that component.

To find the point corresponding to the vector $x = (x_1, x_2, x_3)$, first find the point P corresponding to the subvector (x_1, x_2) in the 2-dimensional plane determined by the x_1 and x_2 axes (called the x_1 x_2-**coordinate plane**) just as in the 2-dimensional case discussed earlier. If $x_3 = 0$, P is the point corresponding to the vector x. If $x_3 \neq 0$, at P draw the line parallel to the x_3-axis and move a distance of $|x_3|$ in x_3-axis units along this line in the positive or negative direction according to whether x_3 is positive or negative, leading to the point corresponding to the vector x. See Figure 3.3.

Now to find the vector $(\bar{x}_1, \bar{x}_2, \bar{x}_3)$ corresponding to a point Q in space: if Q is already a point in the x_1 x_2-coordinate plane, $\bar{x}_3 = 0$, and find the vector $(\bar{x}_1, \bar{x}_2)$ by applying the 2-dimensional procedure discussed earlier in the x_1 x_2-coordinate plane. If Q is not in the x_1 x_2-coordinate plane, drop the perpendicular from Q to the x_1 x_2-coordinate plane, and measure

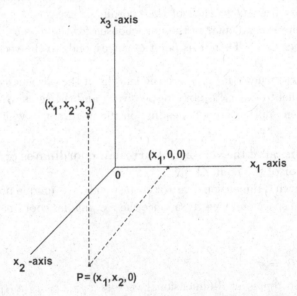

Figure 3.3 The Cartesian (rectangular) coordinate system for 3-dimensional space. All the angles are 90° ($\pi/2$ measured in radians). 3-dimensional vectors (x_1, x_2, x_3) are represented as points in the 3-dimensional space.

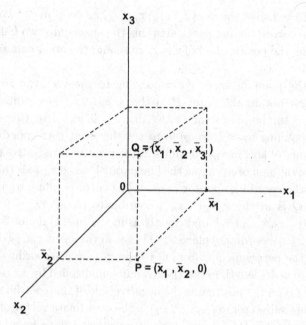

Figure 3.4 P is the foot of the perpendicular from Q to the x_1 x_2-coordinate plane, and $\bar{x}_3$ is the length of QP multiplied by ± 1. All angles are 90° ($\pi/2$ in radians).

its length in x_3-axis units. $\bar{x}_3$ is this length multiplied by ± 1 depending on whether this perpendicular is on the positive or negative side of the x_3-axis in relation to the x_1 x_2-coordinate plane. If the foot of this perpendicular is P, then find the subvector $(\bar{x}_1, \bar{x}_2)$ as the coordinate vector of P in the x_1 x_2-coordinate plane as in the 2-dimensional case discussed earlier. See Figure 3.4.

Here again there is a one-to-one correspondence between 3-dimensional vectors and points in 3-dimensional space. The vector cooresponding to a point in 3-space is called its coordinate vector under the coordinate axes system set up.

The Case $n > 3$

For $n > 3$, the procedure can be visualized to be the same, even though we cannot carry it out physically, since we only live in a 3-dimensional world. We select the coordinate axes to be n mutually perpendicular lines through a point in space selected as the origin representing the 0-vector, and select the positive and negative sides of the origin on each axis, and a scale for measuring lengths on it.

For each $j = 1$ to n, the $x_1 x_2 \ldots x_{j-1} x_{j+1} \ldots x_n$-coordinate plane determined by the corresponding axes partitions the space into two half-spaces, one containing the positive half of the x_j-axis, and the other containing the negative half.

To get the point in space corresponding to the vector $x = (x_1, \ldots, x_{n-1}, x_n)$, first obtain the point P in the $x_1 x_2 \ldots x_{n-1}$-coordinate plane corresponding to the subvector $(x_1, \ldots, x_{n-1})$; if $x_n = 0$, this P is the point corresponding to x. If $x_n \neq 0$, to get the point corresponding to x, move a distance of $|x_n|$ in x_n-axis units in the positive or negative direction depending on the sign of x_n, along the line parallel to the x_n-axis through P.

To find the coordinate vector $(\bar{x}_1, \ldots, \bar{x}_n)$ corresponding to a point Q in space, if Q is in the $x_1 x_2 \ldots x_{n-1}$-coordinate plane, $\bar{x}_n = 0$ and the subvector $(\bar{x}_1, \ldots, \bar{x}_{n-1})$ is found exactly as in the $(n-1)$ dimensional case in the $x_1 \ldots x_{n-1}$-coordinate plane. If Q is not in the $x_1 \ldots x_{n-1}$-coordinate plane, drop the perpendicular from Q to the $x_1 \ldots x_{n-1}$-coordinate plane, and take $\bar{x}_n$ to be its length in the x_n-axis units, multiplied by ± 1 depending on whether Q is on the positive or the negative side of this coordinate plane. The remaining subvector $(\bar{x}_1, \ldots, \bar{x}_{n-1})$ is the coordinate vector of the foot of this perpendicular P in the $x_1 \ldots x_{n-1}$-coordinate plane found as in the $(n-1)$-dimensional case.

With this, the one-to-one correspondence between n-dimensional vectors and points in n-dimensional space holds.

Setting up the coordinate axes and representing each point in space by its coordinate vector makes it possible to use algebraic notation and procedures for the description and study of geometric objects. This area of mathematics is called **analytical geometry**.

So, in the sequel, we will refer to vectors $x \in R^n$ as also **points in R^n**; and the various components in the vector $x = (x_1, \ldots, x_n)$ as its **coordinates**.

In the geometric representation of an n-dimensional vector by a point in n-dimensional space, there is no distinction made between row vectors and column vectors. We will assume that the vectors corresponding to points in n-dimensional space are either all row vectors, or all column vectors, which will be clear from the context.

Exercises

3.1.1: Set up the 2-dimensional Cartesian coordinate system and plot the points: $(1, 2)$, $(-1, -2)$, $(1, -2)$, $(-1, 2)$, $(0, 3)$, $(-3, 0)$, $(0, 2)$, on it.

3.2 Translating the Origin to the Point a, Translation of a Set of Points

Suppose we have set up the Cartesian coordinate system for R^n with a point O selected as the origin represented by the 0-vector. Let $x = (x_1, \ldots, x_n)^T$, $a = (a_1, \ldots, a_n)^T$ be the coordinate vectors corresponding to points P, Q respectively in space WRT this coordinate system.

Now if we change the origin from O to Q, moving each coordinate axis from the present position to the new position parallel to itself, we get a different Cartesian coordinate system which is parallel to the old, in the new variables y. This operation is called **translating the origin to the point** Q represented by the coordinate vector a in the x-space. See Figure 3.5.

The coordinate vector corresponding to the point P WRT the new coordinate system will be $y = x - a$ in the new variables y.

Thus if we translate the origin to a, it has the effect of adding the vector $-a$ to the coordinate vector of every point.

Let T be a set of points $x \in R^n$ in the original x-space which is the set of feasible solutions of a given system of constraints (A) in the original variables x. When we translate the origin in the x-space to $\bar{x}$, in terms of the new coordinate vector y, the set T is the set of feasible solutions of

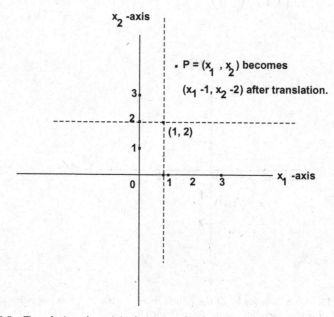

Figure 3.5   Translating the origin from 0 to $(1, 2)$ in the x-space. The new coordinate system in the new variables y is shown in dotted lines. This changes the coordinates of (x_1, x_2) in the original space to $y = (x_1 - 1, x_2 - 2)$.

the system (A') in the new variables y, where (A') is obtained from (A) by substituting $x = y + \bar{x}$.

Another concept that is commonly used is that of translating a given set S of points in R^n to a given point $\bar{x} \in R^n$. Let $S \subset R^n$ be a set of points, and $\bar{x} \in R^n$ a point. Then the **translate** of the set S to the point $\bar{x}$ is the set of points $\{y + \bar{x} : y \in S\}$; which is denoted by the symbol $\bar{x} + S$ sometimes. It is the set obtained by adding $\bar{x}$ to every point in S. The process of obtaining $\bar{x} + S$ from S is called **translation** of S to $\bar{x}$.

Example 1:

Let

$$S = \left\{ \begin{pmatrix} 1 \\ 2 \end{pmatrix}, \begin{pmatrix} -1 \\ 1 \end{pmatrix}, \begin{pmatrix} -1 \\ 2 \end{pmatrix} \right\} \subset R^2; \quad \bar{x} = \begin{pmatrix} -1 \\ -1 \end{pmatrix}.$$

Then $\bar{x} + S$ = translate of S to $\bar{x}$ is:

$$\left\{ \begin{pmatrix} 1 \\ 2 \end{pmatrix} + \begin{pmatrix} -1 \\ -1 \end{pmatrix}, \begin{pmatrix} -1 \\ 1 \end{pmatrix} + \begin{pmatrix} -1 \\ -1 \end{pmatrix}, \begin{pmatrix} -1 \\ 2 \end{pmatrix} + \begin{pmatrix} -1 \\ -1 \end{pmatrix} \right\}$$

$$= \left\{ \begin{pmatrix} 0 \\ 1 \end{pmatrix}, \begin{pmatrix} -2 \\ 0 \end{pmatrix}, \begin{pmatrix} -2 \\ 1 \end{pmatrix} \right\}.$$

See Figure 3.6.

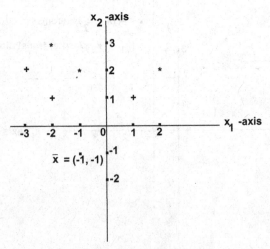

Figure 3.6 The set S consists of all the points marked with "$*$". $\bar{x}$ is the point marked with ".". $\bar{x} + S$ is the set of all points marked with "$+$".

Exercises

3.2.1: We are given the following sets of vectors. Write the set of coordinate vectors of points in these sets when the origin is translated to $(1, 2, -3)^T$.

$$S_1 = \left\{ \begin{pmatrix} 3 \\ 9 \\ 8 \end{pmatrix}, \begin{pmatrix} 3 \\ 9 \\ 8 \end{pmatrix}, \begin{pmatrix} -2 \\ 0 \\ 0 \end{pmatrix}, \begin{pmatrix} 0 \\ 8 \\ -18 \end{pmatrix} \right\}$$

$$S_2 = \left\{ \begin{pmatrix} 0 \\ 0 \\ 0 \end{pmatrix}, \begin{pmatrix} 1 \\ 2 \\ -3 \end{pmatrix}, \begin{pmatrix} -1 \\ -2 \\ 3 \end{pmatrix}, \begin{pmatrix} 2 \\ 4 \\ -6 \end{pmatrix}, \begin{pmatrix} 1 \\ 1 \\ 1 \end{pmatrix} \right\}.$$

3.2.2: Consider the following sets: $S = \{x = (x_1, x_2)^T : 0 \leq x_1 \leq 1, 0 \leq x_2 \leq 1\}$, $T = \{x + (-1, 2)^T : x \in S\}$. To which point should the origin be translated to make the set S into the set T?

3.2.3: In the following pairs S, T check whether T is a translate of S.

$$S = \left\{ \begin{pmatrix} 1 \\ 1 \\ 2 \end{pmatrix}, \begin{pmatrix} 2 \\ 4 \\ -7 \end{pmatrix}, \begin{pmatrix} 0 \\ 3 \\ 8 \end{pmatrix}, \begin{pmatrix} 2 \\ 2 \\ 2 \end{pmatrix} \right\};$$

$$T = \left\{ \begin{pmatrix} 1 \\ 3 \\ -9 \end{pmatrix}, \begin{pmatrix} 1 \\ 1 \\ 0 \end{pmatrix}, \begin{pmatrix} 0 \\ 0 \\ 0 \end{pmatrix}, \begin{pmatrix} -1 \\ 2 \\ 6 \end{pmatrix} \right\}$$

$$S = \left\{ \begin{pmatrix} 2 \\ -3 \end{pmatrix}, \begin{pmatrix} 4 \\ 6 \end{pmatrix}, \begin{pmatrix} -7 \\ -8 \end{pmatrix}, \begin{pmatrix} 8 \\ 4 \end{pmatrix} \right\};$$

$$T = \left\{ \begin{pmatrix} 0 \\ -5 \end{pmatrix}, \begin{pmatrix} 2 \\ 4 \end{pmatrix}, \begin{pmatrix} -9 \\ -10 \end{pmatrix}, \begin{pmatrix} 6 \\ 3 \end{pmatrix} \right\}.$$

3.3   Parametric Representation of Subsets of R^n

For any $n \geq 2$, one way of representing a subset of points in R^n is the **parametric representation**.

In parametric representation of any subset S of points in R^n, the coordinates of a general point in the subset S are expressed as functions of one or more parameters in the form

$$x = \begin{pmatrix} x_1 \\ x_2 \\ \vdots \\ x_j \\ \vdots \\ x_n \end{pmatrix} = \begin{pmatrix} f_1(\alpha_1, \ldots, \alpha_r) \\ f_2(\alpha_1, \ldots, \alpha_r) \\ \vdots \\ f_j(\alpha_1, \ldots, \alpha_r) \\ \vdots \\ f_n(\alpha_1, \ldots, \alpha_r) \end{pmatrix} \qquad (3.1)$$

In this representation, the jth coordinate of a general point in S is expressed as the function $f_j(\alpha_1, \ldots, \alpha_r)$ of r parameters $\alpha_1, \ldots, \alpha_r$, for $j = 1$ to n; where each parameter can take a real value subject to some constraints which will be specified.

If there are no constraints on the parameters $\alpha_1, \ldots, \alpha_r$, by giving all possible real values to each of them in (3.1), we should get all the points in the set S. Thus in this case

$$S = \{(f_1(\alpha_1, \ldots, \alpha_r), \ldots, f_n(\alpha_1, \ldots, \alpha_r))^T :$$
$$\alpha_1, \ldots, \alpha_r \quad \text{take all possible real values}\}.$$

Constraints on the parameters $\alpha_1, \ldots, \alpha_r$ may be simple, such as: $\alpha_t \geq 0$ for all $t = 1$ to r; or $0 \leq \alpha_t \leq 1$ for all $t = 1$ to r, etc. By giving all possible real values to the parameters satisfying these constraints, we should get all points in the set S. Then in this case,

$$S = \{(f_1(\alpha_1, \ldots, \alpha_r), \ldots, f_n(\alpha_1, \ldots, \alpha_r)) : \alpha_1, \ldots, \alpha_r$$
take all possible real values subject to the constraints on them$\}$.

In (3.1), if $r = 1$, i.e., if there is only one parameter in the representation, then the set S usually corresponds to the set of points on a **one dimensional curve** in R^n. If $r = 2$ (two parameters), then the set S usually corresponds to a set of points on a two dimensional surface in R^n.

We will see examples of parametric representations of some subsets in R^n in the following sections. For us, the most important subsets are of course straight lines, half-lines, and line segments, for which we give the definitions here.

The expression in (3.1) represents a **straight line** in R^n for any $n \geq 2$, if $r = 1$ and all the functions $f_1, \ldots, f_n$ are affine functions (i.e., polynomials of degree one or constants) with the parameter appearing with a nonzero coefficient in the expression for at least one x_j, and with no constraints on the value of the parameter. In addition, if the parameter, α_1 say, is required to satisfy an inequality constraint (i.e., either $\alpha_1 \geq k$ or $\alpha_1 \leq k$ for some constant k, this type of constraint is usually transformed into and stated in the form of nonnegativity restriction on the parameter) it represents a **half-line** in R^n. On the other hand if the parameter α_1 is required to lie between two specified bounds (i.e., in the form $k_1 \leq \alpha_1 \leq k_2$ for specified $k_1 < k_2$, this type of restriction is usually transformed into and stated in the form $0 \leq$ parameter ≤ 1) it represents a **line segment** in R^n. We discuss each of these subsets in detail in the following sections.

3.4 Straight Lines in R^n

The general parametric representation of a straight line in R^n involves only one parameter, and is of the form

$$\{x = (a + \alpha b) : \text{parameter } \alpha \text{ taking all real values}\}$$

where a, b are given vectors in R^n and $b \neq 0$.

Notice that if $b = 0$, $x = a$ for all values of α, the above set contains only one point a and not a straight line. That's why in the above representation b is required to be nonzero.

$\alpha = 0$ gives $x = a$, and $\alpha = 1$ gives $x = a + b$. So, both $a, a + b$ are on the straight line. Since two points in any space uniquely determine a straight line, the above straight line is the one joining a and $a + b$.

Here are examples of parametric representations of straight lines in spaces of various dimensions (α is the only parameter in all these examples).

$$L_1 = \left\{ \begin{pmatrix} x_1 \\ x_2 \end{pmatrix} = \begin{pmatrix} 3 - 4\alpha \\ 8 + 2\alpha \end{pmatrix} : \quad \alpha \text{ takes all real values} \right\}$$

$$L_2 = \left\{ \begin{pmatrix} x_1 \\ x_2 \end{pmatrix} = \begin{pmatrix} 2 - \alpha \\ -3 \end{pmatrix} : \quad \alpha \text{ takes all real values} \right\}$$

$$L_3 = \left\{ \begin{pmatrix} x_1 \\ x_2 \\ x_3 \\ x_4 \\ x_5 \end{pmatrix} = \begin{pmatrix} -6 - 4\alpha \\ 3\alpha \\ 0 \\ 3 \\ 8 - 7\alpha \end{pmatrix} : \quad \alpha \text{ takes all real values} \right\}.$$

By giving the parameter the values 0 and 1 respectively, we get two distinct points on the straight line, and the straight line itself is the unique one joining these two points. For example on the straight line L_3 in R^5 given above, these two points are

$$(-6, 0, 0, 3, 8)^T \quad \text{and} \quad (-10, 3, 0, 3, 1)^T$$

and L_3 is the unique straight line joining these two points.

When a parametric representation of a straight line L in R^n is given in terms of the parameter α, using it, it is very easy to check whether a given point $\bar{x} \in R^n$ is contained on L or not. All we need to do is check whether the general point on L expressed in terms of α becomes $\bar{x}$ for some real value of α. If it does, then $\bar{x} \in L$, otherwise $\bar{x} \notin L$. Here are some examples.

Example 1:

Check whether $\bar{x} = (114, -90, 0, 3, 218)^T$ is contained on the straight line L_3 in R^5 given above. For this we need to check whether there is a value of α which will make

$$
\begin{pmatrix} -6 - 4\alpha \\ 3\alpha \\ 0 \\ 3 \\ 8 - 7\alpha \end{pmatrix} = \begin{pmatrix} 114 \\ -90 \\ 0 \\ 3 \\ 218 \end{pmatrix}.
$$

From the first component of these two vectors, we see that $-6 - 4\alpha = 114$ iff $\alpha = -30$. We verify that when we substitute $\alpha = -30$, the equation holds for all components. Hence $\bar{x} \in L_3$.

Now consider the point $x^2 = (-10, 3, 0, 5, 1)^T$. We see that the equation

$$
\begin{pmatrix} -6 - 4\alpha \\ 3\alpha \\ 0 \\ 3 \\ 8 - 7\alpha \end{pmatrix} = \begin{pmatrix} -10 \\ 3 \\ 0 \\ 5 \\ 1 \end{pmatrix}
$$

can never be satisfied for any α, since the fourth components in the two vectors are not equal. So, $x^2 \notin L_3$. Similarly $x^3 = (-10, 3, 0, 3, 15)^T \notin L_3$ as there is no value of α which will satisfy

$$
\begin{pmatrix} -6 - 4\alpha \\ 3\alpha \\ 0 \\ 3 \\ 8 - 7\alpha \end{pmatrix} = \begin{pmatrix} -10 \\ 3 \\ 0 \\ 3 \\ 15 \end{pmatrix}.
$$

Example 2:

Consider the straight line $\{x = x(\alpha) = (-1 + 2\alpha, 2 - \alpha)^T : \alpha \text{ real}\}$. This is the straight line joining the points $x(0) = (-1, 2)^T$ and $x(1) = (1, 1)^T$ in R^2. See Figure 3.7

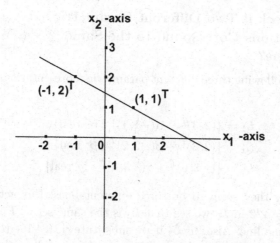

Figure 3.7   A straight line in R^2.

Given Two Distinct Points in R^n, How to Obtain the Parametric Representation of the Straight Line Joining Them?

Suppose $a = (a_1, \ldots, a_n)^T$ and $c = (c_1, \ldots, c_n)^T$ are two distinct points in R^n. The parametric representation of the unique straight line joining them, with parameter α, is

$$x = x(\alpha) = a + \alpha b$$

where $b = c - a$. Since a, c are distinct, $b = c - a \neq 0$. So the above parametric representation does represent a straight line. Also, verify $x(0) = a$, and $x(1) = a + b = a + c - a = c$. So both a, c are on this straight line, hence it is the straight line joining them.

Example 3:

Let

$$a = \begin{pmatrix} -1 \\ 3 \end{pmatrix}, \quad c = \begin{pmatrix} 2 \\ -1 \end{pmatrix}, \quad b = c - a = \begin{pmatrix} 3 \\ -4 \end{pmatrix}.$$

Hence the straight line in R^2 joining a, c is given by

$$\left\{ x = x(\alpha) = \begin{pmatrix} -1 + 3\alpha \\ 3 - 4\alpha \end{pmatrix} : \quad \alpha \text{ takes all real values} \right\}.$$

How to Check if Two Different Parametric Representations Correspond to the Same Straight Line?

Consider the following three different parametric representations of straight lines:

$$L_1 = \{(2,3)^T + \alpha(-1,1)^T : \alpha \text{ real}\}$$
$$L_2 = \{(2,3)^T + \beta(-2,2)^T : \beta \text{ real}\}$$
$$L_3 = \{(-1,6)^T + \gamma(-3,3)^T : \gamma \text{ real}\}.$$

Outwardly they seem to be three different lines. But actually if we substitute $\beta = \alpha/2$ in L_2 we see that L_2 is the same set as L_1, so L_1 and L_2 are the same line. Also, in L_3 if we substitute $\gamma = -1 + (\alpha/3)$ we get the representation for L_1, so L_1 and L_3 are the same line. So, all of the representations above correspond to the same straight line in R^2.

In the same way, given two parametric representations in R^n:

$$U = \{x = a + \alpha b : \alpha \text{ real}\}$$
$$V = \{x = c + \beta d : \beta \text{ real}\}$$

they both correspond to the same straight line in R^n if the following conditions are satisfied:

(i) d is equal to a scalar multiple of b
(ii) $a - c$ is a scalar multiple of b.

If either of these conditions does not hold, the straight lines U, V are different.

Example 4:

Consider the representations for L_1, L_2, L_3 given above. We verify that every pair of these representations correspond to the same straight line in R^2.

Example 5:

Consider the representations

$$L_4 = \{x = (2,3,-7)^T + \alpha(0,1,-1)^T : \alpha \text{ real}\}$$
$$L_5 = \{x = (2,3,-7)^T + \beta(0,6,-1)^T : \beta \text{ real}\}.$$

Since the coefficient vectors of the parameters in these two representations are $(0, 1, -1)^T$, $(0, 6, -1)^T$ and these two vectors are not scalar multiples of each other, L_4, L_5 are different straight lines.

Example 6:

Consider the representations

$$L_6 = \{x = (4, 1, 5)^T + \alpha(1, 2, 3)^T : \alpha \text{ real}\}$$
$$L_7 = \{x = (5, 3, 7)^T + \beta(2, 4, 6)^T : \beta \text{ real}\}.$$

Here the coefficient vectors of the parameters in the two representations $(1, 2, 3)^T$, $(2, 4, 6)^T$ are scalar multiples of each other, so condition (i) is satisfied. When we put $\alpha, \beta = 0$, the points obtained on L_6, L_7 are $(4, 1, 5)^T$, $(5, 3, 7)^T$ and their difference $(1, 2, 2)^T$ is not a scalar multiple of the coefficient vector of the parameter. Hence condition (ii) is violated, so L_6, L_7 are different lines.

How to Obtain a Linear Equation Representation of a Straight Line in R^n Given Its Parametric Representation

A straight line in R^n is the set of solutions of a linearly independent system of $n - 1$ linear equations in n variables. It is the affine hull of any two distinct points in it. Its linear equation representation can be obtained from its parametric representation using the algorithm discussed in Section 3.11.

Exercises

3.4.1: Check whether the straight lines represented by the following pairs of representations are the same or different (in (i), (ii) they are the same; in (iii), (iv) they are different).

(i) $\{x = (0, -7, 2, -6) + \alpha(-2, 1, 2, 4) : \alpha \text{ real}\}$; $\{x = (-6, -4, 8, 6) + \beta(-6, 3, 6, 12) : \beta \text{ real}\}$.

(ii) $\{x = (18, -12, 3, 0) + \alpha(0, 2, 1, -7) : \alpha \text{ real}\}$; $\{x = (18, -12, 3, 0) + \beta(0, 1, 1/2, -7/2) : \beta \text{ real}\}$.

(iii) $\{x = (8, 9 - 5, 4) + \alpha(12, 18, -21, 9) : \alpha \text{ real}\}$; $\{x = (20, 27, -26, 12) + \beta(4, 6, -7, 3) : \beta \text{ real}\}$.

(iv) $\{x = (1,1,1,1,1) + \alpha(10,3,-7,4,2) : \alpha \text{ real}\}$; $\{x = (1,1,1,1,1) + \beta(30,9,-21,11,6) : \beta \text{ real}\}$.

3.5 Half-Lines and Rays in R^n

The general parametric representation of a half-line is of the form $\{x = a + \alpha b : \alpha \geq k\}$ or $\{x = a + \beta b : \beta \leq k\}$ for some given vectors a, b in R^n with $b \neq 0$ and constant k. The first can be expressed as $\{x = (a+kb)+\gamma b : \gamma \geq 0\}$, and the second can be expressed as $\{x = (a+kb)+\delta(-b) : \delta \geq 0\}$. For this reason, we take the general parametric representation for a half-line to be

$$\{x = x(\alpha) = (x_1(\alpha), \ldots, x_n(\alpha))^T : \alpha \geq 0\}$$

where $x(\alpha) = a + \alpha b$ for some $a, b \in R^n$ with $b \neq 0$. It is the subset of the straight line $\{x = x(\alpha) : \alpha \text{ takes all real values}\}$ corresponding to nonnegative values of the parameter, that's the reason for the name. This half-line is said to begin at $a = (a_1, \ldots, a_n)^T$ corresponding to $\alpha = 0$, and said to have the **direction vector** $b = (b_1, \ldots, b_n)^T$. The direction vector for any half-line is always a nonzero vector.

Example 1:

Consider the half-line in R^4

$$L^+ = \{x = x(\alpha) = (6-\alpha, -3+\alpha, 4, 7-2\alpha)^T : \alpha \geq 0\}.$$

Let

$$x^1 = (4,-1,4,3)^T, \quad x^2 = (8,-5,4,11)^T, \quad x^3 = (5,2,4,9)^T.$$

Determine which of these points are in L^+. For this we need to find a nonnegative value for the parameter α that will make $x(\alpha)$ equal to each of these points. For x^1 this equation is

$$\begin{pmatrix} 6-\alpha \\ -3+\alpha \\ 4 \\ 7-2\alpha \end{pmatrix} = \begin{pmatrix} 4 \\ -1 \\ 4 \\ 3 \end{pmatrix}$$

and we verify that $\alpha = 2 > 0$ satisfies this equation. So, $x^1 \in L^+$.

For x^2, the equation is

$$\begin{pmatrix} 6 - \alpha \\ -3 + \alpha \\ 4 \\ 7 - 2\alpha \end{pmatrix} = \begin{pmatrix} 8 \\ -5 \\ 4 \\ 11 \end{pmatrix}$$

and $\alpha = -2$ satisfies this equation. Since $-2 \not\geq 0$, $x^2 \notin L^+$.
For x^3, we have the equation

$$\begin{pmatrix} 6 - \alpha \\ -3 + \alpha \\ 4 \\ 7 - 2\alpha \end{pmatrix} = \begin{pmatrix} 5 \\ 2 \\ 4 \\ 9 \end{pmatrix}.$$

This cannot be satisfied by any α, hence $x^3 \notin L^+$.

Example 2:

Consider the half-line $\{x = x(\alpha) = (1 + 2\alpha, -2 + 3\alpha)^T : \alpha \geq 0\}$ in R^2. It begins at $x(0) = (1, -2)^T$, and $x(1) = (3, 1)^T$ is a point on it. See Figure 3.8. Its direction vector is $x(1) - x(0) = (2, 3)^T$.

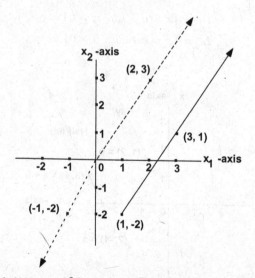

Figure 3.8 The half-line in R^2 from $(1, -2)^T$ in the direction of $(2, 3)^T$ is the solid half-line parallel to the ray of $(2, 3)^T$ which is the dashed half-line through the origin in the positive quadrant. The ray of $(-1, -2)^T$ is the dotted half-line through the origin in the negative quadrant.

A half-line which begins at 0 is called a **ray**. In R^n, for every $a \in R^n$, $a \neq 0$, there is a unique ray, $\{x = \alpha a : \alpha \geq 0\}$. a is called the **direction** or the **direction vector** of this ray.

Example 3:

Let $a = (-1, -2)^T \in R^2$. Its ray is shown in Figure 3.8 as the dotted half-line. Since $(-10, -20)^T = 10(-1, -2)^T$ and the scalar multiple $10 > 0$, the point $(-10, -20)^T$ is contained on this ray.

$(10, 20)^T = -10(-1, -2)^T$, and since the scalar multiplier $-10 \not\geq 0$, the point $(10, 20)^T$ is not contained on the ray of $(-1, -2)^T$.

For $a \in R^n, a \neq 0$, verify that the rays of a, and γa for any $\gamma > 0$, are the same. So, the points $a, \gamma a$ for any $\gamma > 0$, correspond to the same direction.

In R^1, there are only two directions (positive direction, negative direction). For any $n \geq 2$, there are an infinite number of directions in R^n.

Let $a, b \in R^n$, $b \neq 0$. Then the half-line $\{x = a + \alpha b : \alpha \geq 0\}$ in R^n beginning at a, is **parallel** to the ray of b. It is the translate of the ray of b to a. The point $a + \alpha b$ on this half-line is known as the point obtained by moving from a in the direction of b a **step length** of α. See Figure 3.9.

Two parametric representations

$$L_1 = \{x = a + \alpha b : \alpha \geq 0\}$$
$$L_2 = \{x = c + \beta d : \beta \geq 0\}$$

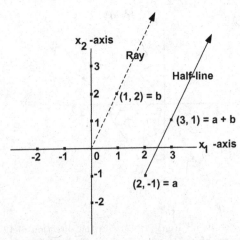

Figure 3.9   A half-line through $(2, -1)^T$ parallel to the ray of $(1, 2)^T$ in R^2. The point $(3, 1)^T = (2, -1)^T + (1, 2)^T$ is obtained by moving from $(2, -1)^T$ a step length of 1 in the direction of $(1, 2)^T$.

represent the same half-line iff (i) starting points are the same, i.e., $a = c$, and (ii) the directions b, d are positive scalar multiples of each other.

For example $\{(2, 1, 3) + \alpha(1, 2, 4) : \alpha \geq 0\}$ and $\{(3, 3, 7) + \beta(1, 2, 4) : \beta \geq 0\}$ are not the same half-line because their starting points are different. In fact the 2nd half-line is a subset of the first.

The half-lines $L_1 = \{(-1, 2, 1) + \alpha(-2, 4, 2) : \alpha \geq 0\}$ and $L_2 = \{(-1, 2, 1) + \beta(-1, 2, 1) : \beta \geq 0\}$ are the same half line. However, as they are represented, the points obtained by moving from the starting point a fixed step length (say 1 unit) in L_1, L_2 are not the same because the direction vector of L_2 is half the direction vector of L_1.

The half-lines $\{(10, 3) + \alpha(-1, 2) : \alpha \geq 0\}$, $\{(10, 3) + \beta(1, -2) : \beta \geq 0\}$ are not the same because their direction vectors are not positive multiples of each other. In fact they form the two half-lines of a straight line broken at the point $(10, 3)$ which is the starting point of both these half-lines.

3.6 Line Segments in R^n

A general line segment in R^n is of the form $\{x = a' + \beta b' : \ell \leq \beta \leq u\}$ where a', b' are given vectors in R^n with $b' \neq 0$, and the bounds on the parameter satisfy $\ell < u$. By defining $a = a' + \ell b'$, $b = (u - \ell)b'$ this can be expressed as $\{x = x(\alpha) = a + \alpha b : 0 \leq \alpha \leq 1\}$, where $b \neq 0$. For this reason, the parametric representation for a line segment is usually stated in this form.

This line segment joins $x(0) = a$ to $x(1) = a + b$, and is the portion of the straight line joining these two points that lies between them. $x(0), x(1)$ are the two end points of this line segment.

Example 1:

Consider the line segment $\{x = x(\alpha) = (6 - \alpha, -3 + \alpha, 4 + 2\alpha, 7 - 2\alpha)^T : 0 \leq \alpha \leq 1\}$. Let $x^1 = (5.5, -2.5, 5, 6)^T$, $x^2 = (4, -1, 8, 3)^T$.

This line segment joins $x(0) = (6, -3, 4, 7)^T$ and $x(1) = (5, -2, 6, 5)^T$ in R^4. To check whether x^1 is on this line segment, we consider the equation

$$\begin{pmatrix} 6 - \alpha \\ -3 + \alpha \\ 4 + 2\alpha \\ 7 - 2\alpha \end{pmatrix} = \begin{pmatrix} 5.5 \\ -2.5 \\ 5 \\ 6 \end{pmatrix}$$

which has the unique solution $\alpha = 0.5$, and since $0 < 0.5 < 1$, x^1 is on this line segment.

Similarly, to check whether x^2 is on this line segment, we consider the equation

$$\begin{pmatrix} 6 - \alpha \\ -3 + \alpha \\ 4 + 2\alpha \\ 7 - 2\alpha \end{pmatrix} = \begin{pmatrix} 4 \\ -1 \\ 8 \\ 3 \end{pmatrix}$$

which has a solution $\alpha = 2$. Since 2 is not between 0 and 1, x^2 is not on this line segment.

Example 2:

Consider the line segment $\{x = x(\alpha) = (1 + 2\alpha, -2 + 3\alpha)^T : 0 \le \alpha \le 1\}$ in R^2. This line segment joins $x(0) = (1, -2)^T$ and $x(1) = (3, 1)^T$. See Figure 3.10.

Suppose $a = (a_1, \ldots, a_n)^T$ and $c = (c_1, \ldots, c_n)^T$ are two distinct points in R^n. The parametric representation (with parameter α) of the line segment joining them is $\{x = x(\alpha) = a + \alpha b : 0 \le \alpha \le 1\}$, where $b = c - a \ne 0$ because $a \ne c$. We verify that $x(0) = a$, and $x(1) = a + b = a + c - a = c$.

Example 3:

Let $a = (1, -2)^T$ and $c = (3, 1)^T$. Then $b = c - a = (2, 3)^T$. Then the line segment joining a, c is $\{x = x(\alpha) = (1 + 2\alpha, -2 + 3\alpha)^T : 0 \le \alpha \le 1\}$ pictured in Figure 3.10 above.

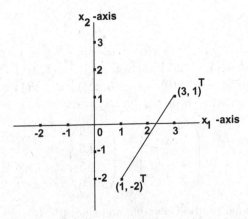

Figure 3.10   The line segment joining $(1, -2)^T$ and $(3, 1)^T$ in R^2.

Given two parametric representations for line segments, they represent the same line segment iff the sets of end points for the line segments in the two representations are the same.

Exercises

3.6.1: (i) Let L be the straight line $\{x = x(\alpha) = (7 - 2\alpha, 8 + 3\alpha, -4, -2 - 3\alpha) : \alpha$ takes all real values$\}$. Check whether the following points are on L or not. (a) (Yes) $(-13, 38, -4, -32)$, $(19, -10, -4, 16)$, $(7, 8, -4, -2)$; (b) (No) $(5, 14, -4, -8)$, $(1, 17, -4, -5)$.

(ii) Express the straight line joining the following pairs of points in parametric representation: $\{(1, 8, -6), (2, 6, -10)\}$, $\{(2, 1, 4, 0), (-3, 0, -2, -8)\}$, $\{(2, 2, 2), (1, 1, 1)\}$, $\{0, (2, 4)\}$. Draw the straight line joining the last pair of points on the 2-dimensional Cartesian plane.

(iii) Consider the half-line $\{x = x(\alpha) = (6 - \alpha, -7 + \alpha, 8 - 2\alpha, 4 + \alpha) : \alpha \geq 0\}$. Mention which of the following points are on this half-line and which are not: (a) (Yes) $(-4, 3, -12, 14)$, $(6, -7, 8, 4)$; (b) (No) $(8, -9, 12, 2)$, $(5, -6, 4, 6)$, $(6, -7, 6, 5)$.

(iv) Mention which of the following points are contained on the ray of $(-5, -4, -1, -2)$ and which are not: (a) (Yes) $(-25, -20, -5, -10)$; (b) (No) $(5, 4, 1, 2)$, $(-10, -8, 2, 4)$.

(v) Mention which of the following points are contained on the line segment $\{x = x(\alpha) = (3 - \alpha, 4 + \alpha, 6 + 2\alpha, -7 - 4\alpha) : 0 \leq \alpha \leq 1\}$, and which are not: (a) (Yes) $(2, 5, 8, -11)$, $(5/2, 9/2, 7, -9)$, $(11/4, 17/4, 13/2, -8)$; (b) (No) $(1, 6, 10, -15)$, $(5, 2, 2, 1)$.

3.7 Straight Lines through the Origin

Let $a = (a_1, \ldots, a_n)^T \neq 0$ be a point in R^n. Then the straight line

$$\{x = x(\alpha) = \alpha a : \alpha \quad \text{takes all real values}\}$$

is the straight line joining 0 and a, because $x(0) = 0$ and $x(1) = a$. This straight line contains the ray of a.

Example 1:

The ray of $(1, -1)^T \in R^2$ and the straight line joining 0 and $(1, -1)^T$ are shown in Figure 3.11.

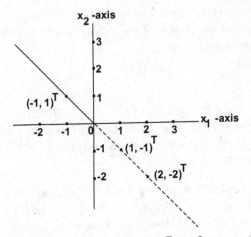

Figure 3.11   The straight line joining 0 and $(1, -1)^T$ in R^2. The dashed portion of this straight line is the ray of $(1, -1)^T$.

Example 2:

Consider the following points in R^4.

$$a = \begin{pmatrix} 1 \\ -1 \\ 2 \\ -3 \end{pmatrix}, \quad a^1 = \begin{pmatrix} -15 \\ 15 \\ -30 \\ 45 \end{pmatrix}, \quad a^2 = \begin{pmatrix} 30 \\ -30 \\ 60 \\ -90 \end{pmatrix}, \quad a^3 = \begin{pmatrix} 15 \\ -15 \\ 40 \\ -60 \end{pmatrix}.$$

We have

Straight line joining 0 and $a = \{(\alpha, -\alpha, 2\alpha, -3\alpha)^T : \alpha$ takes

all real values}

Ray of $a = \{(\alpha, -\alpha, 2\alpha, -3\alpha)^T : \alpha \geq 0\}.$

a^1 is obtained from $(\alpha, -\alpha, 2\alpha, -3\alpha)^T$ when $\alpha = -15$. Since this value of $\alpha < 0$, a^1 is not on the ray of a, but is on the straight line joining 0 and a.

a^2 is obtained from $(\alpha, -\alpha, 2\alpha, -3\alpha)^T$ when $\alpha = 30$. Since this value of $\alpha > 0$, a^2 is on the ray of a.

There is no value of α that will make $a^3 = (\alpha, -\alpha, 2\alpha, -3\alpha)^T$. So, a^3 is not on the straight line joining 0 and a.

Exercises

3.7.1: Let $a = (-1, -2)^T$. Plot the ray of a on the two dimensional Cartesian plane, and check whether $(-6, -12)^T, (7, 14)^T, (-8, -17)^T$ are on the ray of a, or on the straight line joining 0 and a.

3.8 Geometric Interpretation of Addition of Two Vectors, Parallelogram Law

Let

$$a = \begin{pmatrix} 1 \\ 2 \end{pmatrix}, \quad b = \begin{pmatrix} 2 \\ 1 \end{pmatrix}; \quad a + b = \begin{pmatrix} 1 \\ 2 \end{pmatrix} + \begin{pmatrix} 2 \\ 1 \end{pmatrix} = \begin{pmatrix} 3 \\ 3 \end{pmatrix}$$

i.e., $a + b$ is the fourth vertex of the parallelogram whose other three vertices are $0, a, b$. See Figure 3.12.

This is also true in R^n for any $n \geq 2$. Given two distinct nonzero vectors $a, b \in R^n$, the two line segments $[0, a], [0, b]$ are the two sides of a unique two dimensional parallelogram whose fourth vertex is $a + b$. This gives the geometric interpretation for the sum of two vectors, and is known as the **parallelogram law of addition of two vectors.**

Exercises

3.8.1: Plot the points $(1, 2)$, $(2, 1)$, $(3, 3) = (1, 2) + (2, 1)$ on the 2-dimensional Cartesian plane and verify that these three points and 0 form the vertices of a parallelogram.

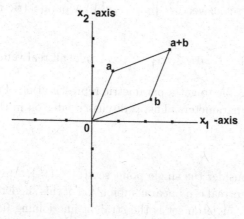

Figure 3.12   Parallelogram law of addition of two vectors.

3.9   Linear Combinations, Linear Hulls or Spans, Subspaces

A subset $S \subset R^n$ is said to be a **subspace of** R^n if it satisfies the following property:

> for any two vectors $a, b \in S$, all linear combinations of a, b (any vector of the form $\alpha a + \beta b$, for any real values of α, β) also belongs to S.

Since the origin, 0, is obtained as the zero linear combination of any pair of vectors, 0 is contained in every subspace of R^n. Stating it another way, a subset of R^n not containing 0, cannot be a subspace of R^n.

Examples of subspaces of R^n are the various coordinate planes of various dimensions.

Linear Combinations of Vectors

Let $\{a^1, \dots, a^k\}$ be a set of k vectors in R^n. By our convention, either all these are column vectors, or all these are row vectors; we will treat the case where they are all column vectors. As defined in Section 1.5, a **linear combination of** $\{a^1, \dots, a^k\}$ is a vector of the form

$$\alpha_1 a^1 + \cdots + \alpha_k a^k$$

where $\alpha_1, \dots, \alpha_k$, the **coefficients** of the vectors $a^1, \dots, a^k$ in this linear combination, can be any real numers.

The set of all such linear combinations of $\{a^1, \dots, a^k\}$ is a subspace of R^n called the **linear hull** or the **linear span of**, or the **subspace spanned by** the set of vectors $\{a^1, \dots, a^k\}$. A parametric representation of this subspace is

$$\{\alpha_1 a^1 + \cdots + \alpha_k a^k : \alpha_1, \dots, \alpha_k \quad \text{take all real values}\}$$

It may be possible to get a parametric representation of the same subspace using fewer parameters. This is discussed later on in this section.

Example 1:

Let $n = 2$, and consider the single point set $\{(-1, 1)^T\}$. Any vector of the form $(-\alpha, \alpha)^T$ for α real is a linear combination of this singleton set. So, the linear hull of this singleton set is the straight line joining 0 and $(-1, 1)^T$. See Figure 3.13.

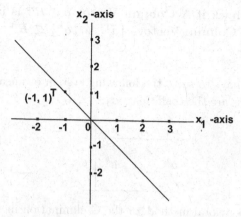

Figure 3.13   The subspace spanned by $(-1, 1)^T$.

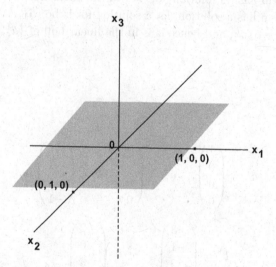

Figure 3.14   The linear hull of $\{(1, 0, 0)^T, (0, 1, 0)^T\}$ in R^3 is the x_1x_2-coordinate plane.

Example 2:

Let $n = 3$, and consider the set of vectors in R^3

$$\left\{ \begin{pmatrix} 1 \\ 0 \\ 0 \end{pmatrix}, \begin{pmatrix} 0 \\ 1 \\ 0 \end{pmatrix} \right\}.$$

A linear combination of this set is a vector of the form $(\alpha_1, \alpha_2, 0)^T$, i.e., a vector in the x_1x_2-coordinate plane in R^3. See Figure 3.14.

Algorithm to Check if A Column Vector $b \in R^n$ is in the Linear Hull of a Set of Column Vectors $\{a^1, \ldots, a^k\} \subset R^n$

BEGIN

For this, you have to solve the following system of linear equations in which the variables are the coefficients $\alpha_1, \ldots, \alpha_k$.

α_1	$\ldots$	α_k	
a^1	$\ldots$	a^k	b

Either the GJ pivotal method, or the G elimination method discussed earlier (Sections 1.16, 1.20) can be used to solve this system.

If the system has no solution, b is not in the linear hull of $\{a^1, \ldots, a^k\}$.

If the system has a solution, let a solution for it be $(\bar{\alpha}_1, \ldots, \bar{\alpha}_k)$. Then $b = \bar{\alpha}_1 a^1 + \cdots + \bar{\alpha}_k a^k$, and hence it is in the linear hull of $\{a^1, \ldots, a^k\}$.

END

Example 3:

Let

$$a^1 = \begin{pmatrix} 1 \\ 0 \\ 0 \\ 0 \end{pmatrix}, \quad a^2 = \begin{pmatrix} 0 \\ 1 \\ 0 \\ 0 \end{pmatrix} \quad a^3 = \begin{pmatrix} 1 \\ -1 \\ 1 \\ -1 \end{pmatrix}$$

$$b^1 = \begin{pmatrix} 3 \\ -1 \\ 2 \\ -2 \end{pmatrix}, \quad b^2 = \begin{pmatrix} 2 \\ 0 \\ 3 \\ -4 \end{pmatrix}.$$

To check whether b^1 is in the linear hull of $\{a^1, a^2, a^3\}$ we solve the following system of equations by the GJ method. Since the first two column vectors in this tableau are already unit vectors, we can select α_1, α_2 as basic variables in rows 1, 2 directly. Then we select the column of α_3 as the pivot column to perform a GJ pivot step in Row 3. The column headed "BV" stores the basic variables selected in the rows. "PC", "PR" indicate the pivot column, pivot row respectively for the pivot step, and the pivot element common to them is boxed. "RC" indicates a "redundant constraint discovered, which is to be eliminated". "IE" indicates an "inconsistent equation" detected in the algorithm.

BV	α_1	α_2	α_3		
α_1	1	0	1	3	
α_2	0	1	-1	-1	
	0	0	$\boxed{1}$	2	PR
	0	0	-1	-2	
			PC		
α_1	1	0	0	1	
α_2	0	1	0	1	
α_3	0	0	1	2	
	0	0	0	0	RC

So, the system has the unique solution $(\alpha_1, \alpha_2, \alpha_3) = (1, 1, 2)$, i.e., $b^1 = a^1 + a^2 + 2a^3$.

So b^1 is in the linear hull of $\{a^1, a^2, a^3\}$.

To check whether b^2 is in the linear hull of $\{a^1, a^2, a^3\}$ we need to solve the following system linear equations.

BV	α_1	α_2	α_3		
α_1	1	0	1	2	
α_2	0	1	-1	0	
	0	0	$\boxed{1}$	3	PR
	0	0	-1	-4	
			PC↑		
α_1	1	0	0	-1	
α_2	0	1	0	3	
α_3	0	0	1	3	
	0	0	0	-1	IE

From the last row of the final tableau, we see that this system has no solution. So, b^2 is not in the linear hull of $\{a^1, a^2, a^3\}$.

Parametric Representation of the Linear Hull of a Set of Vectors Using the Smallest Number of Parameters

Earlier in this section, we discussed the parametric representation of the linear hull of $\{a^1, \ldots, a^k\} \subset R^n$ as

$$L = \{\alpha_1 a^1 + \cdots + \alpha_k a^k : \alpha_1, \ldots, \alpha_k \quad \text{take all real values}\}$$

involving k parameters, that comes directly from its definition. However, it may not be necessary to include all the vectors $a^1, \ldots, a^k$ in this parametric representation to get L. As an example, let

$$a^1 = \begin{pmatrix} 1 \\ 0 \\ 0 \end{pmatrix}, \quad a^2 = \begin{pmatrix} 0 \\ 1 \\ 0 \end{pmatrix}, \quad a^3 = \begin{pmatrix} 1 \\ 1 \\ 0 \end{pmatrix}, \quad a^4 = \begin{pmatrix} 1 \\ -1 \\ 0 \end{pmatrix}.$$

Here a^3, a^4 are themselves linear combinations of a^1, a^2. Hence linear hull of $\{a^1, a^2, a^3, a^4\}$ = linear hull of $\{a^1, a^2\}$.

In the same way, given $\{a^1, \ldots, a^k\} \subset R^n$, if one of the vectors in this set, say a^k, is itself a linear combination of the remaining vectors in the set, then

Linear hull of $\{a^1, \ldots, a^k\}$ = Linear hull of $\{a^1, \ldots, a^{k-1}\}$.

So, to write the parametric representation of $\{a^1, \ldots, a^k\}$, we can omit the vector a^k from this set. The same process can be repeated with the remaining set $\{a^1, \ldots, a^{k-1}\}$.

In the example given above, we have $a^3 = a^1 + a^2$; or

$$-a^1 - a^2 + a^3 = 0.$$

An expression like this, which expresses the 0-vector as a linear combination of vectors in a set, with at least one nonzero coefficient, is called a **linear dependence relation** for that set of vectors.

Whenever there is a linear dependence relation for a set of vectors, that relation can be used to express one of the vectors in the set as a linear combination of the others, and in this case the set of vectors is said to be **linearly dependent**.

On the contrary, if there is no linear dependence relation for a set of vectors, that set is said to be **linearly independent**.

Given a set of vectors $\Gamma = \{a^1, \ldots, a^k\} \subset R^n$, a subset Δ of Γ is said to be a **maximally linearly independent subset of Γ** if it satisfies the following two conditions:

  (a) Δ must be linearly independent, and
  (b) including any other vector from $\Gamma \backslash \Delta$ in Δ makes it linearly dependent.

All maximal linearly independent subsets of a set of vectors $\Gamma \subset R^n$ always contain the same number of vectors in them, this number is called the **rank of the set of vectors** Γ.

The concepts of linear independence, linear dependence of sets of vectors will be discussed at greater length in Chapter 4. For now, all we need is the fact that L = linear hull of Γ is the same as the linear hull of any maximally linearly independent subset of vectors of Γ, and that expressing it this way leads to a parametric representation for L using the smallest number of parameters.

Representing the Linear Hull of a Set of Vectors by a System of Homogeneous Linear Equations

We have the following important result.

Result 3.9.1: Representing the linear hull of a given set of vectors as the solution set of a homogeneous system of linear equations: Let $\Gamma = \{a^1, \ldots, a^k\} \subset R^n$ where, for $t = 1$ to k

$$a^t = \begin{pmatrix} a_1^t \\ \vdots \\ a_n^t \end{pmatrix}$$

and let L = linear hull of Γ. Then there exists a matrix B of order $s \times n$ such that

$$L = \text{linear hull of } \Gamma = \{x : Bx = 0\}$$

where $s \leq n$; and s is the smallest number of homogeneous linear equations required to represent L this way, i.e., the s equations in $Bx = 0$ are linearly independent. (Defined in Section 1.9, it means that none of these equations can be obtained as a linear combination of the others). In fact $s = n - rank$ of the set Γ as explained above.

We will now discuss an algorithm for deriving these equations in $Bx = 0$ given $\{a^1, \ldots, a^k\}$.

Algorithm to Find Representations for $L =$ Linear Hull of $\{a^1, \ldots, a^k\}$ from R^n

BEGIN

Step 1: Let $(b_1, \ldots, b_n) \neq 0$ denote the coefficients in a general homogeneous equation

$$b_1 y_1 + \cdots + b_n y_n = 0 \tag{1}$$

satisfied by each of the vectors $a^t = \left(a_1^t, \ldots, a_n^t\right)^T$. So, the unknown coefficients $b_1, \ldots, b_n$ satisfy the following k equations expressed in detached coefficient tableau form.

Original Tableau

b_1	b_2	$\ldots$	b_n	
a_1^1	a_2^1	$\ldots$	a_n^1	0
a_1^2	a_2^2	$\ldots$	a_n^2	0
$\vdots$	$\vdots$	$\vdots$	$\vdots$	$\vdots$
a_1^k	a_2^k	$\ldots$	a_n^k	0

Notice that the tth row in this tableau is the vector a^t written as a row vector, i.e., $(a^t)^T$, for $t = 1$ to k. The coefficients $b_1, b_2, \ldots, b_n$ in the homogeneous equation (1) are the variables in this system.

Step 2: Now apply the GJ or the G pivotal method (Sections 1.16, 1.23) to find the general solution of this homogeneous system. In this process, the method may discover some constraints in the tableau as being redundant and eliminate them. In the final tableau obtained under this method, a basic variable is selected in each remaining row. Rearrange the variables and their column vectors in the final tableau so that the basic variables appear first in their proper order, followed by nonbasic variables if any. Let r denote the number of rows in the final tableau; and suppose the basic variables are $b_{i_1}, \ldots, b_{i_r}$ in that order; and the nonbasic variables are $b_{j_1}, \ldots, b_{j_{n-r}}$. Let the final tableau be

Final tableau after rearrangement of variables

Basic Vars.	b_{i_1}	b_{i_2}	$\ldots$	b_{i_r}	b_{j_1}	$\ldots$	$b_{j_{n-r}}$	
b_{i_1}	1	0	$\ldots$	0	$\bar{a}_{1,j_1}$	$\ldots$	$\bar{a}_{1,j_{n-r}}$	0
b_{i_2}	0	1	$\ldots$	0	$\bar{a}_{2,j_1}$	$\ldots$	$\bar{a}_{2,j_{n-r}}$	0
$\vdots$	$\vdots$	$\vdots$	$\vdots$	$\vdots$	$\vdots$		$\vdots$	$\vdots$
b_{i_r}	0	0	$\ldots$	1	$\bar{a}_{r,j_1}$	$\ldots$	$\bar{a}_{r,j_{n-r}}$	0

This will be the final tableau if the GJ method is used. If the G method is used instead, the portion of the tableau corresponding to the basic variables will be an upper triangular matrix instead of the unit matrix.

Case 1: If $r = n$, i.e., there are no nonbasic variables in the final tableau; then $(b_1, \ldots, b_n) = 0$ is the unique solution for the homogeneous system in the tableau. In this case, the only equation of the form (1) satisfied by all of the vectors $a^1, \ldots, a^k$ is the one in which all of $b_1, \ldots, b_n$ are 0. In this case the linear hull of $\{a^1, \ldots, a^k\}$ is all of R^n.

Case 2: Suppose $r < n$, i.e., there is at least one nonbasic variable in the final tableau. Then the number of equations s required to represent the linear hull of $\{a^1, \ldots, a^k\}$ is $n - r$. One of these equations corresponds to each nonbasic variable in the final tableau.

For each nonbasic variable b_t in the final tableau, $t = j_1, \ldots, j_{n-r}$, obtain the unique solution of the homogeneous system given by

$b_t = 1$, and all other nonbasic variables $= 0$.

The values of the basic variables in this solution can be obtained from the final tableau by substituting the above values for the nonbasic variables. Write this solution as a row vector. When you put each of these row vectors one below the other, we get a matrix of order $s \times n$ (where $s = n - r$). Call this matrix B. Then

the linear hull of $\{a^1, \ldots, a^k\} = \{x : Bx = 0\}$.

So, $Bx = 0$ gives a linear equation representation for the linear hull of $\{a^1, \ldots, a^k\}$.

Therefore, typically, the linear hull of a set of vectors can be expressed as the set of solutions of a homogeneous system of linear equations.

The number r of equations in the final tableau, is called the **rank** of the set of given vectors $\{a^1, \ldots, a^k\}$. The concept of rank will be discussed at greater length in Chapter 4. This rank is also the **dimension** of the linear hull of $\{a^1, \ldots, a^k\}$.

In the original tableau, the tth row corresponds to the vector a^t for $t = 1$ to k. In this method, suppose pivot steps are carried out from top row to bottom row in that order. When we come to row t, if we find that it is "$0 = 0$" in the current tableau at that time, clearly there is a linear dependence relation among $\{a^1, \ldots, a^t\}$ from which we can get an expression for a^t as a linear combination of $\{a^1, \ldots, a^{t-1}\}$. This linear dependence relation can actually be obtained by introducing the memory matrix as discussed in Section 1.12 at the beginning of computation.

Thus if we number the rows in the original tableau as 1 to k from top to bottom, and keep this numbering in all subsequent tableaus obtained in the method, the rows that remain in the final tableau (i.e., those not eliminated as redundant constraints during the algorithm) correspond to vectors in a maximal linearly independent subset of $\{a^1, \ldots, a^k\}$. Suppose the numbers of the rows in the final tableau are $p_1, \ldots, p_r$. Then

$$\{a^{p_1}, \ldots, a^{p_r}\}$$

is a maximal linearly independent subset of $\{a^1, \ldots, a^k\}$.

Then a parametric representation of the linear hull of $\{a^1, \ldots, a^k\}$ involving the smallest number of parameters is

$L = $ Linear hull of $\{a^1, \ldots, a^k\}$

$\quad = \{x = \alpha_1 a^{p_1} + \alpha_2 a^{p_2} + \cdots \alpha_r a^{p_r} : \quad \alpha_1, \ldots, \alpha_r \quad$ take all possible

real values$\}$

END

Example 4:

Let

$$a^1 = \begin{pmatrix} 1 \\ -1 \\ 0 \\ 1 \\ 1 \end{pmatrix}, \quad a^2 = \begin{pmatrix} 2 \\ -2 \\ 0 \\ 2 \\ 2 \end{pmatrix}, \quad a^3 = \begin{pmatrix} -1 \\ 0 \\ 1 \\ -1 \\ -1 \end{pmatrix}$$

$$a^4 = \begin{pmatrix} 0 \\ -1 \\ 1 \\ 0 \\ 0 \end{pmatrix}, \quad a^5 = \begin{pmatrix} 1 \\ -2 \\ 1 \\ 1 \\ 1 \end{pmatrix}.$$

To get the linear constraint representation for the linear hull of $\{a^1, a^2, a^3, a^4, a^5\}$, we start with the following system of homogeneous equations, on which we apply the GJ pivotal method to get the general solution as discussed in Section 1.23. We number the equations to get the parametric representation of this linear hull using the smallest number of parameters. PR (pivot row), PC (pivot column), RC (redundant constraint to be eliminated), BV (basic variable selected in this row) have the same meanings as in Section 1.23, and in each tableau the pivot element for the pivot step carried out in that tableau is boxed.

BV	b_1	b_2	b_3	b_4	b_5	RHS		Eq. no.
	1	−1	0	1	1	0	PR	1
	2	−2	0	2	2	0		2
	−1	0	1	−1	−1	0		3
	0	−1	1	0	0	0		4
	1	−2	1	1	1	0		5
	PC↑							
b_1	1	−1	0	1	1	0		1
	0	0	0	0	0	0	RC	2
	0	−1	1	0	0	0	PR	3
	0	−1	1	0	0	0		4
	0	−1	1	0	0	0		5
		PC↑						
b_1	1	−1	0	1	1	0		1
b_3	0	−1	1	0	0	0		3
	0	0	0	0	0	0	RC	4
	0	0	0	0	0	0	RC	5

Final tableau after rearrangement of
basic variables in their order

BV	b_1	b_3	b_2	b_4	b_5	RHS	Eq. no.
b_1	1	0	−1	1	1	0	1
b_3	0	1	−1	0	0	0	3

There are three nonbasic variables in the final tableau, b_2, b_4, b_5. By fixing one of these nonbasic variables equal to 1, and the others equal to 0, we are lead to the following solutions of this homogeneous system:

$$(b_2, b_4, b_5) = (1, 0, 0) \text{ yields } (b_1, b_2, b_3, b_4, b_5) = (1, 1, 1, 0, 0)$$
$$(b_2, b_4, b_5) = (0, 1, 0) \text{ yields } (b_1, b_2, b_3, b_4, b_5) = (-1, 0, 0, 1, 0)$$
$$(b_2, b_4, b_5) = (0, 0, 1) \text{ yields } (b_1, b_2, b_3, b_4, b_5) = (-1, 0, 0, 0, 1).$$

Letting

$$B = \begin{pmatrix} 1 & 1 & 1 & 0 & 0 \\ -1 & 0 & 0 & 1 & 0 \\ -1 & 0 & 0 & 0 & 1 \end{pmatrix}, \quad x = (x_1, x_2, x_3, x_4, x_5)^T$$

the linear hull of $\{a^1, a^2, a^3, a^4, a^5\}$ is $\{x : Bx = 0\}$.

It can be verified that each of a^1 to a^5 satisfies all these equations, and so every linear combination of a^1 to a^5 also does.

Since there are only two equations left in the final tableau, the rank of $\{a^1, a^2, a^3, a^4, a^5\}$ is 2, which is also the dimension of their linear hull in R^5.

The original row numbers corresponding to those in the final tableau are 1 and 3. Therefore, the corresponding set of original vectors $\{a^1, a^3\}$ forms a maximal linearly independent subset of $\{a^1, \ldots, a^5\}$, and hence the

linear hull of $\{a^1, \ldots, a^5\} = \{x = \alpha_1 a^1 + \alpha_3 a^3 : \alpha_1, \alpha_3$ take all real values$\}$.

Example 5:

Let

$$a^1 = \begin{pmatrix} 1 \\ 1 \\ 1 \end{pmatrix}, \quad a^2 = \begin{pmatrix} 1 \\ -1 \\ 1 \end{pmatrix}, \quad a^3 = \begin{pmatrix} 1 \\ 1 \\ -1 \end{pmatrix}, \quad a^4 = \begin{pmatrix} 1 \\ 1 \\ 0 \end{pmatrix}.$$

To get the linear constraint representation for the linear hull of $\{a^1, a^2, a^3, a^4, \}$, we start with the following system of homogeneous equations, on which we apply the GJ pivotal method to get the general solution. PR, PC, RC have the same meanings as in the above example. The pivot elements are boxed.

BV	b_1	b_2	b_3	RHS		Eq. no.
	$\boxed{1}$	1	1	0	PR	1
	1	−1	1	0		2
	1	1	−1	0		3
	1	1	0	0		4
	PC↑					
b_1	1	1	1	0		1
	0	$\boxed{-2}$	0	0	PR	2
	0	0	−2	0		3
	0	0	−1	0		4
		PC↑				

BV	b_1	b_2	b_3	RHS		Eq. no.
b_1	1	0	1	0		1
b_2	0	1	0	0		2
	0	0	$\boxed{-2}$	0	PR	3
	0	0	-1	0		4
			PC↑			
b_1	1	0	0	0		1
b_2	0	1	0	0		2
b_3	0	0	1	0		3
	0	0	0	0	RC	4
		Final tableau				
b_1	1	0	0	0		1
b_2	0	1	0	0		2
b_3	0	0	1	0		3

The final tableau has no nonbasic variables, hence $(b_1, b_2, b_3) = 0$ is the unique solution of this homogeneous system. Therefore, the linear hull of $\{a^1, a^2, a^3, a^4\}$ here is R^3.

Also, the original row numbers for the rows in the final tableau are $1, 2, 3$. Therefore the set of corresponding vectors, $\{a^1, a^2, a^3\}$ is a maximal linearly independent subset of $\{a^1, a^2, a^3, a^4\}$, and a parametric representation of the linear hull of this set using the smallest number of parameters is

R^3 = linear hull of $\{a^1, a^2, a^3, a^4\} = \{x = \alpha_1 a^1 + \alpha_2 a^2 + \alpha_3 a^3 : \alpha_1, \alpha_2, \alpha_3$ take all real values$\}$.

Exercises

3.9.1: (i) Let $A_{.1}$ to $A_{.5}$ be the column vectors of A given below.

(a) (Yes) Are b^1, b^2 in the linear hull of $\{A_{.1}, \ldots, A_{.5}\}$? If so, find the actual linear combination for each of them.

(b) (No) Are b^3, b^4 in the linear hull of $\{A_{.1}, \ldots, A_{.5}\}$?

$$A = \begin{pmatrix} 1 & -1 & 1 & 0 & -1 \\ -1 & 2 & 2 & 2 & 1 \\ -1 & 4 & 8 & 6 & 1 \\ 0 & 2 & 3 & 2 & -1 \end{pmatrix}, \quad b^1 = \begin{pmatrix} -2 \\ 3 \\ 5 \\ -1 \end{pmatrix}, \quad b^2 = \begin{pmatrix} 3 & -2 & 0 & 2 \end{pmatrix},$$

$$b^3 = \begin{pmatrix} -1 & -3 & -9 & 2 \end{pmatrix}, \quad b^4 = \begin{pmatrix} 2 & 4 & 13 & 3 \end{pmatrix}$$

(ii) Let L represent the linear hull of the set of row vectors of the following matrix. Express L as the set of solutions of a system of homogeneous linear equations. How many equations in this system? ((a) none, (b) one, (c) three, (d) three).

$$(a) \begin{pmatrix} 1 & -1 & -1 & 2 \\ -1 & 2 & 0 & 3 \\ 0 & -1 & 1 & -4 \\ 0 & -1 & 0 & 0 \\ -1 & 1 & 0 & 4 \end{pmatrix}, \quad (b) \begin{pmatrix} -2 & 3 & -1 & 0 \\ 1 & -4 & 1 & 2 \\ 2 & 1 & 0 & -3 \\ 1 & 0 & 0 & -1 \\ -1 & 3 & -1 & -1 \end{pmatrix},$$

$$(c) \begin{pmatrix} -3 & 1 & -4 & 5 \\ -7 & 2 & 2 & -1 \\ 4 & -1 & -6 & 6 \\ -13 & 4 & -6 & 9 \\ -17 & 5 & 0 & 3 \end{pmatrix}, \quad (d) \begin{pmatrix} 2 & 3 & 3 & 2 \\ 12 & 18 & 18 & 12 \\ 8 & 12 & 12 & 8 \\ 10 & 15 & 15 & 10 \end{pmatrix}.$$

3.9.2: Find a linear constraint representation of the linear hull of: $\{(2,4,6,-8,0)^T\}$.

3.10   Column Space, Row Space, and Null Space of a Matrix

Let A be a given matrix of order $m \times n$. So, the set of row vectors of A is $\{A_{1.}, \ldots, A_{m.}\}$, and its set of column vectors is $\{A_{.1}, \ldots, A_{.n}\}$.

The **row space of** A, the space spanned by the set of row vectors of A, is the linear hull of $\{A_{1.}, \ldots, A_{m.}\}$. A general point in the row space of A, treated as a row vector, is $\alpha_1 A_{1.} + \cdots + \alpha_m A_{m.}$, for any real numbers $\alpha_1, \ldots, \alpha_m$; i.e., it is $(\alpha_1, \ldots, \alpha_m)A$. So, the row space of A viewed as a set of row vectors is $\{yA : y \in R^m\}$.

The **column space of** A, the space spanned by the set of column vectors of A, is the linear hull of $\{A_{.1}, \ldots, A_{.n}\}$. A general point in the column space of A, treated as a column vector, is $\beta_1 A_{.1} + \cdots + \beta_n A_{.n}$, for any real numbers $\beta_1, \ldots, \beta_n$; i.e., it is $A\beta$ where $\beta = (\beta_1, \ldots, \beta_n)^T$. In mathematics books, the column space of A is also called the **range space of** A, or the **image space of** A, or the **range of** A, and denoted by $R(A)$. It is viewed as a set of column vectors, i.e., $R(A) = \{Ax : x \in R^n\}$.

When the row space of A is viewed as a set of column vectors by transposing each vector in it, it is $R(A^T)$ in the above notation.

The **null space of** A, denoted by $N(A)$, is the set of all solutions of the homogeneous system of linear equations $Ax = 0$. So, $N(A) = \{x \in R^n : Ax = 0\}$, and it is a set of column vectors.

3.11 Affine Combinations, Affine Hulls or Spans, Affinespaces

Let $\{a^1, \ldots, a^k\}$ be a set of k vectors in R^n. By our convention, either all these are column vectors, or all these are row vectors. We will treat the case where they are all column vectors. As defined in Section 1.5, an **affine combination** of $\{a^1, \ldots, a^k\}$ is a vector of the form

$$\alpha_1 a^1 + \cdots + \alpha_k a^k$$

where $\alpha_1, \ldots, \alpha_k$, the **coefficients** of the vectors $a^1, \ldots, a^k$ in this linear combination, are real numers satisfying the **affine combination condition**

$$\alpha_1 + \cdots + \alpha_k = 1.$$

The set of all such linear combinations of $\{a^1, \ldots, a^k\}$ is called the **affine hull** or the **affine span** or the **flat** or the **affine space** of the set of vectors $\{a^1, \ldots, a^k\}$. See Figure 3.16.

From the affine combination condition $\alpha_1 + \cdots + \alpha_k = 1$, we have the affine combination

$$\alpha_1 a^1 + \cdots + \alpha_k a^k = (1 - \alpha_2 - \cdots - \alpha_k)a^1 + \alpha_2 a^2 + \cdots + \alpha_k a^k$$
$$= a^1 + [\alpha_2(a^2 - a^1) + \cdots + \alpha_k(a^k - a^1)].$$

Let $\alpha_2(a^2 - a^1) + \cdots + \alpha_k(a^k - a^1) = y$. y is a linear combination of $\{(a^2 - a^1), \ldots, (a^k - a^1)\}$, and as $\alpha_2, \ldots, \alpha_k$ assume all real values, y varies over all of the linear hull of $\{(a^2 - a^1), \ldots, (a^k - a^1)\}$. Therefore, the affine hull of $\{a^1, \ldots, a^k\}$ is

$$a^1 + (\text{linear hull of } \{(a^2 - a^1), \ldots, (a^k - a^1)\})$$

a translate of the linear hull of $\{(a^2 - a^1), \ldots, (a^k - a^1)\}$ to a^1.

A Straight Line Is the Affine Hull of Any Two Points On It

The importance of affine hulls stems from the fact that the straight line joining any two points in R^n is the affine hull of those two points. Consider two distinct points $a = (a_1, \ldots, a_n)^T$, $c = (c_1, \ldots, c_n)^T$ in R^n. From Section 3.4, we know that the parametric representation of the straight line joining a, c is $\{x = a + \alpha(c - a) : \alpha \text{ takes all real values}\}$. So, the general point on this straight line is $a + \alpha(c - a) = (1 - \alpha)a + \alpha c$, an affine combination of $\{a, c\}$.

Thus a straight line is the affine hull of any two distinct points on it. See Figure 3.15.

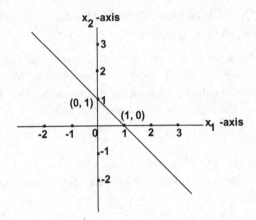

Figure 3.15   The Straight Line Joining $(1,0)^T$, $(0,1)^T$ is their affine hull.

Example 1:

Let

$$a^1 = \begin{pmatrix} 2 \\ -1 \\ 1 \end{pmatrix}, \quad a^2 = \begin{pmatrix} 1 \\ 2 \\ -1 \end{pmatrix}, \quad a^3 = \begin{pmatrix} -2 \\ 11 \\ -7 \end{pmatrix}, \quad a^4 = \begin{pmatrix} 2 \\ -1 \\ 0 \end{pmatrix}.$$

The parametric representation of the straight line joining a^1, a^2 is

$$\{x = a^1 + \alpha(a^2 - a^1) = (2 - \alpha, -1 + 3\alpha, 1 - 2\alpha)^T : \alpha \text{ takes all real values}\}$$

which is the same as $\{x = (1 - \alpha)a^1 + \alpha a^2 : \alpha \text{ takes all real values}\}$.

How to Check if a^{k+1} is in the Affine Hull of $\{a^1, \ldots, a^k\}$?

a^{k+1} is in the affine hull of $\{a^1, \ldots, a^k\}$ iff the following system of equations

$$\alpha_1 a^1 + \cdots + \alpha_k a^k = a^{k+1}$$
$$\alpha_1 + \cdots + \alpha_k = 1$$

in the parameters $\alpha_1, \ldots, \alpha_k$, has a solution. This system can be solved by the GJ pivotal method (Section 1.16) or the G elimination method (Section 1.20). If this system has no solution, a^{k+1} is not in the affine hull of $\{a^1, \ldots, a^k\}$. If the system has the solution $(\bar{\alpha}_1, \ldots, \bar{\alpha}_k)$ then a^{k+1} is the affine combination $\bar{\alpha}_1 a^1 + \cdots + \bar{\alpha}_k a^k$.

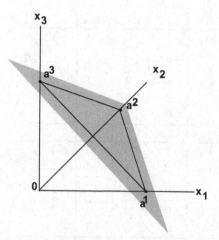

Figure 3.16   The affine hull of $\{a^1, a^2, a^3\}$ in R^3 is the shaded two dimensional plane containing these three points.

Example 2:

Let

$$a^1 = \begin{pmatrix} 1 \\ 0 \\ 0 \\ 0 \end{pmatrix}, \quad a^2 = \begin{pmatrix} 0 \\ 1 \\ 0 \\ 0 \end{pmatrix}, \quad a^3 = \begin{pmatrix} 1 \\ -1 \\ 1 \\ -1 \end{pmatrix}, \quad a^4 = \begin{pmatrix} 3 \\ -1 \\ 2 \\ -2 \end{pmatrix}, \quad a^5 = \begin{pmatrix} 2 \\ -4 \\ 3 \\ -3 \end{pmatrix}.$$

To check whether a^4 is in the affine hull of $\{a^1, a^2, a^3\}$, we solve the following system in the top tableau by the GJ pivotal method. BV (basic variable selected in row), PR (pivot row), PC (pivot column), RC (redundant constraint to be deleted), IE (inconsistent equation identified) have the same meanings as before, and the pivot element in each step is boxed.

BV	α_1	α_2	α_3		
	$\boxed{1}$	0	1	3	PR
	0	1	-1	-1	
	0	0	1	2	
	0	0	-1	-2	
	1	1	1	1	
	PC↑				

BV	α_1	α_2	α_3		
α_1	1	0	1	3	
	0	[1]	−1	−1	PR
	0	0	1	2	
	0	0	−1	−2	
	0	1	0	−2	
		PC↑			
α_1	1	0	1	3	
α_2	0	1	−1	−1	
	0	0	[1]	2	PR
	0	0	−1	−2	
	0	0	1	−1	
		PC↑			
α_1	1	0	0	1	
α_2	0	1	0	1	
α_3	0	0	1	2	
	0	0	0	0	RC
	0	0	0	−3	IE

Since an inconsistent equation is discovered, the system has no solution, therefore a^4 is not in the affine hull of $\{a^1, a^2, a^3\}$.

To check whether a^5 is in the affine hull of $\{a^1, a^2, a^3\}$, we solve the following system of equations in the top tableau given below by the GJ pivotal method. The abbreviations have the same meanings as above, and in each step the pivot element is boxed.

BV	α_1	α_2	α_3		
	[1]	0	1	2	PR
	0	1	−1	−4	
	0	0	1	3	
	0	0	−1	−3	
	1	1	1	1	
	PC↑				
α_1	1	0	1	2	
	0	[1]	−1	−4	PR
	0	0	1	3	
	0	0	−1	−3	
	0	1	0	−1	
		PC↑			

BV	α_1	α_2	α_3		
α_1	1	0	1	2	
α_2	0	1	-1	-4	
	0	0	1	3	PR
	0	0	-1	-3	
	0	0	1	3	
			PC↑		
α_1	1	0	0	-1	
α_2	0	1	0	-1	
α_3	0	0	1	3	
	0	0	0	0	RC
	0	0	0	0	RC

This system has the solution $(\bar{\alpha}_1, \bar{\alpha}_2, \bar{\alpha}_3) = (-1, -1, 3)$. So, a^5 is in the affine hull of $\{a^1, a^2, a^3\}$. In fact

$$a^5 = -a^1 - a^2 + 3a^3.$$

Two Ways of Representing the Affine Hull of a Given Set of Vectors

Let $\{a^1, \ldots, a^k\}$ be a given set of column vectors in R^n, and let F denote their affine hull. From the above discussion we know that

$$F = a^1 + L$$

where L = linear hull of $\{(a^2 - a^1), \ldots, (a^k - a^1)\}$.

Find the representation of L as discussed in Section 3.9. If

$$\text{rank}\{(a^2 - a^1), \ldots, (a^k - a^1)\} \quad \text{is } r$$

and $\{(a^{i_1} - a^1), \ldots, (a^{i_r} - a^1)\}$ is a maximal linearly independent subset of $\{(a^2 - a^1), \ldots, (a^k - a^1)\}$, and a parametric representation of L using the smallest number of parameters is

$$L = \{x = \alpha_1(a^{i_1} - a^1) + \cdots + \alpha_r(a^{i_r} - a^1): \quad \alpha_1, \ldots, \alpha_r \quad \text{real}\}.$$

Thus, a parametric representation of F using the smallest number of parameters is

$$F = \{x = a^1 + \alpha_1(a^{i_1} - a^1) + \cdots + \alpha_r(a^{i_r} - a^1):$$
$$\alpha_1, \ldots, \alpha_r \quad \text{take all possible real values}\}.$$

For another way of representing the affine hull F of $\{a^1, \ldots, a^k\}$, let the linear constraint representation of L = linear hull of $\{(a^2 - a^1), \ldots, (a^k - a^1)\}$ obtained as discussed in Section 3.9 be $\{x : Ax = 0\}$. Let

$$Aa^1 = b.$$

Then, in linear constraint representation,

$$F = \{x : Ax = b\}.$$

So, typically, the affine hull of a set of vectors is the set of solutions of a nonhomogeneous system of linear equations.

Example 3:

Let

$$a^1 = \begin{pmatrix} 1 \\ 1 \\ 1 \\ 1 \\ 1 \end{pmatrix}, \quad a^2 = \begin{pmatrix} 2 \\ 0 \\ 1 \\ 2 \\ 2 \end{pmatrix}, \quad a^3 = \begin{pmatrix} 3 \\ -1 \\ 1 \\ 3 \\ 3 \end{pmatrix},$$

$$a^4 = \begin{pmatrix} 0 \\ 1 \\ 2 \\ 0 \\ 0 \end{pmatrix}, \quad a^5 = \begin{pmatrix} 1 \\ 0 \\ 2 \\ 1 \\ 1 \end{pmatrix}, \quad a^6 = \begin{pmatrix} 2 \\ -1 \\ 2 \\ 2 \\ 2 \end{pmatrix}.$$

Then we have

$$\Gamma = \left\{ a^2 - a^1 = \begin{pmatrix} 1 \\ -1 \\ 0 \\ 1 \\ 1 \end{pmatrix}, \quad a^3 - a^1 = \begin{pmatrix} 2 \\ -2 \\ 0 \\ 2 \\ 2 \end{pmatrix}, \quad a^4 - a^1 = \begin{pmatrix} -1 \\ 0 \\ 1 \\ -1 \\ -1 \end{pmatrix}, \right.$$

$$\left. a^5 - a^1 = \begin{pmatrix} 0 \\ -1 \\ 1 \\ 0 \\ 0 \end{pmatrix}, \quad a^6 - a^1 = \begin{pmatrix} 1 \\ -2 \\ 1 \\ 1 \\ 1 \end{pmatrix} \right\}.$$

The linear constraint representation of the linear hull of Γ was derived in Section 3.9 to be $Ax = 0$ (in Section 3.9 the coefficient matrix was denoted by B, here we denote the same by A) where

$$x = (x_1, x_2, x_3, x_4, x_5)^T$$

$$A = \begin{pmatrix} 1 & 1 & 1 & 0 & 0 \\ -1 & 0 & 0 & 1 & 0 \\ -1 & 0 & 0 & 0 & 1 \end{pmatrix}.$$

Verify that $Aa^1 = (3, 0, 0)^T = b$.

Then the linear constraint representation of the affine hull of $\{a^1$ to $a^6\}$ is $Ax = b$ with, A, b given above.

Also, in Section 3.9 we found that a maximal linearly independent subset of Γ is $\{a^2 - a^1, a^3 - a^1\}$. So, a parametric representation of the affine hull of $\{a^1$ to $a^6\}$ using the smallest number of parameters is $\{x = a^1 + \alpha_1(a^2 - a^1) + \alpha_2(a^3 - a^1) : \alpha_1, \alpha_2$ take all real values$\}$.

Direct Algorithm to Find a Linear Equation Representation of the Affine Hull of a Given Set of Vectors Directly Using GJ or G Pivot Steps

Let F be the affine hull of $\{a^1, \ldots, a^k\}$, a given set of column vectors in R^n. So,

$$F = \{x = \alpha_1 a^1 + \alpha_2 a^2 + \cdots + \alpha_k a^k : \alpha_1 + \cdots + \alpha_k = 1\}$$

$$= \{x = a^1 + \beta_1 c^1 + \cdots + \beta_{k-1} c^{k-1} : \beta_1, \ldots, \beta_{k-1} \text{ real}\}$$

where $\beta_1 = \alpha_2 - \alpha_1, \ldots, \beta_{k-1} = \alpha_k - \alpha_1$ (and hence as $\alpha_1, \ldots, \alpha_k$ take all real values subject to $\alpha_1 + \cdots + \alpha_k = 1, \beta_1, \ldots, \beta_{k-1}$ take all real values) and $c^1 = a^2 - a^1, \ldots, c^{k-1} = a^{k-1} - a^1$.

This representation of F is like the parametric representation of the general solution to a system of linear equations discussed in Section 1.8, but somewhat more general. In the parametric representation in Section 1.8, the following property holds:

| Special property of prametric representation obtained in GJ, or G methods (Sections 1.16, 1.20) | Each of the parameters is the value of one of the variables x_j. |

For instance, here is a parametric representation from Section 1.8

$$\begin{pmatrix} x_1 \\ x_2 \\ x_3 \end{pmatrix} = \begin{pmatrix} 10 - \bar{x}_3 \\ 20 - \bar{x}_3 \\ \bar{x}_3 \end{pmatrix}, \quad \bar{x}_3 \text{ real parameter.}$$

Here the only parameter $\bar{x}_3$ is the value of the variable x_3 in this representation.

Under this special property, all you need to do to get a linear equation system with this general solution is to replace each parameter in the representation by the variable to which it is equal, and then write the system of equations that comes from the representation.

For instance in the above representation, replacing the parameter $\bar{x}_3$ by x_3 leads to

$$\begin{pmatrix} x_1 \\ x_2 \\ x_3 \end{pmatrix} = \begin{pmatrix} 10 - x_3 \\ 20 - x_3 \\ x_3 \end{pmatrix}$$

or the system: $x_1 + x_3 = 10$; $x_2 + x_3 = 20$ as the linear equation representation.

For another example, we take the parametric representation of the general solution in Example 1, Section 1.16; it is

$$\begin{pmatrix} x_1 \\ x_2 \\ x_3 \\ x_4 \\ x_5 \end{pmatrix} = \begin{pmatrix} \bar{x}_1 \\ -60 - 9\bar{x}_1 - 9\bar{x}_4 \\ -40 + 5\bar{x}_1 - 5\bar{x}_4 \\ \bar{x}_4 \\ 25 - 2\bar{x}_1 + 2\bar{x}_4 \end{pmatrix}.$$

Here the parameters $\bar{x}_4, \bar{x}_1$ are the values of x_4, x_1 in the representation. Replacing $\bar{x}_4$ by x_4, and $\bar{x}_1$ by x_1 leads to the system of equations

$$x_2 = -60 + 9x_1 - 9x_4$$
$$x_3 = -40 + 5x_1 - 5x_4$$
$$x_5 = 25 - 2x_1 + 2x_4.$$

This system is exactly the system of equations from the final tableau obtained under the GJ method. It is equivalent to the original system of equations in this example in the sense that both systems have the same set of solutions.

However this special property may not hold for the general parametric representation. For an example see Example 4 given below in which each of the variables involves at least two parameters with nonzero coefficients.

Algorithm to Find a Linear Equation Representation of $F = \{x = a^1 + \beta_1 c^1 + \cdots + \beta_{k-1} c^{k-1} : \beta_1, \ldots, \beta_{k-1} \text{ real}\}$

BEGIN

Step 1. Write the n equations expressing each of the variables x_j in terms of the parameters, with the terms involving the parameters transferred to the left, and put it in detached coefficient tableau form. It is the original tableau.

Original tableau

x_1	$\ldots$	x_n	β_1	$\ldots$	β_{k-1}	
$I_{.1}$	$\ldots$	$I_{.n}$	$-c^1$	$\ldots$	$-c^{k-1}$	a^1

where $I_{.1}, \ldots, I_{.n}$ are the n column vectors of the unit matrix of order n.

Step 2. Perform GJ or G pivot steps on this tableau in as many rows as possible *using only the columns of the parameters as pivot columns.* Whenever a pivot step is completed with a row as the pivot row, record the parameter associated with the pivot column as the basic parameter in that row.

Terminate the pivot steps when each of the remaining rows in which pivot steps have not been carried out so far has a 0-entry in the column associated with each of the parameters. This is the final tableau in the procedure.

Step 3. If every row in the final tableau has a basic parameter, F is the whole space R^n, terminate.

If there are rows without a basic parameter in the final tableau, each such row must have a 0-entry in the column of every parameter (otherwise Step 2 in incomplete as a pivot step can be carried out in this row). So, the equation corresponding to this row in the final tableau involves only the x_j variables and none of the parameters. Write the equations corresponding to all such rows in a system, this system provides a linear equation representation for F, terminate.

END

Example 4:

Find a linear equation representation for the affine set

$$F = \left\{ \begin{pmatrix} x_1 \\ x_2 \\ x_3 \end{pmatrix} = \begin{pmatrix} 9 - \beta_1 - 2\beta_2 \\ -6 + \beta_1 + 3\beta_2 - \beta_3 \\ -7 - \beta_2 + 2\beta_3 \end{pmatrix} : \beta_1, \beta_2, \beta_3 \quad \text{real} \right\}.$$

We apply the algorithm using GJ pivot steps. The original tableau is the one at the top. Pivot elements are boxed. The PR, pivot row is the row with the pivot element; and the PC, pivot column is the column with the pivot element. BP stands for "Basic Parameter selected in this row".

BP	x_1	x_2	x_3	β_1	β_2	β_3	
	1	0	0	$\boxed{1}$	2	0	9
	0	1	0	-1	-3	1	-6
	0	0	1	0	1	-2	-7
β_1	1	0	0	1	2	0	9
	1	1	0	0	$\boxed{-1}$	1	3
	0	0	1	0	1	-2	-7
β_1	3	2	0	1	0	2	15
β_2	-1	-1	0	0	1	-1	-3
	1	1	1	0	0	$\boxed{-1}$	-4
Final tableau							
β_1	5	4	2	1	0	0	7
β_2	-2	-2	-1	0	1	0	1
β_3	-1	-1	-1	0	0	1	4

Since every row in the final tableau has a basic parameter, $F = R^3$, the whole space in this example.

Example 5:

Find a linear equation representation for the affine hull of $\{a^1, a^2, a^3, a^4, a^5\}$ given below.

$$a^1 = \begin{pmatrix} 1 \\ 1 \\ 1 \\ 1 \end{pmatrix}, \quad a^2 = \begin{pmatrix} 2 \\ 1 \\ 2 \\ 2 \end{pmatrix}, \quad a^3 = \begin{pmatrix} 3 \\ 1 \\ 3 \\ 3 \end{pmatrix},$$

$$a^4 = \begin{pmatrix} 1 \\ 2 \\ 1 \\ 1 \end{pmatrix}, \quad a^5 = \begin{pmatrix} 2 \\ 2 \\ 2 \\ 2 \end{pmatrix}.$$

So, $F = \{x = a^1 + \beta_1 c^1 + \beta_2 c^2 + \beta_3 c^3 + \beta_4 c^4 : \beta_1$ to β_4 take all real values$\}$, where c^1 to c^4 are given below.

$$c^1 = a^2 - a^1 = \begin{pmatrix} 1 \\ 0 \\ 1 \\ 1 \end{pmatrix}, \quad c^2 = a^3 - a^1 = \begin{pmatrix} 2 \\ 0 \\ 2 \\ 2 \end{pmatrix}$$

$$c^3 = a^4 - a^1 = \begin{pmatrix} 0 \\ 1 \\ 0 \\ 0 \end{pmatrix}, \quad c^4 = a^5 - a^1 = \begin{pmatrix} 1 \\ 1 \\ 1 \\ 1 \end{pmatrix}.$$

We apply the algorithm using GJ pivot steps and the terminology of the previous example.

BP	x_1	x_2	x_3	x_4	β_1	β_2	β_3	β_4	
	1	0	0	0	-1	-2	0	-1	1
	0	1	0	0	0	0	-1	-1	1
	0	0	1	0	-1	-2	0	-1	1
	0	0	0	1	-1	-2	0	-1	1
β_1	-1	0	0	0	1	2	0	1	-1
	0	1	0	0	0	0	-1	-1	1
	-1	0	1	0	0	0	0	0	0
	-1	0	0	1	0	0	0	0	0
β_1	-1	0	0	0	1	2	0	1	-1
β_3	0	-1	0	0	0	0	1	1	-1
	-1	0	1	0	0	0	0	0	0
	-1	0	0	1	0	0	0	0	0

In rows 3, 4 in which pivot steps have not been carried out yet, all the parameter columns have 0 entries, hence the pivot steps are terminated.

The system of equations corresponding to rows 3, 4 in the final tableau which do not have a basic parameter is

$$-x_1 + x_4 = 0$$
$$-x_1 + x_5 = 0$$

which provides a linear equation representation for F.

Also, from earlier discussion we see that the column vectors of the nonbasic parameters β_2, β_4 are linear combinations of column vectors of β_1, β_3. This implies that

$$\{\beta_1 c^1 + \cdots + \beta_4 c^4 : \beta_1, \ldots, \beta_4 \text{ real}\} = \{\beta_1 c^1 + \beta_3 c^3 : \beta_1, \beta_3 \text{ real}\}.$$

So, a parametric representation of F using the smallest number of parameters is $\{a^1 + \beta_1 c^1 + \beta_3 c^3 : \beta_1, \beta_3 \text{ take all real values}\}$. This implies that $F = $ affine hull of $\{a^1, a^1 + c^1, a^1 + c^3\} = $ affine hull of $\{a^1, a^2, a^4\}$.

Example 6:

Find a linear equation representation for the affine hull of $\{a^1 = (-3, 4, 0, -6)^T, a^2 = (-1, 3, -3, -4)^T\}$ which is a straight line L in R^4.

So, $L = \{x = (x_1, x_2, x_3, x_4)^T = (-3, 4, 0, -6)^T + \alpha(2, -1, -3, 2)^T : \alpha$ takes all real values$\}$.

To get the linear equation representation of L we apply the algorithm using GJ pivot steps and the terminology of the previous examples.

BP	x_1	x_2	x_3	x_4	α	
	1	0	0	0	-2	-3
	0	1	0	0	$\boxed{1}$	4
	0	0	1	0	3	0
	0	0	0	1	-2	6
	1	2	0	0	0	5
α	0	1	0	0	1	4
	0	-3	1	0	0	-12
	0	2	0	1	0	14

Now the only parameter α in this representation, is eliminated from rows 1, 3, and 4. The equations corresponding to these rows form the following system of three equations in four unknowns, which provides a linear equation representation for the straight line L.

$$x_1 + 2x_2 = 5$$
$$-3x_2 + x_3 = -12$$
$$2x_2 + x_4 = 14.$$

Example 7:

Find a linear constraint representation for the half-line $L^+ = \{x = (-3, 4, 0, -6)^T + \alpha(2, -1, -3, 2)^T : \alpha \geq 0\}$.

Let L be the straight line containing L^+. In Example 6 above we found the linear equation representation for L.

But in L^+ we have the restriction $\alpha \geq 0$. From the 2nd row in the final tableau in Example 6 we get the equation $x_2 + \alpha = 4$, or $\alpha = 4 - x_2$ expressing the parameter α in terms of the variables x_j on L. So the restriction $\alpha \geq 0$ is equivalent to the inequality constraint $4 - x_2 \geq 0$ or $x_2 \leq 4$. When we augment the linear equation representation for L with this additional constraint, we get the following system which is the linear constraint representation for L^+.

$$x_1 + 2x_2 = 5$$
$$-3x_2 + x_3 = -12$$
$$2x_2 + x_4 = 14$$
$$x_2 \leq 4.$$

Exercises

3.11.1: Let $A_{.1}$ to $A_{.5}$ be the column vectors of A given below.

(i) (Yes) Are b^1, b^2 in the affine hull of $\{A_{.1}, \ldots, A_{.5}\}$? If so, find the actual affine combination for each of them.

(ii) (No) Are b^3, b^4 in the affine hull of $\{A_{.1}, \ldots, A_{.5}\}$?

$$A = \begin{pmatrix} 1 & -1 & 1 & 0 & -1 \\ -1 & 2 & 2 & 2 & 1 \\ -1 & 4 & 8 & 6 & 1 \\ 0 & 2 & 3 & 2 & -1 \end{pmatrix}, \quad b^1 = \begin{pmatrix} 5 \\ -5 \\ -5 \\ 3 \end{pmatrix}, \quad b^2 = \begin{pmatrix} 4 & -3 & -1 & 2 \end{pmatrix},$$

$$b^3 = \begin{pmatrix} 3 & 2 & 12 & -8 \end{pmatrix}, \quad b^4 = \begin{pmatrix} -4 & -2 & -14 & 10 \end{pmatrix}$$

3.11.2: Find the linear equation representation of the affine hull of the set of row vectors of the following matrices.

$$
\text{(i)} \quad
\begin{pmatrix}
-3 & 1 & -4 & 5 \\
-7 & 2 & 2 & -1 \\
4 & -1 & -6 & 6 \\
-13 & 4 & -6 & 9 \\
-17 & 5 & 0 & 3
\end{pmatrix},
\quad \text{(ii)} \quad
\begin{pmatrix}
2 & 3 & 3 & 2 \\
12 & 18 & 18 & 12 \\
8 & 12 & 12 & 8 \\
10 & 15 & 15 & 10
\end{pmatrix}
$$

3.11.3: Obtain the linear equation representations of the following straight lines given in parametric form: $\{x = (-3,7,9,-13)^T + \alpha(-2,-7,1,-2)^T : \alpha \text{ real}\}$, $\{x = (2,3) + \alpha(1,-2) : \alpha \text{ real}\}$.

3.12   Convex Combinations, Convex Hulls

Let a^1, a^2 be two distinct points in R^n. A **convex combination** of a^1, a^2 is

$$
\alpha_1 a^1 + \alpha_2 a^2
$$

where the coefficients α_1, α_2 have to satisfy the **convex combination conditions**

$$
\alpha_1 + \alpha_2 = 1, \quad \text{and} \quad \alpha_1, \alpha_2 \geq 0.
$$

The set of all convex combinations of $\{a^1, a^2\}$ is called the **convex hull** of $\{a^1, a^2\}$, and it is the **line segment joining** a^1 and a^2. See Figure 3.17. Its parametric representation is

$$
\{x = \alpha_1 a^1 + \alpha_2 a^2 : \alpha_1 + \alpha_2 = 1, \alpha_1, \alpha_2 \geq 0\}
$$

$$
\text{or} \quad \{x = \alpha a^1 + (1-\alpha)a^2 : 0 \leq \alpha \leq 1\}.
$$

Example 1:

Let

$$
a^1 = \begin{pmatrix} 2 \\ 2 \\ 2 \\ 2 \end{pmatrix}, \quad
a^2 = \begin{pmatrix} 6 \\ 6 \\ 6 \\ 6 \end{pmatrix}, \quad
a^3 = \begin{pmatrix} 3 \\ 3 \\ 3 \\ 3 \end{pmatrix}, \quad
a^4 = \begin{pmatrix} 10 \\ 10 \\ 10 \\ 10 \end{pmatrix}, \quad
a^5 = \begin{pmatrix} 3 \\ 4 \\ 2 \\ 5 \end{pmatrix}.
$$

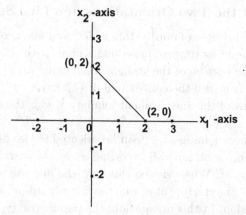

Figure 3.17   The convex hull of any two points in any space is the line segment joining them.

To check if a^3 is in the line segment joining a^1, a^2 we solve the following equations for α.

$$\begin{pmatrix} 2\alpha + 6(1 - \alpha) \\ 2\alpha + 6(1 - \alpha) \\ 2\alpha + 6(1 - \alpha) \\ 2\alpha + 6(1 - \alpha) \end{pmatrix} = \begin{pmatrix} 3 \\ 3 \\ 3 \\ 3 \end{pmatrix}$$

which yields $\alpha = 0.25$. Since $0 \le 0.25 \le 1$, we conclude that a^3 is in the line segment joining a^1, a^2.

To check if a^4 is in the line segment joining a^1, a^2 we solve

$$\begin{pmatrix} 2\alpha + 6(1 - \alpha) \\ 2\alpha + 6(1 - \alpha) \\ 2\alpha + 6(1 - \alpha) \\ 2\alpha + 6(1 - \alpha) \end{pmatrix} = \begin{pmatrix} 10 \\ 10 \\ 10 \\ 10 \end{pmatrix}$$

which yields $\alpha = -1$. Since -1 is not between 0 and 1, we conclude that a^4 is on the straight line joining a^1, a^2 but not on the line segment joining them.

To check if a^5 is in the line segment joining a^1, a^2 we solve

$$\begin{pmatrix} 2\alpha + 6(1 - \alpha) \\ 2\alpha + 6(1 - \alpha) \\ 2\alpha + 6(1 - \alpha) \\ 2\alpha + 6(1 - \alpha) \end{pmatrix} = \begin{pmatrix} 3 \\ 4 \\ 2 \\ 5 \end{pmatrix}$$

which has no solution. So, a^5 is not even on the straight line joining a^1, a^2.

Midpoint, and the Two Orientations of a Line Segment

Let a^1, a^2 be two distinct column vectors in R^n, and also geometrically the two points corresponding to them in n-dimensional space. The line segment joining a^1, a^2 is the portion of the straight line joining a^1, a^2 lying between the points a^1, a^2. It is also the convex hull of $\{a^1, a^2\}$.

The **midpoint** of this line segment joining a^1, a^2 is the convex combination $\frac{1}{2}a^1 + \frac{1}{2}a^2$.

The line segment joining a^1, a^2 can be oriented in two different ways.

One orientation is obtained by treating a^1 as the starting point, and then move towards a^2. When viewed this way, the line segment becomes a portion of the half-line starting at a^1 and moving towards a^2 and continuing in the same direction. In this orientation, the parametric representation of the line segment is $\{a^1 + \alpha(a^2 - a^1) : 0 \leq \alpha \leq 1\}$. In this representation $\alpha = 0$ corresponds to the starting point a^1; and $\alpha = 1$ gives a^2, the end point in the line segment in this orientation. $a^1 + \alpha(a^2 - a^1)$ is the point on the line segment that is a fraction α away from a^1 towards a^2 for $0 \leq \alpha \leq 1$. Verify that $\alpha = 1/2$ gives the midpoint of the line segment.

The 2nd orientation is obtained by starting at a^2 and moving towards a^1, under this orientation the representation of the line segment is $\{a^2 + \alpha(a^1 - a^2) : 0 \leq \alpha \leq 1\}$; $a^2 + \alpha(a^1 - a^2)$ is the point on the line segment that is a fraction α away from a^2 towards a^1 for $0 \leq \alpha \leq 1$.

Example 2:

Let $a^1 = (1, 0, -2, 3)^T$, $a^2 = (-2, 2, 1, -7)^T$. So, $a^2 - a^1 = (-3, 2, 3, -10)^T$, and $a^1 - a^2 = (3, -2, -3, 10)^T$.

The midpoint of the line segment joining a^1, a^2 is $\frac{1}{2}a^1 + \frac{1}{2}a^2 = (-1/2, 1, -1/2, -2)^T$.

The orientation of this line segment from a^1 towards a^2 is $\{(1 - 3\alpha, 2\alpha, -2 + 3\alpha, 3 - 10\alpha)^T : 0 \leq \alpha \leq 1\}$. Taking $\alpha = 2/3$, we see that $(-1, 4/3, 0, -11/3)^T$ is the point on this line segment that is two-thirds of the way away from a^1 towards a^2.

The other orientation of this line segment is $\{(-2 + 3\alpha, 2 - 2\alpha, 1 - 3\alpha, -7 + 10\alpha)^T : 0 \leq \alpha \leq 1\}$.

Convex Combinations of a General Set of k Points

Let $\{a^1, \ldots, a^k\}$ be a set of k column vectors in R^n where k is finite. A **convex combination** of $\{a^1, \ldots, a^k\}$ is

$$\alpha_1 a^1 + \cdots + \alpha_k a^k$$

where the coefficients $\alpha_1, \ldots, \alpha_k$ have to satisfy the **convex combination conditions**

$$\alpha_1 + \cdots + \alpha_k = 1, \quad \text{and all of} \quad \alpha_1, \ldots, \alpha_k \geq 0.$$

The set of all convex combinations of $\{a^1, \ldots, a^k\}$ is called the **convex hull** of $\{a^1, \ldots, a^k\}$, it is $\{x = \alpha_1 a^1 + \cdots + \alpha_k a^k : \alpha_1 + \cdots + \alpha_k = 1, \alpha_1, \ldots, \alpha_k \geq 0\}$, it is also known as a **convex polytope** or a **bounded convex polyhedron**. So, the convex hull of a set of points is a subset of its affine hull, which is itself a subset of its linear hull. See Figure 3.18.

How to Check if a^{k+1} is in the Convex Hull of $\{a^1, \ldots, a^k\}$?

a^{k+1} is in the convex hull of $\{a^1, \ldots, a^k\}$ iff the following system of linear constraints

$$\alpha_1 a^1 + \cdots + \alpha_k a^k = a^{k+1}$$
$$\alpha_1 + \cdots + \alpha_k = 1$$
$$\alpha_1, \ldots, \alpha_k \geq 0$$

in the parameters $\alpha_1, \ldots, \alpha_k$, has a feasible solution. Since this system involves inequality constraints (nonnegativity constraints on the variables), to solve it we need to use linear programming techniques, which are outside the scope of this book. We can ignore the nonnegativity constraints on the variables, and solve the remaining system which is a system of equations, by the GJ pivotal method, or the G elimination method. This could terminate in 3 possible ways discussed below.

1. It may yield a solution which satisfies the nonnegativity constraints. In this case we conclude that a^{k+1} is in the convex hull of $\{a^1, \ldots, a^k\}$.
2. It may conclude that the system of equations itself is inconsistent, i.e., has no solution. In this case we conclude that a^{k+1} is not in the convex hull of $\{a^1, \ldots, a^k\}$.
3. The method may terminate with a solution to the system of equations $\bar{\alpha} = (\bar{\alpha}_1, \ldots, \bar{\alpha}_k)$ where some $\bar{\alpha}_t < 0$. In this case, if this solution is the unique solution of the system of equations, we are able to conclude that the system of equations has no nonnegative solution, and therefore that a^{k+1} is not in the convex hull of $\{a^1, \ldots, a^k\}$. On the other hand if the system of equations has multiple solutions, this method has been inconclusive. To resolve the question, we need to check whether

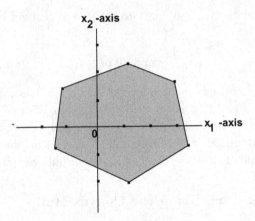

Figure 3.18   A convex polytope, the convex hull of its 6 corner points in R^2.

there is a nonnegative solution to the system for which we need linear programming techniques.

Example 3:

Let

$$
a^1 = \begin{pmatrix} 1 \\ 0 \\ 0 \\ 0 \\ 1 \end{pmatrix}, \ a^2 = \begin{pmatrix} 0 \\ 1 \\ 0 \\ 0 \\ 1 \end{pmatrix}, \ a^3 = \begin{pmatrix} 1 \\ -1 \\ 1 \\ -1 \\ 1 \end{pmatrix}, \ a^4 = \begin{pmatrix} 2 \\ -4 \\ 3 \\ -3 \\ 1 \end{pmatrix}, \ a^5 = \begin{pmatrix} 3/4 \\ -1/4 \\ 1/2 \\ -1/2 \\ 1 \end{pmatrix}.
$$

To check if a^4 is in the convex hull of $\{a^1, a^2, a^3\}$ we solve the following system of equations (after ignoring the nonnegativity constraints) by the GJ pivotal method. PR (pivot row), PC (pivot column), RC (redundant constraint to be deleted), BV (basic variable selected in row) have the same meanings as before, and the pivot element is boxed in each step.

BV	α_1	α_2	α_3		
	$\boxed{1}$	0	1	2	PR
	0	1	-1	-4	
	0	0	1	3	
	0	0	-1	-3	
	1	1	1	1	
	PC↑				

BV	α_1	α_2	α_3		
α_1	1	0	1	2	
	0	$\boxed{1}$	-1	-4	PR
	0	0	1	3	
	0	0	-1	-3	
	0	1	0	-1	
		PC↑			
α_1	1	0	1	2	
α_2	0	1	-1	-4	
	0	0	$\boxed{1}$	3	PR
	0	0	-1	-3	
	0	0	1	3	
		PC↑			
α_1	1	0	0	-1	
α_2	0	1	0	-1	
α_3	0	0	1	3	
	0	0	0	0	RC
	0	0	0	0	RC

The system of equations in this case has a unique solution $(\alpha_1, \alpha_2, \alpha_3) = (-1, -1, 3)$ which violates the nonnegativity constraints. So, we are able to conclude that the system has no nonnegative solution, and therefore a^4 is not in the convex hull of $\{a^1, a^2, a^3\}$.

To check whether a^5 is in the convex hull of $\{a^1, a^2, a^3\}$, we apply the GJ method on the following system of equations.

BV	α_1	α_2	α_3		
	$\boxed{1}$	0	1	3/4	PR
	0	1	-1	$-1/4$	
	0	0	1	1/2	
	0	0	-1	$-1/2$	
	1	1	1	1	
	PC↑				
α_1	1	0	1	3/4	
	0	$\boxed{1}$	-1	$-1/4$	PR
	0	0	1	1/2	
	0	0	-1	$-1/2$	
	0	1	0	1/4	
		PC↑			

BV	α_1	α_2	α_3	
α_1	1	0	1	3/4
α_2	0	1	$-1/4$	-4
	0	0	$\boxed{1}$	1/2 PR
	0	0	-1	$-1/2$
	0	0	1	1/2
			PC↑	
α_1	1	0	0	1/4
α_2	0	1	0	1/4
α_3	0	0	1	1/2
	0	0	0	0 RC
	0	0	0	0 RC

So, the method yields the solution $(\alpha_1, \alpha_2, \alpha_3) = (1/4, 1/4, 1/2)$ which satisfies the nonnegativity constraints. Hence a^5 is in the convex hull of $\{a^1, a^2, a^3\}$. In fact $a^5 = (1/4)a^1 + (1/4)a^2 + (1/2)a^3$.

How to Represent the Convex Hull of a Given Set of Points through a System of Linear Constraints?

The convex hull of a given set of points $\{a^1, \ldots, a^k\}$ in R^n can be represented through a system of linear constraints, but this system would involve some linear inequalities. Typically the number of linear inequalities in this representation grows exponentially with k. Deriving these inequalities requires linear programming techniques which are outside the scope of this book. However, as the number of the inequalities in the representation is typically very very large, even though linear programming techniques are available to generate them, those techniques are rarely used directly in their entirety in practice. However, implementations of these techniques which generate these constraints one at a time are of great importance in the area of Combinatorial Optimization.

Exercises

3.12.1: Check whether the following vectors are in the convex hull of the set of column vectors of the matrix A given below: (a) (Yes) b^1, b^2; (b) (No) b^3, b^4.

$$A = \begin{pmatrix} 3 & 0 & 9 \\ 6 & -3 & 0 \\ 9 & 9 & -9 \\ 12 & -6 & 3 \end{pmatrix}, \quad b^1 = \begin{pmatrix} 4 \\ 1 \\ 3 \\ 9 \end{pmatrix}, \quad b^2 = \begin{pmatrix} 5 & 4 & 3 & 9 \end{pmatrix},$$

$$b^3 = \begin{pmatrix} 15 & 9 & 18 & 21 \end{pmatrix}, \quad b^4 = \begin{pmatrix} 6 & 15 & 9 & 30 \end{pmatrix}.$$

3.12.2: For each pair of points given below, find: (i) the midpoint of the line segment joining them, (ii) parametric representation of the line segment oriented with each point in the pair as the starting point, and directed towards the other point in the pair, and (iii) the point on the line segment that is two-thirds of the way from the starting point towards the other end point in each of the parametric representations above.

$$\{(1, -1, 0, 2)^T, (-3, 4, -7, -9)^T\}; \{(0, -3, 4, -6, 7)^T, (6, 0, 3, -4, -2)^T\};$$
$$\{(1, -3, -1)^T, (-2, -1, -4)^T\}.$$

3.13 Nonnegative Combinations, Pos Cones

A **nonnegative linear combination** of a given set of vectors $\{a^1, \ldots, a^k\}$ in R^n is a vector of the form

$$\alpha_1 a^1 + \cdots + \alpha_k a^k \quad \text{where all the coefficients} \quad \alpha_1, \ldots, \alpha_k \geq 0.$$

The set of all nonnegative linear combinations of $\{a^1, \ldots, a^k\}$ is $\{x = \alpha_1 a^1 + \cdots + \alpha_k a^k : \alpha_1, \ldots, \alpha_k \geq 0\}$. It is called the **nonnegative hull** or the **pos cone** of $\{a^1, \ldots, a^k\}$ and is denoted by $\mathrm{Pos}\{a^1, \ldots, a^k\}$. It is a **cone**.

Example 1:

Here we show how to draw a pos cone in R^2. Let

$$\Gamma = \left\{ a^1 = \begin{pmatrix} 1 \\ 2 \end{pmatrix}, \quad a^2 = \begin{pmatrix} 2 \\ -3 \end{pmatrix}, \quad a^3 = \begin{pmatrix} 1 \\ 0 \end{pmatrix}, \quad a^4 = \begin{pmatrix} 1 \\ -1 \end{pmatrix} \right\}.$$

To draw $\mathrm{Pos}(\Gamma)$, plot each of the points in Γ, and draw the ray of each of these points (the half-line joining the origin to the point and continuing in that direction indefinitely). The Pos cone of Γ is the smallest angle at the origin containing all these rays. See Figure 3.19.

In higher dimensional spaces, even though we cannot physically draw the pos cone of a set of points, we can visualize it as the smallest solid angle at the origin containing the rays of all the points in the set.

To check whether a point a^{k+1} is in $\mathrm{Pos}\{a^1, \ldots, a^k\}$, we need to solve the following system of equations in nonnegative variables.

$$\alpha_1 a^1 + \cdots + \alpha_k a^k = a^{k+1}$$
$$\alpha_1, \ldots, \alpha_k \geq 0.$$

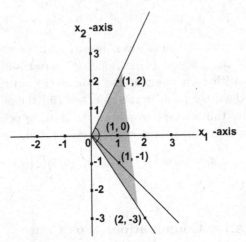

Figure 3.19 The cone Pos(Γ) in R^2 is the smallest angle at the origin containing the rays of all the points in Γ.

It is not always possible to get a feasible solution to this system, or conclude that it has no feasible solution, using the GJ pivotal method or the G elimination method applied to the equality constraints only, ignoring the nonnegativity requirements. To conclusively solve this system, linear programming techniques which are outside the scope of this book are needed.

Pos$\{a^1, \ldots, a^k\}$ can be represented through a system of linear constraints. It leads to a **homogeneous system of linear inequalities** of the form $Ax \geq 0$. To derive the constraints in this representation requires linear programming techniques not discussed in this book. Unfortunately, the number of constraints in such a representation typically grows exponentially with k, so this linear constraint representation is rarely used in practice in its entirety, but implementations of these techniques that generate these constraints one at a time are of great importance in the area of Combinatorial Optimization.

Summary of Various Types of Combinations of a Set of Vectors $\Gamma = \{a^1, \ldots, a^k\} \subset R^n$

Linear combination of Γ: Any point of the form $\alpha_1 a^1 + \cdots + \alpha_k a^k$ where $\alpha_1, \ldots, \alpha_k$ can take all possible real values. Set of all these combinations is the linear hull of Γ. It is a subspace of R^n. If it is not R^n, it is the set of solutions of a homogeneous system of linear equations of the form

$Ax = 0$ where A is of order $m \times n$, with $1 \le m \le n$. Given Γ, this linear constraint representation of its linear hull can be found efficiently.

Affine combination of Γ: Any point of the form $\alpha_1 a^1 + \cdots + \alpha_k a^k$ where $\alpha_1 + \cdots + \alpha_k = 1$. Set of all these combinations is the affine hull of Γ, it is the translate of a subspace of R^n. If it is not R^n, it is the set of solutions of a nonhomogeneous system of linear equations of the form $Ax = b$ where A is of order $m \times n$, with $1 \le m \le n$. Given Γ, this linear constraint representation of its affine hull can be found efficiently.

Convex combination of Γ: Any point of the form $\alpha_1 a^1 + \cdots + \alpha_k a^k$ where $\alpha_1 + \cdots + \alpha_k = 1$, and $\alpha_1, \ldots, \alpha_k \ge 0$. Set of all these combinations is the convex hull of Γ. It is a convex polytope in R^n. It is the set of feasible solutions of a system of nonhomogeneous linear inequalities of the form $Ax \ge b$ where A is of order $m \times n$, with m typically growing exponentially in k. Given Γ, this linear constraint representation of its convex hull can be derived using linear programming techniques.

Nonnegative linear combination of Γ: Any point of the form $\alpha_1 a^1 + \cdots + \alpha_k a^k$ where $\alpha_1, \ldots, \alpha_k \ge 0$. Set of all these combinations is the pos cone of Γ. It is a cone in R^n. If it is not R^n, it is the set of solutions of a homogeneous system of linear inequalities of the form $Ax \ge 0$.

3.14 Hyperplanes

For $n \ge 2$, a **hyperplane** in R^n is defined to be the set of all vectors $x = (x_1, \ldots, x_n)^T$ which satisfy a single **nontrivial linear equation**, i.e., one of the form

$$a_1 x_1 + \cdots + a_n x_n = b_1 \quad \text{where } (a_1, \ldots, a_n) \ne 0 \ (b_1 \text{ could be 0}).$$

So, a nontrivial linear equation is one in which at least one of the variables appears with a nonzero coefficient. In R^n, each hyperplane has dimension $n-1$. So, in R^2 each straight line is a hyperplane and vice versa. See Figure 3.20 for a straight line in R^2 (hyperplane in R^2).

We have seen earlier the parametric representation for straight lines in R^n for any $n \ge 2$, which is the most convenient way for representing straight lines.

An alternate representation for a straight line in R^2 only, is through a single linear equation. This works in R^2 because every straight line in R^2 is a hyperplane and vice versa. For $n \ge 3$, in R^n a hyperplane has dimension $n-1 \ge 2$, and is not a straight line. **That's why a single linear equation does not represent a straight line in spaces of dimension ≥ 3.**

We will now include figures of some hyperplanes in R^2 and R^3.

Figure 3.21 to 3.23: In R^3 the three coordinate hyperplanes are given by "$x_3 = 0$" (contains the x_1 and x_2 coordinate axes, Figure 3.21), "$x_1 = 0$" (contains the x_2 and x_3 coordinate axes, Figure 3.22), and "$x_2 = 0$" (contains the x_1 and x_3 coordinate axes). You can visualize these coordinate

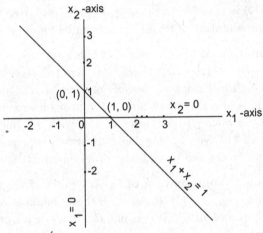

Figure 3.20 In R^2, "$x_2 = 0$" (x_1-axis), "$x_1 = 0$" (x_2-axis) are the two coordinate hyperplanes. We also show the hyperplane (straight line) represented by "$x_1 + x_2 = 1$.

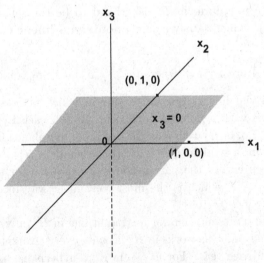

Figure 3.21 The x_1, x_2-coordinate hyperplane represented by the equation $x_3 = 0$ in R^3.

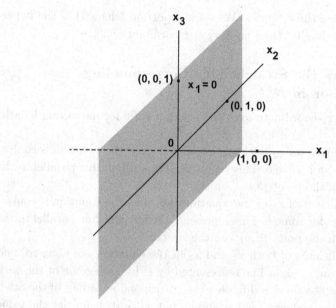

Figure 3.22   The x_2, x_3-coordinate hyperplane represented by the equation $x_1 = 0$ in R^3.

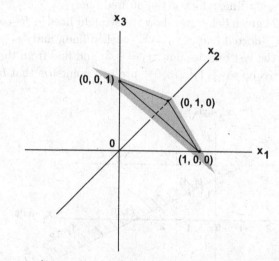

Figure 3.23   Hyperplane represented by $x_1 + x_2 + x_3 = 1$ in R^3.

hyperplanes from these figures. We show a portion (shaded) of the hyperplane corresponding to "$x_1 + x_2 + x_3 = 1$" in Figure 3.23.

How to Draw the Straight Line Corresponding to an Equation in R^2?

Set up the x_1, x_2-coordinate axes and select a scale for measuring lengths on each axes.

If the coefficient of x_2 in the equation is 0, then the equation is equivalent to "$x_1 = b$" for some b. This represents the straight line parallel to the x_2-axis through the point $(b, 0)^T$ on the x_1-axis.

If the coefficient of x_1 in the equation is 0, then the equation is equivalent to "$x_2 = b$" for some b. This represents the straight line parallel to the x_1-axis through the point $(0, b)^T$ on the x_2-axis.

If the coefficients of both x_1 and x_2 in the equation are nonzero, find two points on the straight line represented by it by giving one of the variables two different values and finding the corresponding values of the other variable from the equation. For example, put $x_1 = 0, 1$ and let the value of x_2 from the equation be respectively b_0, b_1. Then $(0, b_0)^T$, $(1, b_1)^T$ are points on the straight line represented by it. Join the two points $(0, b_0)^T$, $(1, b_1)^T$ by a straight line, which is the desired one.

In Figure 3.24 given below, we show the straight lines in R^2 corresponding to "$x_1 = 1.5$" (dotted line), "$x_2 = 2$" (dashed line), and "$x_1 + 2x_2 = 3$" (solid line). For the last one, we put "$x_1 = 1, 3$, and find from the equation $x_2 = 1, 0$ respectively; so $(1, 1)^T$, $(3, 0)^T$ are two points on that line.

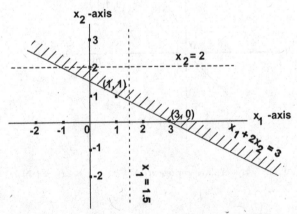

Figure 3.24   The straight lines in R^2 represented by the equations $x_2 = 2$, $x_1 = 1.5$, and $x_1 + 2x_2 = 3$, and the halfspace represented by $x_1 + 2x_2 \geq 3$.

Exercises

3.14.1: Draw the straight lines in R^2 corresponding to each of the following equations.

$$\text{(i)} \ -3x_1 = -9 \qquad \text{(iii)} \ 4x_2 = 12$$
$$\text{(ii)} \ x_1 + 2x_2 = 5 \qquad \text{(iv)} \ 3x_1 - x_2 = 9$$

3.15 Halfspaces

For any $n \geq 2$, a hyperplane in R^n divides the space into two halfspaces, each of which consists of all the points on one side of the hyperplane. Look at the hyperplane defined by the equation "$x_1 + x_2 = 2$" in R^2 in Figure 3.25, it divides R^2 into two halfspaces, one above it (the lined side), one below it. Points in the halfspace above this line satisfy the constraint "$x_1 + x_2 \geq 2$". Points in the halfspace below this line, containing the origin 0, satisfy the constraint "$x_1 + x_2 \leq 2$".

In any space a straight line always has dimension 1. One can verify that a straight line in R^3 does not divide R^3 into two halfspaces because it does not have enough dimension. To divide R^3 into two halfspaces you need a plane of dimension 2.

In the same way, a hyperplane in R^n represented by the equation

$$a_1 x_1 + \cdots + a_n x_n = a_0$$

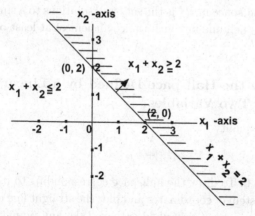

Figure 3.25 A hyperplane in R^2 (Straight line) divides the space into two halfspaces. The thick straight line represents "$x_1 + x_2 = 2$". The inequality "$x_1 + x_2 \geq 2$" represents the halfspace of all points in R^2 lying above (i.e., on the marked side of) this line.

where $(a_1, \ldots, a_n) \neq 0$ divides the space into two halfspaces, one of which is

$$\{x : a_1 x_1 + \cdots + a_n x_n \geq a_0\}$$

and the other is

$$\{x : a_1 x_1 + \cdots + a_n x_n \leq a_0\}.$$

The hyperplane represented by the equation "$a_1 x_1 + \cdots + a_n x_n = a_0$" is itself common to both these halfspaces. The above inequalities are said to be **closed linear inequalities** (because they both allow the possibility of the two sides of the inequality being equal), and the halfspaces represented by them are called **closed halfspaces**. Each closed halfspace includes the hyperplane defining it. The intersection of the two halfspaces given above is the hyperplane defining them. The single equation $a_1 x_1 + \cdots + a_n x_n = a_0$ is equivalent to the pair of inequalities

$$a_1 x_1 + \cdots + a_n x_n \geq a_0$$
$$a_1 x_1 + \cdots + a_n x_n \leq a_0.$$

This explains the fact that a hyperplane in R^n is the intersection of the two closed halfspaces defined by it.

The strict inequalities

$$a_1 x_1 + \cdots + a_n x_n > a_0 \quad \text{or} \quad a_1 x_1 + \cdots + a_n x_n < a_0$$

are said to be **open linear inequalities**, they represent **open halfspaces**. An open halfspace does not contain the hyperplane defining it.

Thus each halfspace in R^n is the set of all solutions to a single nontrivial linear inequality (i.e., one in which the coefficient of at least one variable is nonzero).

How to Draw the Halfspace Defined by a Linear Inequality in Two Variables?

Consider the inequality

$$a_1 x_1 + a_2 x_2 \geq a_0$$

where $(a_1, a_2) \neq 0$. To draw the halfspace corresponding to it in R^2, set up the Cartesian system of coordinates and plot the straight line corresponding to $a_1 x_1 + a_2 x_2 + = a_0$ as described earlier. Take any point $\bar{x} = (\bar{x}_1, \bar{x}_2)^T$ in one of the open halfspaces into which this straight line divides R^2. If $a_1 \bar{x}_1 + a_2 \bar{x}_2 > a_0$, the halfspace containing $\bar{x}$ is the one corresponding

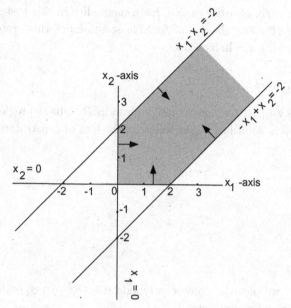

Figure 3.26   Four halfspaces in R^2, and the unbounded convex polyhedron that is their intersection.

to $a_1 x_1 + a_2 x_2 \geq a_0$. On the other hand if $a_1 \bar{x}_1 + a_2 \bar{x}_2 < a_0$, the other halfspace not containing $\bar{x}$ is the one corresponding to $a_1 x_1 + a_2 x_2 \geq a_0$.

Example 1:

In Figure 3.26 we draw the halfspaces corresponding to the four inequalities given below and mark each of them by an arrow on the straight line defining it.

$$x_1 \geq 0$$
$$x_2 \geq 0$$
$$x_1 - x_2 \geq -2$$
$$-x_1 + x_2 \leq -2.$$

3.16   Convex Polyhedra

A **convex polyhedron** in R^n is the intersection of a finite number of halfspaces, i.e., it is the set of feasible solutions of a system of linear inequalities of the form

$$Ax \geq b$$

where A is a matrix of order $m \times n$. Each inequality in this system defines a halfspace in R^n; and the set of feasible solutions of this system is the intersection of all these halfspaces.

Example 1:

In Figure 3.26 we plot the convex polyhedron in R^2 (shaded region) that is the set of feasible solutions of the following system of constraints

$$x_1 \geq 0$$
$$x_2 \geq 0$$
$$x_1 - x_2 \geq -2$$
$$-x_1 + x_2 \leq -2.$$

Example 2:

In Figure 3.27 we plot the convex polyhedron in R^2 corresponding to the following system of constraints

$$x_1 + x_2 \leq 2$$
$$-x_1 + x_2 \geq -2$$
$$-x_1 - x_2 \leq 2$$
$$x_1 - x_2 \geq -2.$$

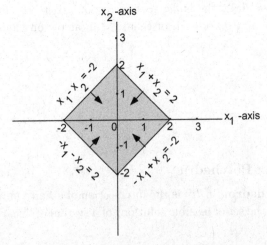

Figure 3.27   A convex polytope (bounded convex polyhedron) in R^2.

A bounded convex polyhedron is called a **convex polytope**.

The set of feasible solutions of a homogeneous system of linear inequalities, i.e., a system of the form

$$Ax \geq 0$$

is said to be a **convex polyhedral cone**. As an example, consider the system

$$2x_1 + x_2 \geq 0$$
$$x_1 + 3x_2 \geq 0$$

Its set of feasible solutions is the polyhedral cone (marked by an angle sign) in Figure 3.28.

Exercises

3.16.1: Set up the 2-dimensional Cartesian coordinate system and draw the convex polyhedra represented by the following systems of constraints. Mention whether each of them is a polytope or an unbounded polyhedron.

(i) (Unbounded) (a) $-5 \leq -x_1 + x_2 \leq 4$, $x_1 + x_2 \geq 3$, $x_1 + 2x_2 \geq 4$, $x_1, x_2 \geq 0$.

(b) $x_1, x_2 \geq 0$, $x_1 \leq 50$, $x_2 \leq 80$, $-2x_1 + x_2 \leq 60$, $3x_1 + x_2 \geq 30$, $x_1 + 3x_2 \geq 30$.

(c) $-2x_1 + x_2 \leq 0$, $x_1 - 2x_2 \leq 0$.

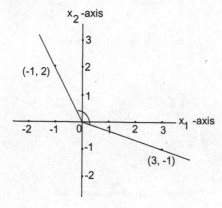

Figure 3.28   A polyhedral cone in R^2.

(ii) (Bounded) (d) $-4 \leq x_1 \leq 4$, $-4 \leq x_2 \leq 4$, $-6 \leq -x_1 + x_2 \leq 6$, $-6 \leq x_1 + x_2 \leq 6$.

(e) $x_1, x_2 \geq 0$, $x_1 + x_2 \geq 10$, $x_1 + 4x_2 \leq -20$.

3.17   Convex and Nonconvex Sets

A subset $K \subset R^n$ is said to be a convex set if it satisfies the following condition

$$\text{for every pair of points} \quad x^1, x^2 \in K;$$

$$\alpha x^1 + (1 - \alpha)x^2 \in K \quad \text{for all } 0 \leq \alpha \leq 1$$

i.e., for every pair of points x^1, x^2 in K, the entire line segment joining x^1, x^2 must be in K.

Figure 3.29 has examples of convex sets in R^2; and Figure 3.30 has examples of nonconvex sets.

Some examples of convex sets in R^n are

all hyperplanes, and halfspaces, convex polyhedra, convex polyhedral cones, and balls.

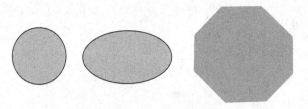

Figure 3.29   Convex sets.

Figure 3.30   Nonconvex sets.

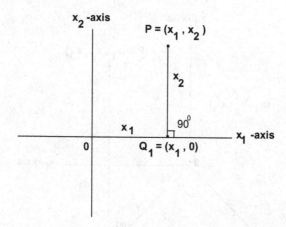

Figure 3.31   Euclidean distance function in R^2 explained.

3.18   Euclidean Distance, and Angles

The Euclidean Distance Between Two Points in R^n

Consider the point $P = (x_1, x_2)^T$ in R^2. Let $Q_1 = (x_1, 0)^T$. Then from Figure 3.31 we see that 0, Q_1, P form a right angled triangle. The famous **Pythagoras theorem** states that in any right angled triangle the square of the diagonal is the sum of the squares of the two sides.

Applying this theorem we see that

$$\text{distance between } 0 \text{ and } P = (x_1, x_2)^T \text{ in } R^2 = \sqrt{x_1^2 + x_2^2}.$$

Now consider the point $P = (x_1, x_2, x_3)^T$ in R^3, see Figure 3.32. Let the point $Q_2 = (x_1, x_2, 0)^T$ in the x_1, x_2-coordinate plane for R^3. From the above argument we know that $0Q_2 = $ the length of the line segment $0Q_2$ is $\sqrt{x_1^2 + x_2^2}$. Also, 0, Q_2, P form a right angled triangle in the two dimensional plane of R^3 containing the rays of P and Q_2. Applying the Pythagoras theorem to this right angled triangle, we have

$$\text{distance between } 0 \text{ and } P = (x_1, x_2, x_3)^T \text{ in } R^3 = \sqrt{(0Q_2)^2 + x_3^2}$$

$$= \sqrt{x_1^2 + x_2^2 + x_3^2}.$$

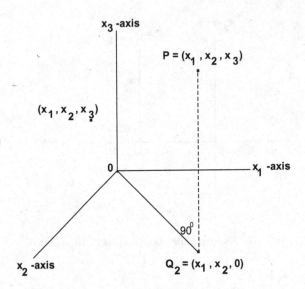

Figure 3.32   Eoclidean distance function in R^3 explained.

Continuing this argument the same way, we get the following definition of distance in R^n

distance between 0 and $P = (x_1, \ldots, x_n)^T$ in R^n

$$= \sqrt{x_1^2 + \cdots + x_n^2} = \sqrt{x^T x}$$

The distance between 0 and $x \in R^n$ given above is usually denoted by the symbol $\|x\|$, and is known as the **Euclidean norm** of the vector x.

To find the distance between two points $L = (y_1, \ldots, y_n)^T$ and $P = (x_1, \ldots, x_n)^T$ in R^n, we translate the origin to L, this has no effect on distances. After the translation L becomes 0 and P becomes $(x_1 - y_1, \ldots, x_n - y_n)^T$. From this and the above, we conclude that

distance between $x = (x_1, \ldots, x_n)^T$ and $y = (y_1, \ldots, y_n)^T$ in R^n

$$= \sqrt{(x_1 - y_1)^2 + \cdots + (x_n - y_n)^2} = \sqrt{(x - y)^T (x - y)}.$$

The distance function given above is known as the **Euclidean distance**. This is the most commonly used distance measure.

The Euclidean norm satisfies the following properties:

(i) $\|x\| \geq 0$ for all $x \in R^n$; and $\|x\| = 0$ only if $x = 0$.

(ii) $\|\lambda x\| = |\lambda| \|x\|$ for all $x \in R^n$, and all real numbers λ.

(iii) **Triangle inequality:** For all $x, y \in R^n$, $\|x + y\| \leq \|x\| + \|y\|$.

(iv) **Cauchy-Schwartz inequality:** For all $x, y \in R^n$, $|\langle x, y \rangle| \leq (\|x\|)(\|y\|)$ where $\langle x, y \rangle$ is the dot product of the two vectors x, y defined in Section 1.5 (if x, y are both column vectors, $\langle x, y \rangle = x^T y$). Also, $|\langle x, y \rangle| = (\|x\|)(\|y\|)$ iff one of the two vectors among x, y is a scalar multiple of the other.

Using the parallelogram law of addition of vectors, it can be seen that the triangle inequality is the same as the geometric fact that for any $n \geq 2$, the sum of the lengths of two sides of a two dimensional triangle in R^n is $\geq$ the length of the third side. To see this let x, y be two nonzero vectors in R^n. Figure 3.33 shows the two dimensional subspace of R^n containing the rays of x and y. Look at the triangle with vertices $0, x$ and $x + y$. The lengths of the sides of this triangle are $\|x\|$ (side joining 0 and x), $\|y\|$ (side joining $x, x + y$), and $\|x + y\|$ (side joining 0 and $x + y$); and so the triangle inequality $\|x\| + \|y\| \geq \|x + y\|$ is an expression of this geometric fact.

Mathematicians have constructed many other functions which can be used as distance functions (these distance functions do not satisfy the Pythagoras theorem), and they study properties of R^n under different distance measures. In this book we will only use the Euclidean distance function defined above.

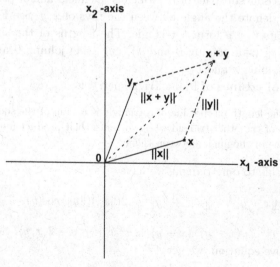

Figure 3.33   Interpretation of triangle inequality using the parallelogram law of addition of vectors.

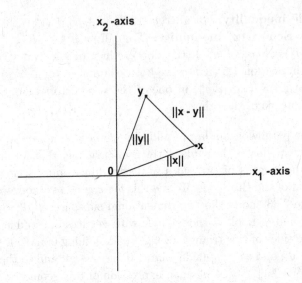

Figure 3.34 Angle between the rays of two nonzero vectors in R^n.

Relationship Between the Dot Product of Two Vectors and the Angle Between their Rays

Let x, y be two nonzero vectors in R^n. First assume that neither of these vectors is a scalar multiple of the other. Then the rays of x, y are contained in a two dimensional subspace of R^n which is the linear hull of $\{x, y\}$. In this subspace let θ denote the angle between the rays of x, y. See Figure 3.34. The three points $0, x, y$ form a triangle. The lengths of the sides of this triangle are: $\|x\|$ (side joining 0 and x), $\|y\|$ (side joining 0 and y), and $\|x - y\|$ (side joining x and y).

The **Law of cosines** of Plane Trignometry, is:

Square of the length of one side of a triangle is = sum of the squares of the lengths of the other two sides − two times their product multiplied by the cosine of the angle between them.

Applying this to our triangle, we have

$$\|x - y\|^2 = \|x\|^2 + \|y\|^2 - 2(\|x\|)(\|y\|)\cos(\theta).$$

But $\|x - y\|^2 = (x - y)^T(x - y) = x^T x + y^T y - 2\langle x, y\rangle$. Substituting this in the above equation, we get

$$\cos(\theta) = \langle x, y\rangle / ((\|x\|)(\|y\|)).$$

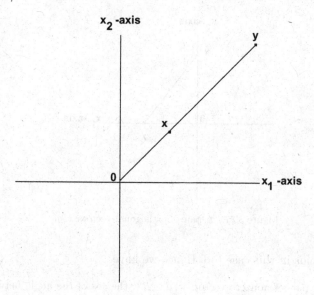

Figure 3.35   Angle between the rays of x, y is 0 when x is a positive multiple of y.

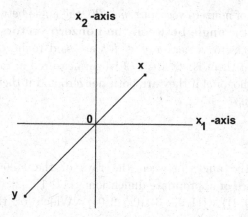

Figure 3.36   Angle between the rays of x, y is 180° when x is a negative multiple of y.

If x is a positive multiple of y (see Figure 3.35), then the rays of x and y are the same, so the angle between these rays is 0, and since $\cos(0) = 1$, the above equation holds.

If x is a negative multiple of y (see Figure 3.36), then the rays of x, y are the two halfs of a straight line, i.e., the angle θ between these rays is 180° (or π if measured in radians) and since $\cos(180°) = -1$, the above

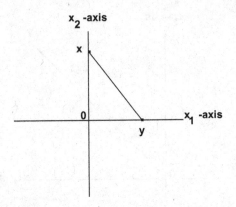

Figure 3.37   A pair of orthogonal vectors x, y.

equation holds in this case too. Hence we have

> For any pair of nonzero vectors $x, y \in R^n$, the cos of the angle between their rays is = the dot product of x and y divided by the product of the norms of x and y.

For any pair of nonzero vectors $x, y \in R^n$, the angle between their rays is also known as the **angle between the nonzero vectors** x, y.

A pair of nonzero vectors $x, y \in R^n$ are said to be **orthogonal** if the angle between them is $90°$ (or $\pi/2$ when measured in radians). Hence $x, y \in R^n$ are orthogonal if they are both nonzero and if their dot product is 0. See Figure 3.37.

Exercises

3.18.1:   Find the angle between the rays of the following pairs of points in the space of appropriate dimension: $\{(1, -1, 1, 2), (0, 1, -2, -2)\}$, $\{(1, 2, -3), (-2, 0, 4)\}$, $\{(1, 0, -3), (0, -2, 0)\}$. Which of these pairs is orthogonal?

3.19   The Determinant Interpreted as a Volume Function

Consider a matrix B of order 2×2

$$B = \begin{pmatrix} b_{11} & b_{12} \\ b_{21} & b_{22} \end{pmatrix}$$

and let $B_1., B_2.$ be its row vectors (we can carry this discussion in terms of column vectors too, we are doing it in terms of row vectors). Let

$$f(B_1., B_2.) = \text{area of the parallelogram with } 0, B_1., B_2. \text{ as three of its}$$
$$\text{vertices in } R^2, \text{ see Figure 3.38.}$$

Then from geometry, we know that the area of a parallelogram is the product of the length of its base and height. From this, we verify that the function $f(B_1., B_2.)$ satisfies the following properties:

(a) $f((1,0),(0,1)) = 1$, since by definition, the area of the unit square is 1.
(b) $f(\lambda B_1., B_2.) = f(B_1., \lambda B_2.) = \lambda f(B_1., B_2.)$ for all $\lambda \geq 0$. See Figure 3.38 where we illustrate this for $\lambda = 2, 3$.
(c) $f(B_1. + B_2., B_2.) = f(B_1., B_2.) = f(B_1., B_2. + B_1.)$. See Figure 3.39 where this property is illustrated.

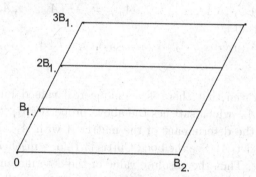

Figure 3.38   For nonzero $x, y \in R^2$, $f(x, y)$ is the area of the parallelogram with $0, x, y$ as three of its vertices. You can verify that $f(2B_1., B_2.) = 2f(B_1., B_2.)$ and $f(3B_1., B_2.) = 3f(B_1., B_2.)$.

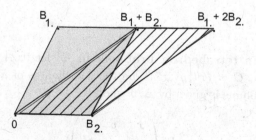

Figure 3.39   For nonzero $x, y \in R^2$, $f(x, y)$ is the area of the parallelogram with $0, x, y$ as three of its vertices. You can verify that $f(B_1. + B_2., B_2.) = f(B_1., B_2.)$.

It turns out that there is a unique real valued function of the row vectors B_1, B_2. that satisfies the three properties (a), (b), (c) mentioned above, it is $\det(B)$ = the determinant of the matrix B with B_1, B_2. as its row vectors. Also, the area function defined above $f(B_1, B_2.) = |\det(B)|$.

In the same way let $A = (a_{ij})$ be a square matrix of order n and let $A_1, \ldots, A_n$. be its row vectors. Define

$V(A_1, \ldots, A_n.) = n$-dimensional volume (volume in R^n, also called content) of the parallelopiped P with 0 as a vertex and $A_1, \ldots, A_n$. as its adjacent vertices in R^n ($P = \{x : x = \sum_{i=1}^n \lambda_i A_i, 0 \le \lambda_i \le 1$ for all $i\}$).

This volume function $V(A_1, \ldots, A_n.)$ is always nonnegative, it satisfies $V(A_1, \ldots, A_n.) = |F(A_1, \ldots, A_n.)|$ where the function $F(A_1, \ldots, A_n.)$ satisfies the following three properties:

(i) $F(I_1, \ldots, I_n.) = 1$, where $I_1, \ldots, I_n$. are the row vectors of the unit matrix of order n.

(ii) $F(A_1, \ldots, A_{i-1}, \lambda A_i, A_{i+1}, \ldots, A_n.) = \lambda F(A_1, \ldots, A_n.)$ for all λ real, and for all i.

(iii) $F(A_1, \ldots, A_{i-1}, A_i. + A_j, A_{i+1}, \ldots, A_n.) = F(A_1, \ldots, A_n.)$ for all i and $j \ne i$.

It can be proven that there is a unique real valued function of the vectors $A_1, \ldots, A_n$. which satisfies the above properties (i), (ii), (iii), and that function is the determinant of the matrix A with $A_1, \ldots, A_n$. as its row vectors, i.e., $\det(A)$. See the book Curtis [3.1] for a mathematical proof of this statement. Thus the absolute value of the determinant of a square matrix of order n, is the n-dimensional volume of the parallelopiped with 0 as a vertex and the row vectors of the matrix as its adjacent vertices.

This partly explains the major role that determinants play in mathematical research.

Exercises

3.19.1: Euler's tetrahedron problem: In R^3 let 0, $A = (x, y, z)^T$, $B = (x', y', z')^T$, $C = (x'', y'', z'')^T$ be the coordinates of a tetrahedron. Prove that its volume is given by $\Delta/6$ where

$$\Delta = \det \begin{pmatrix} x & y & z \\ x' & y' & z' \\ x'' & y'' & z'' \end{pmatrix}.$$

3.20 Orthogonal Projections (Nearest Points)

The following problem arises in many applications.

> We are given a subset $S \subset R^n$, usually as the set of feasible solutions of a given system of constraints, or sometimes in parametric representation; and a point $\bar{x} \neq S$. We are required to find the **nearest point** in S to $\bar{x}$.

In general this leads to a **quadratic programming problem**, a special version of **nonlinear programming**. Both these topics are outside the scope of this book. However, we will show the solution to this problem when S is a simple set such as a straight line, or a hyperplane, or the set of solutions of a system of linear equations.

Nearest Point When S Is a Straight Line

Consider the case where $S = L = \{x : x = a + \lambda c, \lambda$ takes all real values$\}$, where $a = (a_i)$, $c = (c_i)$ are given column vectors in R^n and $c \neq 0$. L is given in parametric representation. So, $a + \lambda c$ is a general point on L, and the problem of finding the nearest point in L to $\bar{x}$ (note that minimizing the distance is the same as minimizing its square) is:

$$\text{Minimize } f_1(\lambda) = (a + \lambda c - \bar{x})^T (a + \lambda c - \bar{x})$$

$$= \sum_{i=1}^{n} (a_i + \lambda c_i - \bar{x}_i)^2$$

over λ taking all real values. From calculus, we know that the minimum of $f_1(\lambda)$ is attained at a λ which makes the derivative $f_1'(\lambda) = 0$, i.e., satisfies

$$2 \sum_{i=1}^{n} (a_i + \lambda c_i - \bar{x}_i) c_i = 0.$$

Let λ^* denote the solution of this. So,

$$\lambda^* = \frac{\sum_{i=1}^{n} (\bar{x}_i - a_i) c_i}{\sum_{i=1}^{n} c_i^2} = \frac{c^T (\bar{x} - a)}{\|c\|^2}.$$

So, the nearest point in L to $\bar{x}$ is $a + \lambda^* c = x^*$ say. It can be verified that for any point $x = a + \lambda c \in L$,

$$(\bar{x} - x^*)^T (x - x^*) = (\bar{x} - a - \lambda^* c)^T (a + \lambda c - a - \lambda^* c)$$
$$= (\lambda - \lambda^*)(\bar{x} - a - \lambda^* c)^T c$$

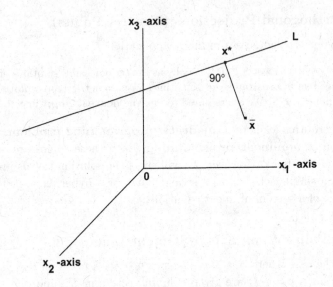

Figure 3.40   The nearest point, x^* in L to $\bar{x}$, is the orthogonal projection of $\bar{x}$ on L.

$$= (\lambda - \lambda^*)((\bar{x} - a)^T c - \lambda^* c^T c)$$
$$= (\lambda - \lambda^*)((\bar{x} - a)^T c - (\bar{x} - a)^T c)$$
$$= 0.$$

See Figure 3.40. So, the nearest point, x^*, on the straight line L to $\bar{x}$ satisfies the property that $(\bar{x} - x^*)$ is orthogonal to $(x - x^*)$ for all $x \in L$. That's why the nearest point x^* is also known as the **orthogonal projection** of $\bar{x}$ on L.

Example 1:

Consider the straight line L and point $\bar{x} \in R^3$ given below.

$$L = \left\{ x : x = \begin{pmatrix} 3 \\ -4 \\ -11 \end{pmatrix} + \lambda \begin{pmatrix} 0 \\ -4 \\ 9 \end{pmatrix}, \ \lambda \text{ real} \right\}, \quad \bar{x} = \begin{pmatrix} 1 \\ 2 \\ -3 \end{pmatrix}.$$

Then from the above, the nearest point in L to $\bar{x}$ is obtained by taking

$$\lambda = \lambda^* = \frac{(0, -4, 9)((1, 2, -3) - (3, -4, -11))^T}{\|(0, -4, 9)^T\|^2} = \frac{48}{25}$$

So, the nearest point to $\bar{x}$ in L in this example is

$$x^* = (3, -4, -11)^T + (48/25)(0, -4, 9)^T = (3, -11.68, 6.28)^T.$$

x^* is the orthogonal projection of $\bar{x}$ on the straight line L.

Nearest Point When S Is a Hyperplane

Consider the case where $S = H = \{x : ax = a_0\}$, where $a = (a_1, \ldots, a_n) \neq 0$; a hyperplane. In this case the problem of finding the nearest point in H to $\bar{x}$ is the quadratic program

$$\text{Minimize} \quad (x - \bar{x})^T (x - \bar{x})$$
$$\text{subject to} \quad ax = a_0.$$

We can solve this problem using the **Lagrange multiplier technique** of calculus. We construct the Lagrangian for this problem, which is

$$L(x, \delta) = (x - \bar{x})^T (x - \bar{x}) - \delta(ax - a_0)$$

where δ is the **Lagrange multiplier** associated with the constraint in the above problem. The optimum solution to the problem, x^*, δ^* is then obtained by solving the system

$$\frac{\partial L(x, \delta)}{\partial x} = 2(x - \bar{x})^T - \delta a = 0$$
$$ax - a_0 = 0.$$

From the first equation we have $x^* = \bar{x} + (\delta^*/2)a^T$. Substituting this in the 2nd equation we get $(\delta^*/2) = -(a\bar{x} - a_0)/\|a\|^2 = \Delta^*$ say. So, the nearest point x^* in H to $\bar{x}$ is given by

$$x^* = \bar{x} + a^T \Delta^*$$

where $\Delta^* = -(a\bar{x} - a_0)/\|a\|^2$. It can be verified that for any $x \in H$

$$(\bar{x} - x^*)^T (x - x^*) = -\Delta^* a(x - \bar{x} - a^T \Delta^*)$$
$$= -\Delta^* (ax - a\bar{x} + (a\bar{x} - a_0))$$
$$= \Delta^* (a_0 - a\bar{x} - (a\bar{x} - a_0)) = 0$$

since $ax = a_0$ for all $x \in H$. Hence $(\bar{x} - x^*)$ is orthogonal to $(x - x^*)$ for all $x \in H$. See Figure 3.41. Here again, the nearest point x^* is therefore known also as the **orthogonal projection** of $\bar{x}$ in H.

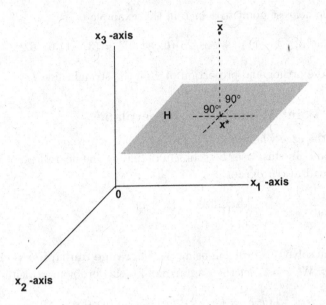

Figure 3.41 x^* is the orthogonal projection (and the nearest point to) $\bar{x}$ in H.

Nearest Point When $S = \{x : Ax = b\}$

Here consider $S = \{x : Ax = b\}$ where A is a matrix of order $m \times n$. Without any loss of generality, assume that there are no redundant equations in the system "$Ax = b$" and that it has at least one solution. The problem of finding the nearest point in S to $\bar{x}$ is the quadratic program

$$\text{Minimize} \quad (x - \bar{x})^T (x - \bar{x})$$
$$\text{subject to} \quad Ax = b.$$

Again we will use the Lagrange multiplier technique to solve this problem. Since there are m constraints, we need to use m Lagrange multipliers here, one for each constraint. Let $\pi = (\pi_1, \ldots, \pi_m)$ be the row vector of Lagrange multipliers. Then the Lagrangian function for this problem is

$$L(x, \pi) = (x - \bar{x})^T (x - \bar{x}) - \pi(Ax - b).$$

The optimum solution, x^*, π^* is then obtained by solving the system

$$\frac{\partial L(x, \pi)}{\partial x} = 2(x - \bar{x})^T - \pi A = 0$$
$$Ax - b = 0.$$

From the first equation we have

$$x^* = \bar{x} + A^T \left(\frac{(\pi^*)^T}{2} \right).$$

Substituting this in the 2nd equation we get

$$A\bar{x} - b + AA^T \left(\frac{(\pi^*)^T}{2} \right) = 0.$$

The fact that the consistent system of equations "$Ax = b$" has no redundant equations implies that the matrix AA^T of order $m \times m$ is nonsingular (see Chapter 4), and hence has an inverse. Therefore from the above equation, we have

$$\frac{(\pi^*)^T}{2} = -(AA^T)^{-1}(A\bar{x} - b)$$

and hence

$$x^* = \bar{x} - A^T(AA^T)^{-1}(A\bar{x} - b).$$

Again it can be verified that $\bar{x} - x^*$ is orthogonal to $x - x^*$ for all $x \in S = \{x : Ax = b\}$. Hence the nearest point x^* is also known as the orthogonal projection of $\bar{x}$ in S. See Figure 3.42.

Example 2:

Consider the set $S \subset R^3$ of solutions to the system of constraints "$Ax = b$" where

$$A = \begin{pmatrix} 0 & 1 & -1 \\ 0 & 1 & 1 \end{pmatrix}, \quad b = \begin{pmatrix} 2 \\ 9 \end{pmatrix}.$$

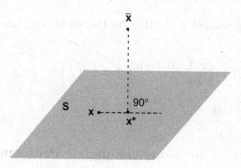

Figure 3.42 x^* is the orthogonal projection of (and the nearest point to) $\bar{x}$ in S.

Let $\bar{x} = (1, 1, 0)^T$. Then

$$A\bar{x} - b = (0, -8)^T$$

$$AA^T = \begin{pmatrix} 3 & 0 \\ 0 & 2 \end{pmatrix}, \quad (AA^T)^{-1} = \begin{pmatrix} 1/3 & 0 \\ 0 & 1/2 \end{pmatrix}$$

$$A^T(AA^T)^{-1}(A\bar{x} - b) = \begin{pmatrix} 1 & 0 \\ 1 & 1 \\ -1 & 1 \end{pmatrix} \begin{pmatrix} 1/3 & 0 \\ 0 & 1/2 \end{pmatrix} \begin{pmatrix} 0 \\ -8 \end{pmatrix} = \begin{pmatrix} 0 \\ -4 \\ -4 \end{pmatrix}.$$

So, the orthogonal projection of $\bar{x}$ in S is

$$x^* = \bar{x} - A^T(AA^T)^{-1}(A\bar{x} - b) = \begin{pmatrix} 1 \\ 1 \\ 0 \end{pmatrix} - \begin{pmatrix} 0 \\ -4 \\ -4 \end{pmatrix} = \begin{pmatrix} 1 \\ 5 \\ 4 \end{pmatrix}.$$

x^* is the nearest point in S to $\bar{x}$.

Exercises

3.20.1: Find the nearest point to $\bar{x}$ on the straight line $L = \{x : x = a + \lambda c, \lambda$ takes all real values$\}$ when

(i) $\bar{x} = (2, -3, 0, 2)$, $a = (1, -3, 2, 0)$, $c = (-1, 1, -1, 1)$.
(ii) $\bar{x} = (-4, 10, -12, -3)$, $a = (2, 1, -3, -6)$, $c = (2, -3, 3, -1)$.
(iii) $\bar{x} = (4, 9, 18, 6)$, $a = (3, -2, 6, 5)$, $c = (-2, 2, 1, 0)$.

3.20.2: Find the nearest point to $\bar{x}$ on the hyperplane $H = \{x : ax = a_0\}$ where

(i) $\bar{x} = (2, 1, 1, 1)$, $a = (1, -1, -1, 1)$, $a_0 = 6$.
(ii) $\bar{x} = (1, -2, -1, -2)$, $a = (-2, 11, 6, -3)$, $a_0 = -24$.
(iii) $\bar{x} = (-1, 0, 3)$, $a = (3, -4, 0)$, $a_0 = 6$.

3.20.3: Find the nearest point to $\bar{x}$ on the set of solutions of the system $Ax = b$ where

(i) $\bar{x} = (2, 3, 1)$, $A = \begin{pmatrix} 1 & 0 & -1 \\ -1 & 1 & 0 \end{pmatrix}$, $b = \begin{pmatrix} 3 \\ -9 \end{pmatrix}$.

(ii) $\bar{x} = (-2, -1, 1)$, $A = \begin{pmatrix} 1 & 2 & 1 \\ 2 & 1 & 0 \end{pmatrix}$, $b = \begin{pmatrix} -3 \\ -5 \end{pmatrix}$.

3.21 Procedures for Some Geometric Problems

Here we describe briefly procedures for common geometric problems that arise in many applications, based on algorithms discussed in earlier sections.

1. Identical, Parallel, Intersecting, Orthogonal, and Nonintersecting Straight Line Pairs

Let $L_1 = \{x = a + \alpha b : \alpha \text{ real}\}$, $L_2 = \{x = c + \beta d : \beta \text{ real}\}$ be two straight lines in R^n given in parametric representation. Assume that all of a, b, c, d are column vectors.

(i) L_1, L_2 are identical, i.e., they are the same straight line if $b, d, a - c$ are all scalar multiples of each other.

When L_1, L_2 are not the same straight lines, but $n = 2$:

(ii) L_1, L_2 are distinct but parallel straight lines in R^2 if b, d are scalar multiples of each other, but $a - c$ is not a scalar multiple of b.

(iii) If b is not a scalar multiple of d, then L_1, L_2 are not parallel lines in R^2 and have a unique point of intersection.

In this case, if the dot product $b^T d = \langle b, d \rangle = 0$, then L_1, L_2 are orthogonal straight lines.

When L_1, L_2 are not the same straight lines, but $n \geq 3$:

(iv) If b, d are scalar multiples of each other, and $(a - c)$ is not a scalar multiple of b, then L_1, L_2 are two parallel lines in R^n both contained in an affine space of dimension 2.

(v) If b, d are not scalar multiples of each other, and rank$\{b, d, a - c\}$ is 2, then L_1, L_2 are not parallel, but are contained in an affine space of dimension 2, and they intersect at a unique point.

In this case L_1, L_2 are an intersecting orthogonal pair of straight lines in R^n if $b^T d = 0$. If $b^T d \neq 0$, they are not orthogonal.

(vi) If b, d are not scalar multiples of each other, and rank$\{b, d, a - c\}$ is 3, then L_1, L_2 are not parallel, but are not contained together in any affine space of dimension 2 in R^n, and do not intersect.

In this case they are a nonintersecting, orthogonal pair of straight lines if $b^T d = 0$. If $b^T d \neq 0$ they are not orthogonal.

The rank of a set of vectors needed in the above conditions is briefly mentioned in Section 3.9, but detailed treatment of rank along with algorithms for computing it are given in Sections 4.2, 4.3.

So, to summarize, for any $n \geq 2$, L_1, L_2 are nonidentical parallel straight lines iff b, d are scalar multiples of each other and $a - c$ is not a scalar multiple of b.

Also, whether they intersect or not, they are an orthogonal pair iff $b^T d = 0$.

2. Angle Between Half-Lines

Given two half-lines in R^n in parametric form, $L_1^+ = \{x = a + \alpha b : \alpha \geq 0\}$, $L_2^+ = \{x = c + \beta d : \beta \geq 0\}$, whether they have a common point or not, the angle between them is defined to be the angle between their directions. If this angle is θ, we therefore have $\cos(\theta) = \langle b, d \rangle / ((\|b\|)(\|d\|))$.

The half-lines L_1^+, L_2^+ are said to make an **acute angle** if $0 \leq \theta \leq 90°$, and an **obtuse angle** if $90° < \theta \leq 180°$. Therefore L_1^+, L_2^+ make an acute angle if $\langle b, d \rangle \geq 0$; and an obtuse angle if $\langle b, d \rangle < 0$.

3. To Find the Equation for a Hyperplane Containing a Given Set of Points in R^n

Let the given set of points be $\Gamma = \{a^1, \ldots, a^k\}$ where each a^t is a column vector. Let $c_1 x_1 + \cdots + c_n x_n = c_0$ be the equation for a hyperplane containing all the points in Γ. We need to find $c = (c_1, \ldots, c_n) \neq 0$ and c_0, these are the variables in this problem. If $0 \in \Gamma$, we know that c_0 must be zero. If $0 \notin \Gamma$ we do not know the value of c_0. The system of equations that c, c_0 satisfies is:

$$
\begin{array}{cc|c}
c & c_0 & \\
\hline
a^1 & -1 & 0 \\
\vdots & \vdots & \vdots \\
a^k & -1 & 0 \\
\end{array}
$$

As in the algorithm discussed in Section 3.9, apply the algorithm of Section 1.23 on this system. This algorithm terminates in three different ways with a final tableau.

(i) **No nonbasic variable in the final tableau:** In this case $(c, c_0) = (0, 0)$ is the only solution to the above system, so there is no hyperplane in R^n that contains all the points in Γ. This happens only if dimension$(\Gamma) = n$.

(ii) **Exactly one nonbasic variable in the final tableau:** Let $(\bar{c}, \bar{c}_0)$ be the solution of the above system obtained by setting the nonbasic variable $= 1$, and finding the values of the basic variables from the final tableau. In this case dimension$(\Gamma) = n - 1$, and $\bar{c}x = \bar{c}_0$ is the equation for the unique hyperplane in R^n containing all the points in Γ.

(iii) **Two or more nonbasic variables in the final tableau:** Let $(\bar{c}^r, \bar{c}_0^r)$ be the solution of the above system obtained by setting the rth nonbasic variable $= 1$, other nonbasic variables $= 0$, and then obtaining the values of the basic variables from the final tableau; for $r = 1$ to k.

In this case dimension $(\Gamma) < (n-1)$, and there are many (in fact an infinite number) hyperplanes containing all the points in Γ. Each equation in the system

$$\bar{c}^1 x = \bar{c}_0^1$$
$$\vdots \quad \vdots$$
$$\bar{c}^k x = \bar{c}_0^k$$

represents a different hyperplane containing all the points in Γ, and any nonzero linear combination of these equations also gives such a hyperplane. The above system of equations together defines the affine hull of Γ. So, this is one way of getting the linear equation representation of the affine hull of a given set of points.

4. Checking Whether a Straight Line Lies on a Hyperplane, or Parallel to it, or Orthogonal to it

Let a, b be column vectors, and c a row vector in R^n. Let c_0 be a real number. Consider the straight line $L = \{x = a + \alpha b : \alpha \text{ real}\}$, and the hyperplane $H = \{x : cx = c_0\}$.

Consider the equation $c(a + \alpha b) = c_0$ in the parameter α.

If $cb \neq 0$, the point on L given by $\alpha = (ca - c_0)/(-cb)$ is the unique point of intersection of L and H. So in this case L and H intersect.

If c, b are scalar multiples of each other, L is orthogonal to H.

If $cb = 0$ and $ca - c_0 \neq 0$, there is no finite value of the parameter α that will satisfy the above equation. In this case, L and H are parallel.

If $cb = 0$, and $ca - c_0 = 0$, the above equation is satisfied by all values of the parameter. In this case, L lies completely in H.

A straight line lies completely on a hyperplane, if they have two common points.

5. Normal Line to a Hyperplane Through a Point on It

Let $cx = c_0$ represent a hyperplane H in R^n, and let $\bar{x}$ be a point on it. So, $c\bar{x} = c_0$.

$\{x = \bar{x} + \alpha d : \alpha \text{ real}\}$ represents a straight line through $\bar{x}$ lying in H if $d \neq 0$ and $cd = 0$.

A stright line through $\bar{x}$ is said to be the **normal line** to H through $\bar{x}$ if it is orthogonal to every straight line through $\bar{x}$ lying in H. So, the normal line to H through $\bar{x}$ is given by the representation $\{x = \bar{x} + \beta c : \beta \text{ real}\}$.

6. Parallel and Orthogonal Hyperplanes

Two hyperplanes in R^n are said to be **parallel** iff they do not intersect. This happens iff the equation for one of them can be obtained from the equation for the other by changing its RHS constant. That's why changing the RHS constant in the equation for a hyperplane is called **moving the hyperplane parallel to itself**.

So, two hyperplanes $H_1 = \{x : cx = c_0\}$, $H_2 = \{x : dx = d_0\}$ are parallel to each other if c, d are scalar multiples of each other, i.e., if there exists a nonzero number δ such that $c = \delta d$, and $c_0 \neq \delta d_0$.

Two hyperplanes are said to be **orthogonal** if their normal lines form an orthogonal pair of straight lines. So, H_1, H_2 are orthogonal iff the dot product $\langle c, d \rangle = 0$.

Exercises

3.21.1: The angle between two nonzero points is defined to be the angle between the rays of those two points. If $a = (1, a_2, -2), b = (3, 1, 2)$ what condition should a_2 satisfy so that the angle between a, b is: (i) $90°$, (ii) $<90°$, (iii) $>90°$.

3.21.2: Formulate the system of equations to find a point $x \in R^3$ that is orthogonal to both the vectors $(1, 1, -1)$ and $(1, -1, 1)$ and has Euclidean norm 1.

3.21.3: Does the straight line through $(1, 2, 3)$ and $(-1, 0, 1)$ intersect the hyperplane defined by the equation $2x_1 + 3x_2 - x_3 = 5$? If so find the point of intersection. Do the same for the same straight line and the hyperplane defined by $6x_1 + 8x_2 - 14x_3 = 15$.

How many points of intersection does the straight line through $(2, 3, 0, 7)$ and $(1, -8, 5, 0)$ have with the hyperplane defined by $x_1 + x_2 - x_3 - x_4 = -2$?

3.21.4: Find a hyperplane in R^4 containing all the points $(-1, 1, 0, 0)$, $(0, 1, -1, 0)$, $(0, 0, -1, 1)$, $(0, 1, 0, 1)$, $(-1, 0, -1, 0)$, $(1, 2, 1, 2)$.

Do all the points $(2, 0, 1)$, $(0, 2, 1)$, $(1, 1, 1)$, $(1, 1, -1)$ lie on a hyperplane in R^3? If not is there a half-space in R^3 that contains all of them? If so find one such half-space.

In R^3 find a hyperplane if possible, or a half-space otherwise, that contains all the points in the set $\{(1, 2, 0), (1, 0, 2), (1, 1, 0), (1, 0, 1), (1, 1, 1),$ $(2, 1, 0), (2, 0, 1), (0, 1, 1)\}$. Can you do this if the requirement is to contain all but one of the points in the set?

3.21.5: Find a hyperplane in R^3 that is orthogonal to both the hyperplanes represented by the equations $x_1 + x_2 - x_3 = 6$, $x_1 - x_2 + x_3 = -2$, and passes through the point $(1, 1, 1)$.

Find the equation of the hyperplane in R^3 that contains the point $(1, -1, 1)$ and contains the entire straight line joining $(2, 1, 1)$ and $(1, 2, 1)$.

3.21.6: Check whether the two straight lines joining the following pairs of points are parallel.

(Yes) $\{\{(1, -1, 0, 7), (2, 1, 3, 6)\}, \{(5, 5, 5, 5), (2, -1, -4, 8)\}\}$, $\{\{(10, 20, 30, 40), (9, 19, 31, 41)\}, \{(40, 40, 40, 40), (30, 30, 50, 50)\}\}$

(No) $\{\{(-6, -7, 0, 3), (-5, -6, 1, 4)\}, \{(14, 13, 20, 23), (34, 33, 40, 43)\}\}$, $\{\{(1, 1, 2), (2, 0, 3)\}, \{(5, 6, 7), (6, 7, 8)\}\}$.

3.21.7: Check whether the following pair of straight lines, each one specified by a pair of points on it, are orthogonal.

(Yes) $\{\{(1, 2, 1, 1), (2, 1, 0, 2)\}, \{(30, 40, 50, 60), (28, 38, 53, 63)\}\}$, $\{\{(10, 10, 10, 10), (11, 9, 10, 10)\}, \{(11, 9, 10, 10), (11, 9, 11, 12)\}\}$.

(No) $\{\{(2, 1, 1, 1), (2, 3, 0, 1)\}, \{(5, 5, 5, 5), (3, 6, 4, 15)\}\}$, $\{\{(2, 1, 2), (3, 3, 5)\}, \{(7, 0, -1), (1, 1, 0)\}\}$.

3.21.8: Check whether the straight line L lies on the hyperplane H in the following.

(Yes) $\{L$ joins $(1, -2, 3)$ and $(2, -1, 2)$, $H = \{x : 2x_1 + 3x_2 + 5x_3 = 11\}\}$, $\{L$ joins $(2, 1, 4, 3)$ and $(3, 0, 3, 4)$, $H = \{x : 2x_2 + 4x_3 + 6x_4 = 36\}\}$.

(No) $\{L$ joins $(2, 1, 5)$ and $(1, 0, 6)$, $H = \{x : 3x_1 + 2x_2 + 4x_3 = 28\}\}$, $\{L$ joins $(10, 15, 20, 25)$ and $(11, 16, 21, 26)$, $H = \{x : 3x_1 + 2x_2 - 2x_3 - 3x_4 = -50\}\}$.

3.21.9: Find the normal lines to the hyperplanes H at points $\bar{x}$ given on them.

$$\bar{x} = (1, -1, -1, 2)^T \text{ on } H = \{x : 2x_1 - 3x_2 + 4x_4 = 13\}.$$
$$\bar{x} = (0, 3, -7)^T \text{ on } H = \{x : x_1 - x_3 = 7\}.$$

3.21.10: Find the equation of a hyperplane containing the point $(5, 0, 3, -3)^T$ for which the normal at that point is the straight line $\{x = (1, 2, -1, -1)^T + \alpha(2, -1, 2, -1)^T : \alpha \text{ real}\}$.

3.21.11: Check whether the following pairs of hyperplanes H_1 and H_2 are parallel.

(Yes) $\{H_1 = \{x : 2x_1 - 6x_2 + x_3 = -10\}, H_2 = \{x : -6x_1 + 18x_2 - 3x_3 = 30\}\}$, $\{H_1 = \{x : 4x_1 - 10x_4 = 6\}, H_2 = \{x : 160x_1 - 400x_4 = 240\}\}$, $\{H_1$ is the hyperplane through the points $\{(1, 1, 1), (2, 0, 1), (-1, 2, 0)\}$, H_2 is the hyperplane through the points $\{(1, 1, 0), (2, 1, 1), (1, 2, 1)\}\}$.

(No) $\{H_1 = \{x : 6x_1 - 7x_2 + 18x_3 - 19x_4 = 6\}, H_2 = \{x : 6x_1 - 7x_2 + 18x_3 - 18x_4 = 6\}\}$, $\{H_1 = \{x : 4x_1 - 2x_2 + 8x_3 - 7x_4 + 20x_5 = 500\}$, $H_2 = \{x : 200x_1 - 100x_2 + 400x_3 - 350x_4 + 990x_5 = 25,000\}\}$, $\{H_1$ is the hyperplane through the points $\{(1,1,1), (1,0,2), (-1,-3,1)\}$, H_2 is the hyperplane through the points $\{(1,1,1), (2,0,2), (1,2,-1)\}\}$.

3.21.12: Find the angle between L_1, L_2 in the following pairs and check whether they make an acute or obtuse angle.

(Acute) $\{L_1$ joins $\{(3,2,-1,2)$ to $(4,4,-2,2)\}$, L_2 joins $\{(5,5,5,5)$ to $(4,7,4,7)\}\}$, $\{L_1$ joins $\{(1,2,2,1)$ to $(2,1,1,2)\}$, L_2 joins $\{(3,1,1,3)$ to $(2,0,2,4)\}\}$.

(Obtuse) $\{L_1$ joins $\{(7,-4,3,2)$ to $(8,-3,4,3)\}$, L_2 joins $\{(9,10, -1,-2)$ to $(8,9,-2,-3)\}\}$, $\{L_1$ joins $\{(0,0)$ to $(1,0)\}$, L_2 joins $\{(0,0)$ to $(-1,0)\}\}$.

3.21.13: Determine whether the straight line L and the hyperplane H in the following pairs are parallel.

(Yes) $\{L$ joins $\{(1,1,1), (2,1,1)\}$, H is the hyperplane through $\{(1,1,0), (1,0,0), (0,1,0)\}\}$, $\{L$ joins $\{(5,2,-2,2), (6,1,-2,2)\}$, H is the hyperplane through $\{(1,1,2,1), (-1,1,1,0), (-1,-1,0,-1), (1,0,-1,-2)\}\}$.

(No) $\{L$ joins $\{(1,1,1,1), (2,2,3,2)\}$, H is the hyperplane through $\{(1,1,0,0), (0,1,1,0), (1,2,1,0), (0,0,1,1)\}\}$, $\{L$ joins $\{(5,2,-2,2), (6,1,-2,2)\}$, H is the hyperplane through $\{(1,0,1,1), (1,1,0,1), (0,0,1,2), (1,-1,1,-1)\}\}$.

3.21.14: Check whether the following pairs of hyperplanes H_1, H_2 are orthogonal.

(Yes) $\{H_1$ is the hyperplane through $\{(1,1,0), (2,0,1), (0,2,1)\}$, H_2 is the hyperplane through $\{(1,0,1), (2,1,1), (1,1,2)\}\}$, $\{H_1$ is the hyperplane through $\{(1,1,1,0), (0,1,1,1), (1,0,0,1), (0,0,1,1)\}$, H_2 is the hyperplane through $\{(0,-1,-1,0), (2,0,0,1), (1,-1,0,0), (1,0,-1,1)\}\}$.

(No) $\{H_1$ is the hyperplane through $\{(1,1,0,0), (2,0,1,0), (2,0,0,1), (1,1,1,1)\}$, H_2 is the hyperplane through $\{(0,1,1,0), (1,2,0,0), (1,0,2,1), (1,1,1,1)\}\}$, $\{H_1$ is the hyperplane defined by the equation $x_1 + x_2 - x_3 = 2$, H_2 is the hyperplane through $\{(-1,-1,0,0), (0,-1,-1,0), -(0,0,-1,-1), (0,-1,-1,0)\}\}$.

3.21.15: Find a hyperplane through the following sets of vectors Γ.

(None) $\Gamma = \{(1,1,0), (1,0,1), (0,1,1), (1,1,1)\}$.

(Unique) $\Gamma = \{(1,1,1,0), (0,1,1,1), (1,0,0,1), (0,0,1,1)\}$.

(Nonunique) $\Gamma = \{(2,0,1), (1,1,2), (3,-1,0)\}$.

3.21.16: Find the equation of the hyperplane through the point $(7, -8, -6)$, parallel to the hyperplane containing the points $(1, 1, 0)$, $(0, 1, 1)$, $(1, 0, 1)$.

3.22 Additional Exercises

3.22.1: The general equation for a circle in the x_1, x_2-coordinate plane is $a_0(x_1^2 + x_2^2) + a_1 x_1 + a_2 x_2 + a_3 = 0$, where $a_0 \neq 0$. Given three points which do not all lie on a straight line, there is a unique circle containing all of them, develop a method for computing the equation of that circle using the algorithm in Section 1.23. Use that method to find the equations of the circles containing all the points in the following sets: $\{(1, 0), (0, 1), (1, 2)\}$, $\{(1, 1), (2, 3), (3, 5)\}$, $\{(2, 2), (2, -2), (4, 0)\}$.

3.22.2: x^r, $r = 1, 2, 3$ are the coordinates of the vertices of a triangle ABC in R^2 in that order. P, Q are the midpoints of the sides AB, AC of this triangle. Prove that the line PQ is parallel to the line BC, and that the length of the line segment PQ is half the length of the side BC of the triangle.

Let D, E be the points on the sides AB, AC of this triangle that are a fixed fraction α of the length of that side away from A, for some $0 < \alpha < 1$. Prove that the line segment DE is parallel to the side BC of the triangle, and that its length is α times the length of the side BC.

3.22.3: x^r, $r = 1, 2, 3, 4$ are the coordinates of the vertices of a quadrilateral ABCD in R^2 in that order. P, Q, R, S are the midpoints of the sides of this quadrilateral. Prove that P, Q, R, S are the vertices of a parallelogram. Also show that the diagonals of this parallelogram intersect at the centroid of the quadrilateral which is $(x^1 + x^2 + x^3 + x^4)/4$.

3.22.4: A, B, C, D, E are the points $(4, 4, 0)$, $(8, 2, 0)$, $(8, 8, 0)$, $(4, 8, 0)$, $(6, 6, 4)$ in R^3 respectively. They are the vertices of a **pyramid** in R^3. The 4 vertices A, B, C, D all lie on a hyperplane which is the base of this pyramid. E lies above this base and is known as the apex of the pyramid.

Let P, Q, R, S be the midpoints of the edges AE, BE, CE, DE of this pyramid. Show that all these four points lie on a hyperplane which is parallel to the base of the pyramid.

3.22.5: Consider a parallelogram in R^2. Set up the axes of coordinates with one of the vertices of the parallelogram as the origin. Let $a = (a_1, a_2)$, $b = (b_1, b_2)$ be the coordinate vectors for the two vertices of the parallelogram adjacent to 0. By parallelogram law, the 4th vertex must be $a + b$.

Using this framework, show that the diagonals of the parallelogram intersect at the point which is the midpoint of both the diagonals.

3.22.6: Let a, b be two nonzero vectors in R^n. Show that the parallelogram $\{x = \alpha a + \beta b : 0 \leq \alpha \leq 1, 0 \leq \beta \leq 1\}$ is a convex set.

3.22.7: **Finding the path of a flying bird:** Axes of coordinates have been set up to trace the path of a bird flying high up in the sky. The x_1, x_2-axes are on the ground, and the x_3-axis is along the vertical altitude. The unit of measurement along each axis is 1 foot. At time 0-minutes, she is at the point $(100, -200, 400)$. She is flying in the direction $(-2, 1, 0)$ at a constant speed of 200 feet per minute. At her altitude, the wind is blowing in the direction of $(2, -3, 0)$ at a constant speed of 1500 feet per minute. Derive a parametric representation of her path, and find her location at time 10 minutes.

Reference

[3.1] C. W. Curtis, *Linear Algebra: An Introductory Approach* (Springer, 1991).

Chapter 4

Numerical Linear Algebra

4.1 Linear Dependence, Independence of Sets of Vectors

Given a set of vectors (all of them column vectors, or all of them row vectors; here we will assume that they are all column vectors) $\Delta = \{a^1, \ldots, a^k\}$ from R^n, it is sometimes important to be able to check whether any of the vectors in this set can be expressed as a linear combination of the others.

In Section 1.9 we encountered this problem in order to check whether a system of linear equations contains a redundant constraint that can be eliminated without effecting the set of solutions. For this, the coefficient vector of one of the equations in the system should be a linear combination of the others. In Section 3.9 we encountered this problem again in the process of developing parametric representations for the linear hull of a given set of vectors using the smallest number of parameters.

Broadly speaking, we will say that the set $\Delta = \{a^1, \ldots, a^k\}$ of vectors is **linearly dependent** if one of the vectors in it can be expressed as a linear combination of the others, **linearly independent** otherwise. For example, if a^1 can be expressed this way as in

$$a^1 = \alpha_2 a^2 + \cdots + \alpha_k a^k.$$

This leads to

$$-a^1 + \alpha_2 a^2 + \cdots + \alpha_k a^k = 0.$$

Now defining $\alpha_1 = -1$, this is

$$\alpha_1 a^1 + \alpha_2 a^2 + \cdots + \alpha_k a^k = 0$$

where $(\alpha_1, \ldots, \alpha_k) \neq 0$ (because $\alpha_1 = -1$, the vector $(\alpha_1, \ldots, \alpha_k) \neq 0$). In keeping with the name that we gave for a similar relation among constraints, we will call a relation like this among vectors in the set Δ a **linear dependence relation** as long as the vector of coefficients in it $(\alpha_1, \ldots, \alpha_k) \neq 0$.

When $k = 1$, i.e., Δ contains only one vector a^1 in it, there is no possibility of expressing a vector in Δ as a linear combination of the others because the only vector in Δ is a^1. In this special case, for technical reasons of consistency, we will define the singleton set Δ to be linearly dependent if the only vector a^1 in it is 0, linearly independent if $a^1 \neq 0$. With this convention, we again see that in this special case also, the set Δ is linearly dependent if there is a linear dependence relation for it (because if $a^1 = 0$, then $\alpha_1 a^1 = 0$ for all $\alpha_1 \neq 0$).

The only other special set to consider is the empty set $\emptyset$, i.e., the set containing no vectors at all. Again for technical reasons of consistency, we define the empty set to be linearly independent. So, here is the mathematical definition.

Definition: The set of vectors $\Delta = \{a^1, \ldots, a^k\}$ in R^n is **linearly dependent** iff it is possible to express the 0-vector as a linear combination of vectors in it with at least one nonzero coefficient; i.e., iff there exists $\alpha = (\alpha_1, \ldots, \alpha_k) \neq 0$ such that

$$\alpha_1 a^1 + \alpha_2 a^2 + \cdots + \alpha_k a^k = 0. \tag{4.1}$$

Since each vector here is from R^n, this is a homogeneous system of n equations in k variables. In this case (4.1) is known as a **linear dependence relation** for Δ. The set Δ is said to be **linearly independent** iff there is no linear dependence relation for it. $\bowtie$

If the set Δ has two or more vectors in it and it is linearly dependent, then from a linear dependence relation for it, we can find an expression for one of the vectors in Δ as a linear combination of the others. For example, in (4.1), since $\alpha = (\alpha_1, \ldots, \alpha_k) \neq 0$, at least one of the α_t must be nonzero. If $\alpha_1 \neq 0$, then from (4.1) we have

$$a^1 = -\frac{1}{\alpha_1}(\alpha_2 a^2 + \cdots + \alpha_k a^k).$$

Example 1:

Let

$$\Delta = \left\{ a^1 = \begin{pmatrix} 1 \\ 0 \\ 0 \end{pmatrix}, \ a^2 = \begin{pmatrix} 0 \\ 1 \\ 0 \end{pmatrix}, \ a^3 = \begin{pmatrix} 0 \\ 0 \\ 1 \end{pmatrix}, \ a^4 = \begin{pmatrix} 5 \\ -6 \\ 0 \end{pmatrix} \right\}.$$

Since $-5a^1 + 6a^2 + a^4 = 0$, Δ is linearly dependent and this equation is a linear dependence relation for it. Also, from this relation we have

$$a^1 = -(6/5)a^2 - (1/5)a^4$$
$$a^2 = (5/6)a^1 - (1/6)a^4$$
$$a^4 = 5a^1 - 6a^2$$

Example 2:

Let

$$\Delta = \left\{ a^1 = \begin{pmatrix} 1 \\ 0 \\ 0 \end{pmatrix}, \ a^2 = \begin{pmatrix} 0 \\ 1 \\ 0 \end{pmatrix}, \ a^3 = \begin{pmatrix} 0 \\ 0 \\ 1 \end{pmatrix} \right\}.$$

The system of equations (4.1) for this set Δ here is given below in detached coefficient form.

α_1	α_2	α_3	
1	0	0	0
0	1	0	0
0	0	1	0

This homogeneous system has the unique solution $\alpha = (\alpha_1, \alpha_2, \alpha_3) = 0$. So, there is no linear dependence relation for this set Δ, hence this set Δ is linearly independent.

Using the same argument, it can be verified that for any $n \geq 2$, the set of n unit vectors in R^n is linearly independent.

We will discuss two algorithms for checking linear dependence or independence of a given set of vectors in Section 4.3. When the set is linearly dependent, these algorithms will always output one or more linear dependence relations for it.

4.2   Rank and Bases for a Set of Vectors

From the definition it is clear that any subset of a linearly independent set of vectors is linearly independent. So, the property of linear independence of a set of vectors is always preserved when vectors are deleted from it, but may be lost if new vectors are included in it. For example, the linearly independent set $\{(1,0,0)^T, (0,1,0)^T, (0,0,1)^T\}$ remains linearly independent when vectors are deleted from it, but becomes linearly dependent when any vector from R^3 is included in it.

Maximal Linearly Independent Subsets

Let $\Gamma \subset R^n$ be a set of vectors. As defined in Section 3.9, a subset $E \subset \Gamma$ is said to be a **maximal linearly independent subset** of γ if E is linearly independent, and either $E = \Gamma$, or $E \cup \{a\}$ is linearly dependent for all $a \in \Gamma \backslash E$.

If $\Gamma = \emptyset$ or $\Gamma = \{0\}$, the only maximal linearly independent subset of Γ is the emptyset.

If Γ has at least one nonzero vector, one procedure for finding a maximal linearly independent subset of Γ is the following simple **augmentation scheme:**

> Start the scheme with the subset consisting of one nonzero vector from Γ. Always maintain the subset linearly independent. Augment it by one new element of Γ at a time, making sure that linear independence holds; until no new vectors from Γ can be included, at which stage the subset becomes a maximal linearly independent subset of Γ.

But this scheme is not computationally efficient. We will discuss an efficient method for finding a maximal linearly independent subset based on pivot operations, in the next section.

Minimal Spanning Subsets

Let $\Gamma \subset R^n$ be a set of vectors. Let $F \subset \Gamma$ satisfy the property that every vector in Γ can be expressed as a linear combination of vectors in F. Then F is said to be a **spanning subset of** Γ (or, to have the property of **spanning** Γ). As an example, let

$$\Gamma_1 = \left\{ a^1 = \begin{pmatrix} 1 \\ 0 \\ 0 \end{pmatrix}, \ a^2 = \begin{pmatrix} -2 \\ 0 \\ 0 \end{pmatrix}, \ a^3 = \begin{pmatrix} 1 \\ 1 \\ 0 \end{pmatrix}, \right.$$

$$\left. a^4 = \begin{pmatrix} -3 \\ -4 \\ 0 \end{pmatrix}, \ a^5 = \begin{pmatrix} 1 \\ 1 \\ -1 \end{pmatrix} \right\}.$$

Then the subset $F_1 = \{a^1, a^3, a^5\}$ is a spanning subset of Γ_1. But $F_2 = \{a^1, a^3, a^4\}$ is not a spanning subset of Γ_1, because a^5 in Γ_1 cannot be expressed as a linear combination of vectors in F_2.

Given $\Gamma \subset R^n$, it is clear from the definition that a spanning subset of Γ retains its spanning property when more vectors are included in it, but may lose the spanning property if vectors are deleted from it.

A subset $F \subset \Gamma$ is said to be a **minimal spanning subset of Γ** if it is a spanning subset of Γ, but none of the proper subsets of F have the spanning property for Γ.

The definition itself suggests a simple deletion scheme for finding a minimal spanning subset of Γ. The scheme maintains a subset of Γ. Initially that subset is Γ itself. In each step, delete one vector from the subset making sure that the remaining subset always has the spanning property for Γ (for this it is enough if the vector to be deleted is in the linear hull of the remaining elements in the subset at that stage). When you reach a stage where no more vectors can be deleted from the subset without losing the spanning property for Γ (this will happen if the subset becomes linearly independent), you have a minimal spanning subset of Γ.

Again this scheme is computationally expensive. We will discuss an efficient method for finding a minimal spanning subset based on row operations in the next section.

The Rank of a Set of Vectors, Bases

We have the following results.

Result 4.2.1: Maximal linear independence and minimal spanning property imply each other: *Every maximal linearly independent subset of a set $\Gamma \subset R^n$, is a minimal spanning subset for Γ and vice versa.*

Result 4.2.2: Rank of a set of vectors: *All minimal spanning subsets of $\Gamma \subset R^n$ have the same cardinality (i.e., same number of elements in it); it is the same as the cardinality of any maximal linearly independent subset of Γ; as mentioned in Section 3.9 earlier, this number is called the **rank** of the set Γ. It is also the dimension of the linear hull of Γ.*

Definition: Every maximal linearly independent subset of Γ or a minimal spanning subset of Γ is called a **basis** for Γ. ⋈

Thus a basis for a set of vectors Γ in R^n is a linearly independent subset satisfying the property that all vectors in the set Γ not in the basis can be represented as linear combinations of vectors in the basis. A set of vectors Γ may have many bases, but every basis for Γ consists of the same number of vectors, this number is the rank of Γ.

When referring to a basis B for a set $\Gamma \subset R^n$, vectors in B are called **basic vectors**, and vectors in Γ not in B are called **nonbasic vectors**. In most applications, the basic vectors themselves are arranged in some order and called the **first basic vector, second basic vector**, etc. Then the word basis refers to this ordered set with the basic vectors arranged in this order.

Let I be the unit matrix of order n. Its set of column vectors $\{I_{.1}, I_{.2}, \ldots, I_{.n}\}$ is a basis for R^n called the **unit basis**. For example, for $n = 3$

$$\left\{ \begin{pmatrix} 1 \\ 0 \\ 0 \end{pmatrix}, \begin{pmatrix} 0 \\ 1 \\ 0 \end{pmatrix}, \begin{pmatrix} 0 \\ 0 \\ 1 \end{pmatrix} \right\}$$

is the unit basis for R^3. Hence we have the following result.

Result 4.2.3: Ranks of subsets of R^n: *The rank of R^n is n. Every basis for R^n always consists of n vectors. And therefore any set of $n + 1$ or more vectors from R^n is always linearly dependent. The rank of any subset of R^n is $\leq n$.*

Result 4.2.4: Basic vectors span each nonbasic vector uniquely: *Every nonbasic vector can be expressed as a linear combination of basic vectors in a unique manner.*

By definition, every nonbasic vector can be expressed as a linear combination of basic vectors. We will now show that this expression is always unique.

Let the basis $B = \{A_1, \ldots, A_r\}$ and let A_{r+1} be a nonbasic vector. Suppose A_{r+1} can be expressed as a linear combination of vectors in B in two different ways, say as

$$A_{r+1} = \sum_{t=1}^{r} \beta_t A_t \quad \text{and as} \quad A_{r+1} = \sum_{t=1}^{r} \gamma_t A_t$$

where $(\beta_1, \ldots, \beta_r) \neq (\gamma_1, \ldots, \gamma_r)$. So, if $\alpha = (\alpha_1, \ldots, \alpha_r) = (\beta_1 - \gamma_1, \ldots, \beta_r - \gamma_r)$ then $\alpha \neq 0$ and we have

$$\sum_{t=1}^{r} \alpha_t A_t = \sum_{t=1}^{r} \beta_t A_t - \sum_{t=1}^{r} \gamma_t A_t = A_{r+1} - A_{r+1} = 0.$$

This is a linear dependence relation for $B = \{A_1, \ldots, A_r\}$, a contradiction since B is a basis. So, the expression for the nonbasic vector A_{r+1} as a linear combination of basic vectors must be unique.

Definition: The vector of coefficients in the expression of a nonbasic vector as a linear combination of basic vectors, arranged in the same order in which the basic vectors are arranged in the basis, is called **the representation of this nonbasic vector in terms of this basis.**

Thus if the basis $B = \{A_1, \ldots, A_r\}$ with the vectors arranged in this order, and the nonbasic vector $A_{r+1} = \sum_{t=1}^{r} \beta_t A_t$, then $(\beta_1, \ldots, \beta_r)$ is the representation of A_{r+1} in terms of the basis B. As an example, let

$$\Gamma_1 = \left\{ a^1 = \begin{pmatrix} 1 \\ 0 \\ 0 \end{pmatrix}, \; a^2 = \begin{pmatrix} -2 \\ 0 \\ 0 \end{pmatrix}, \; a^3 = \begin{pmatrix} 1 \\ 1 \\ 0 \end{pmatrix}, \right.$$

$$\left. a^4 = \begin{pmatrix} -3 \\ -4 \\ 0 \end{pmatrix}, \; a^5 = \begin{pmatrix} 1 \\ 1 \\ -1 \end{pmatrix} \right\}.$$

Let $F_1 = \{a^1, a^3, a^5\}$. F_1 is a basis for the set Γ_1. The nonbasic vector $a^4 = a^1 - 4a^3$, and hence the representation of a^4 in terms of the basis F_1 for Γ_1 is $(1, -4, 0)^T$.

4.3 Two Algorithms to Check Linear Dependence, Find Rank and a Basis for a Given Set of Vectors

Let $\Delta = \{a^1, \ldots, a^k\}$ be a given set of vectors in R^n. Here we discuss two algorithms using pivot steps, for checking whether Δ is linearly dependent, find its rank, a basis, and the representation of each nonbasic vector in terms of that basis. Algorithm 1 is based on using the memory matrix discussed in Section 1.12. Algorithm 2 operates directly on the linear dependence relation for Δ treated as a system of homogeneous linear equations in the coefficients, and tries to find a nonzero solution for it using the method discussed in Section 1.23.

Algorithm 1: Checking Linear Dependence Using the Memory Matrix

In this algorithm, row operations involved in G or GJ pivot steps are carried out. Because it is based on row operations, it is necessary to enter each vector in Δ as a row vector in the original tableau, even though the vectors in Δ may be column vectors.

BEGIN

Step 1: Enter the vectors $a^1, \ldots, a^k$ in Δ in that order, as row vectors in the original tableau (if the vectors in Δ are column vectors, you will enter their transposes as rows, but still call them by the same names).

Set up a memory matrix (initialy unit matrix of order k) with its columns labeled $a^1, \ldots, a^k$ in that order, on the left side of the tableau.

Step 2: Try to perform GJ pivot steps (see Sections 1.15, 1.16), or G pivot steps with row interchanges as necessary (see Sections 1.19, 1.20, these lead to a more efficient algorithm) in each of the tableau in any order.

A pivot step is not carried out with a row r as pivot row only if that row has become the 0-row (i.e., all the entries on the right hand portion of it in the current tableau are zero) when it has come up for consideration as pivot row.

Step 3: If you are only interested in checking whether Δ is linearly dependent and obtain a single linear dependence relation for it when it is, then this algorithm can be terminated when a 0-row shows up in the algorithm for the first time. Then the entries in the memory part (the left hand part) of the 0-row in the final tableau are the coefficients for a linear dependence relation for Δ.

If pivot steps are carried out in every row of the tableau (i.e., no 0-rows in the final tableau, then Δ is linearly independent, its rank is k, and Δ is the basis for itself.

If only $s < k$ pivot steps are performed, then rank of Δ is s, the subset $\{a^i : i \text{ such that the } i\text{-th row in the final tableau is not a 0-row}\}$ (these are the vectors corresponding to rows in which pivot steps are carried out) is a basis for Δ. Also, from the linear dependence relation of each 0-row, you can get the representation of the nonbasic variable corresponding to that row in terms of the basis.

END

Example 1:

Consider the set $\Gamma = \{A_{.1}, \text{ to } A_{.5}\}$ of column vectors in R^6 of the matrix given below. We will find the rank and a basis for this set.

$A_{.1}$	$A_{.2}$	$A_{.3}$	$A_{.4}$	$A_{.5}$
1	2	0	2	1
2	4	0	4	−1
−1	−2	3	1	−1
1	2	1	3	1
0	0	2	2	−2
−2	−4	2	−2	1

We enter each of the vectors $A_{.1}$ to $A_{.5}$ as a row in a original tableau (the top one among those given below), set up the memory matrix, and apply the algorithm. The various tableaus obtained are given below. We carry out G pivot steps, no row interchanges were necessary. PR (pivot row), PC (pivot column) are indicated in each tableau, and the pivot element is boxed. "0-row" indicates 0-rows discovered. In the bottommost tableau, PR, PC are indicated, but since PR there is the last row, and the pivot step is the G pivot step, there is no work to be done. Hence the bottommost tableau is the final tableau.

Memory matrix					Tableau						
$A_{.1}$	$A_{.2}$	$A_{.3}$	$A_{.4}$	$A_{.5}$							
1	0	0	0	0	$\boxed{1}$	2	−1	1	0	−2	PR
0	1	0	0	0	2	4	−2	2	0	−4	
0	0	1	0	0	0	0	3	1	2	2	
0	0	0	1	0	2	4	1	3	2	−2	
0	0	0	0	1	1	−1	−1	1	−2	1	
					PC↑						
1	0	0	0	0	1	2	−1	1	0	−2	
−2	1	0	0	0	0	0	0	0	0	0	0-row
0	0	1	0	0	0	0	3	$\boxed{1}$	2	2	PR
−2	0	0	1	0	0	0	3	1	2	2	
−1	0	0	0	1	0	−3	0	0	−2	3	
								PC↑			
1	0	0	0	0	1	2	−1	1	0	−2	
−2	1	0	0	0	0	0	0	0	0	0	0-row
0	0	1	0	0	0	0	3	1	2	2	
−2	0	−1	1	0	0	0	0	0	0	0	0-row
−1	0	0	0	1	0	−3	0	0	$\boxed{−2}$	3	PR
								PC↑			

The nonzero rows in the right-hand portion of the final (bottommost) tableau are rows 1, 3, 5. Hence $\{A_{.1}, A_{.3}, A_{.5}\}$ is a basis for the original set Γ in this example, and the rank of Γ is 3. From rows 2 and 4 in the final tableau we have the linear dependence relations

$$-2A_{.1} + A_{.2} = 0 \quad \text{and} \quad -2A_{.1} - A_{.3} + A_{.4} = 0.$$
$$\text{So,} \quad A_{.2} = 2A_{.1} \quad \text{and} \quad A_{.4} = 2A_{.1} + A_{.3}.$$

So, the representations of the nonbasic vectors $A_{.2}, A_{.4}$ in terms of the basis $\{A_{.1}, A_{.3}, A_{.5}\}$ are $(2, 0, 0)^T$ and $(2, 1, 0)^T$ respectively.

Example 2:

Consider the set $\Gamma = \{a^1 = (1, 0, -1, 1), a^2 = (-1, -1, 0, 1), a^3 = (0, 1, 1, -1)\}$ of row vectors in R^4. We will find the rank and a basis for this set. We enter each of these vectors as rows in the original tableau (the top one among those given below), set up the memory matrix, and apply the algorithm using GJ pivot steps. The various tableaus obtained are given below. PR (pivot row), PC (pivot column) are indicated in each tableau, and the pivot element is boxed. The bottommost tableau is the final tableau.

a^1	a^2	a^3					
1	0	0	$\boxed{1}$	0	-1	1	PR
0	1	0	-1	-1	0	1	
0	0	1	0	1	1	-1	
			PC↑				
1	0	0	1	0	-1	1	
1	1	0	0	$\boxed{-1}$	-1	2	PR
0	0	1	0	1	1	-1	
			PC↑				
1	0	0	1	0	-1	1	
-1	-1	0	0	1	1	-2	
1	1	1	0	0	0	$\boxed{1}$	PR
					PC↑		
0	-1	-1	1	0	-1	0	
1	1	2	0	1	1	0	
1	1	1	0	0	0	1	

Pivot steps are carried out in every row of the tableau (i.e., no 0-rows in the final tableau), therefore the set Γ here is linearly independent, and its rank is 3.

Algorithm 2: Checking Linear Dependence by Row Reduction of th System of Equations in the Linear Dependence Relation

BEGIN

Step 1: Let $\Delta = \{a^1, \ldots, a^k\} \subset R^n$ be the set of vectors whose linear dependence is to be checked. This algorithm treats the linear dependence relation

$$\alpha_1 a^1 + \cdots + \alpha_k a^k = 0$$

as a system of n homogeneous equations in the unknown coefficients $\alpha_1, \ldots, \alpha_k$; and tries to find nonzero solutions to it by the algorithm discussed in Section 1.23 using GJ pivot steps. The system in detached coefficient tableau form is

Original tableau

α_1	$\cdots$	α_j	$\cdots$	α_k	
a^1	$\cdots$	a^j	$\cdots$	a^k	0

In this tableau, each of the vectors in Δ is entered as a column vector (if it is given originally as a row vector, enter its transpose). For $j = 1$ to k, the vector a^j is the column vector of the variable α_j in the original tableau.

Step 2: In the final tableau obtained at the end of the method if every variable α_j is a basic variable (i.e., if there are no nonbasic variables at this stage), then $\alpha = (\alpha_1, \ldots, \alpha_k) = 0$ is the unique solution for the system. In this case Δ is linearly independent, its rank is k, and a basis for Δ is Δ itself.

On the other hand, if there is at least one nonbasic variable in the final tableau, the system has nonzero solutions, and Δ is linearly dependent. Suppose there are p rows in the final tableau with α_{i_t} as the tth basic variable for $t = 1$ to p. Rearrange the variables and their column vectors in the final tableau so that the basic variables appear in their proper order first, followed by nonbasic variables.

Final tableau after rearrangement of variables

Basic Vars.	α_{i_1}	α_{i_2}	$\cdots$	α_{i_p}	α_{j_1}	$\cdots$	$\alpha_{j_{k-p}}$	
α_{i_1}	1	0	$\cdots$	0	$\bar{a}_{1,j_1}$	$\cdots$	$\bar{a}_{1,j_{k-p}}$	0
α_{i_2}	0	1	$\cdots$	0	$\bar{a}_{2,j_1}$	$\cdots$	$\bar{a}_{2,j_{k-p}}$	0
$\vdots$	$\vdots$	$\vdots$	$\vdots$	$\vdots$	$\vdots$		$\vdots$	$\vdots$
α_{i_p}	0	0	$\cdots$	1	$\bar{a}_{p,j_1}$	$\cdots$	$\bar{a}_{p,j_{k-p}}$	0

Then the rank of Δ is $p =$ the number of pivot steps we were able to carry out on the system.

The basis (maximal linearly independent subset of vectors) of Δ identified by the method is $\{a^{i_1}, \ldots, a^{i_p}\}$ with a^{i_t} as the tth basic vector for $t = 1$ to p. These correspond to columns of the tableau in which pivot steps were carried out during the method.

For each $u = 1$ to $k - p$, by fixing the nonbasic variable $\alpha_{j_u} = 1$, and all other nonbasic variables $= 0$; leads to the nonzero solution

$$\alpha_{i_t} = -\bar{a}_{t,j_u}, \quad \text{for } t = 1 \text{ to } p,$$

$$\alpha_{j_u} = 1, \quad \text{and other nonbasic variables } \alpha_{j_v} = 0 \text{ for } v \neq u$$

for the linear dependence relation for Δ. From this we conclude that the nonbasic vector

$$a^{j_u} = \sum_{t=1}^{p} \bar{a}_{t,j_u} a^{i_t}.$$

Hence, for $u = 1$ to $k - p$, the representation of the nonbasic vector a^{j_u} in terms of the basis $(a_{i_1}, \ldots, a^{i_p})$ is $(\bar{a}_{1,j_u}, \ldots, \bar{a}_{p,j_u})$.

Hence, the column vector of the nonbasic variable α_{j_u} in the final tableau, called its **updated column**, is the representation of its column vector in the original tableau in terms of the basis obtained.

END

Example 3:

Let Δ be the set of column vectors of the matrix given below, with names as indicated there.

$A_{.1}$	$A_{.2}$	$A_{.3}$	$A_{.4}$	$A_{.5}$
1	2	0	2	1
2	4	0	4	-1
-1	-2	3	1	-1
1	2	1	3	1
0	0	2	2	-2
-2	-4	2	-2	1

We will find a basis for Δ using the above algorithm. The detached coefficient original tableau for the linear dependence relation is the first tableau given below. We apply the GJ pivotal method on it, showing the BV (basic variables in each row), PR (pivot row), PC (pivot column), and enclosing the pivot element in a box in each step. RC indicates redundant constraint to be eliminated.

BV	α_1	α_2	α_3	α_4	α_5		
	$\boxed{1}$	2	0	2	1	0	PR
	2	4	0	4	−1	0	
	−1	−2	3	1	−1	0	
	1	2	1	3	1	0	
	0	0	2	2	−2	0	
	−2	−4	2	−2	1	0	
	PC↑						
α_1	1	2	0	2	1	0	
	0	0	0	0	$\boxed{-3}$	0	PR
	0	0	3	3	0	0	
	0	0	1	1	0	0	
	0	0	2	2	−2	0	
	0	0	2	2	3	0	
					PC↑		
α_1	1	2	0	2	0	0	
α_5	0	0	0	0	1	0	
	0	0	$\boxed{3}$	3	0	0	PR
	0	0	1	1	0	0	
	0	0	2	2	0	0	
	0	0	2	2	0	0	
			PC↑				
α_1	1	2	0	2	0	0	
α_5	0	0	0	0	1	0	
α_3	0	0	1	1	0	0	
	0	0	0	0	0	0	RC
	0	0	0	0	0	0	RC
	0	0	0	0	0	0	RC
α_1	1	2	0	2	0	0	
α_5	0	0	0	0	1	0	
α_3	0	0	1	1	0	0	

Final tableau with variables rearranged

BV	α_1	α_5	α_3	α_2	α_4	
α_1	1	0	0	2	2	0
α_5	0	1	0	0	0	0
α_3	0	0	1	0	1	0

The basic variables are $(\alpha_1, \alpha_5, \alpha_3)$ in that order; and α_2, α_4 are the nonbasic variables. So, Δ is linearly dependent, its rank is 3, and $\{A_{.1}, A_{.5}, A_{.3}\}$ is a basis (maximal linearly independent subset) for it.

Fixing the nonbasics $\alpha_2 = 1$, $\alpha_4 = 0$ leads to the solution

$$(\alpha_1, \alpha_5, \alpha_3, \alpha_2, \alpha_4) = (-2, 0, 0, 1, 0)$$

to the linear dependence relation. Therefore

$$-2A_{.1} + A_{.2} = 0 \quad \text{or} \quad A_{.2} = 2A_{.1}.$$

Similarly fixing the nonbasics $\alpha_4 = 1$, $\alpha_2 = 0$ leads to the solution

$$(\alpha_1, \alpha_5, \alpha_3, \alpha_2, \alpha_4) = (-2, 0, -1, 0, 1)$$

to the linear dependence relation. Therefore

$$-2A_{.1} - A_{.3} + A_{.4} = 0 \quad \text{or} \quad A_{.4} = 2A_{.1} + A_{.3}.$$

Hence the representations of the nonbasic vectors $A_{.2}, A_{.4}$ in terms of the basis $(A_{.1}, A_{.5}, A_{.3})$ are $(2, 0, 0)^T$ and $(2, 0, 1)^T$ respectively; and we verify that these are the column vectors of the corresponding variables α_2, α_4 in the final tableau.

4.4 How to Get an Alternate Basis by Introducing a Nonbasic Vector into A Basis?

Let $\Delta = \{a^1, \ldots, a^k\}$ be a set of vectors in R^n whose rank is $r < k$, and let $B = \{a^1, \ldots, a^r\}$ be a basis for it.

In some applications, it is important to find out whether there are other subsets of Δ which are also bases for it; and if so, how one can obtain them starting with the known basis B. A procedure called **introducing a nonbasic vector into a basis** can be used to do this. We will now discuss this procedure.

Any nonbasic vector which is nonzero can be introduced into a basis. For the operation of introducing the nonbasic vector $a^{r+1} \neq 0$ into the basis B, a^{r+1} is called **the entering vector**.

By Result 4.2.2 of Section 4.2, every basis for Δ always contains the same number of vectors. So, when a^{r+1} is introduced into B, one of the present basic vectors in B must be deleted from it, this basic vector is called the **dropping basic vector** in this operation.

The answer to the question: *which basic vector in B can the entering vector a^{r+1} replace?* is contained in the representation of a^{r+1} in terms of B. Suppose this representation is $(\alpha_1, \ldots, \alpha_r)$, i.e.,

$$a^{r+1} = \alpha_1 a^1 + \cdots + \alpha_r a^r.$$

Then a^{r+1} can replace any of the basic vectors a^j for which $\alpha_j \neq 0$, to lead to a new basis for Δ.

Example 1:

As an example, consider the set $\Gamma = \{A_{.1}, \text{ to } A_{.5}\}$ of column vectors in R^6 of the matrix given below, from the examples discussed in Section 4.3.

$A_{.1}$	$A_{.2}$	$A_{.3}$	$A_{.4}$	$A_{.5}$
1	2	0	2	1
2	4	0	4	-1
-1	-2	3	1	-1
1	2	1	3	1
0	0	2	2	-2
-2	-4	2	-2	1

In Section 4.3, we found the basis $B_1 = \{A_{.1}, A_{.3}, A_{.5}\}$ for this set Δ, and that the representations of the nonbasic vectors in terms of this basis are:

$A_{.2}$ representation is $(2, 0, 0)$; i.e., $A_{.2} = 2A_{.1}$

$A_{.4}$ representation is $(2, 1, 0)$; i.e., $A_{.4} = 2A_{.1} + A_{.3}$.

Therefore, the nonbasic vector $A_{.2}$ can only replace the basic vector $A_{.1}$ from the basis B_1 (because $A_{.1}$ is the only basic vector with a nonzero coefficient in the representation of $A_{.2}$), leading to another basis $B_2 = \{A_{.2}, A_{.3}, A_{.5}\}$ for Δ.

This implies that if we replace any of the basic vectors $A_{.3}, A_{.5}$ from the basis B_1 by $A_{.2}$, we will get subsets of vectors which are linearly dependent, and hence are not bases. In fact, we can easily verify that the subsets $\{A_{.1}, A_{.2}, A_{.5}\}, \{A_{.1}, A_{.3}, A_{.2}\}$ are both linearly dependent because $-2A_{.1} + A_{.2} = 0$ is a linear dependence relation for them.

In the same way, we conclude that the nonbasic vector $A_{.4}$ can replace either of the basic vectors $A_{.1}, A_{.3}$ from the basis B_1, but not $A_{.5}$. Hence, $B_3 = \{A_{.4}, A_{.3}, A_{.5}\}, B_4 = \{A_{.1}, A_{.4}, A_{.5}\}$ are also basic vectors for Δ.

Exercises

4.4.1: Find the rank of the following set of vectors Γ, a basis for this set, representation of each nonbasic vector in terms of this basis; and determine for each nonbasic vector which basic vectors it can replace to lead to another basis: (i) $\Gamma = \{(1,0,-1,2,1), (-1,1,2,-1,0), (3,-1, -4,5,2), (1,1,1,0,2), (1,2,2,1,3)\}$ (ii) $\Gamma = \{(1,1,-1,2,1), (2,-1,2,1,2), (4,1,0,5,4), (0,0,1,0,0), (7,1,2,8,7)\}$.

4.4.2: Show that the following sets of vectors are linearly independent: i $\{(0,1,-1,0,1), (1,-1,1,1,1), (1,1,1,2,-1), (0,0,1,3,2), (0,0,-1,-1,-1)\}$ (ii) $\{(2,-2,2,1), (-2,3,-3,-1), (1,-2,1,0), (0,2,2,2)\}$ (iii) $\{(3,1,4), (7,2,6), (-2,0,5)\}$.

4.4.3: If the set $\{A_{.1}, A_{.2}, A_{.3}\}$ is linearly independent, show that the set $\{A_{.2} + A_{.3}, A_{.3} + A_{.1}, A_{.1} + A_{.2}\}$ must also be linearly independent. Is the converse also correct?

Let $y^1 = A_{.1} + A_{.2} + A_{.3}$, $y^2 = A_{.1} + \alpha A_{.2}$, $y^3 = A_{.2} + \beta A_{.3}$. If $\{A_{.1}, A_{.2}, A_{.3}\}$ is linearly independent, what condition should α, β satisfy so that $\{y^1, y^2, y^3\}$ is also linearly independent?

4.4.4: Given a linearly independent set of vectors $\{A_{.1}, A_{.2}, \ldots, A_{.m}\}$ in R^n, and another vector $A_{.m+1}$ in R^n show that $\{A_{.1}, A_{.2}, \ldots, A_{.m}, A_{.m+1}\}$ is linearly independent iff $A_{.m+1}$ is not in the linear hull of $\{A_{.1}, A_{.2}, \ldots, A_{.m}\}$.

4.5 The Rank of a Matrix

Let $A = (a_{ij})$ be a matrix of order $m \times n$. In our notation, $\{A_{1.}, \ldots, A_{m.}\}$ is the set of row vectors of A. The rank of this set is called the **row rank** of matrix A.

$\{A_{.1}, \ldots, A_{.n}\}$ is the set of column vectors of A. The rank of this set of vectors is called the **column rank** of matrix A.

Result 4.5.1: Row rank and column rank are the same: *For any matrix A, its row rank is always equal to its column rank. Their common value is called the **rank** of matrix A. If A is of order $m \times n$ and rank r, then $r \leq m$ and $r \leq n$.*

Historical note on the rank of a matrix: The term *rank of a matrix A* was introduced in the 1870s by the German mathematician F. G. Frobinius, as the order of the largest order square submatrix of A with a nonzero determinant. His definition of rank turns out to be equivalent to the number of vectors in any maximal linearly independent subset of column (or row) vectors of A, or a minimal spanning subset of these vectors.

The rank of a matrix A can be computed by applying either of the algorithms discussed in Section 4.3 to find the rank of the set of row (or column) vectors of A.

Example 1: The Row Reduction Method for finding the rank of a matrix, and bases for its column space using GJ pivot steps.

This is exactly Algorithm 2 discussed in Section 4.3. Let A be the matrix whose rank we want to find. Put A in a tableau form, and label its column vectors by their names $A_{.1}, A_{.2}$, etc. in a row at the top. Try to perform a GJ pivot step in each row of this tableau, one after the other, say from top to bottom (any other order would be OK too). When a pivot step is performed in a row i, if the pivot column for that pivot step is $A_{.j}$, then record $A_{.j}$ as the basic column vector selected in this step in row i. If a zero row is encountered, move to the next row, and continue the same way until each row is either the zero row or had a pivot step performed in it. Then rank of A is the total number of pivot steps performed in this algorithm, and the final tableau contains a basis for the column space of A, and the representation of each nonbasic column in terms of this basis.

We will apply this method on the following matrix A of order 6×5. For each pivot step PR, PC indicate the pivot row, column respectively and the pivot element is enclosed in a box.

$$
A = \begin{pmatrix}
1 & 2 & 0 & 2 & 1 \\
2 & 4 & 0 & 4 & -1 \\
-1 & -2 & 3 & 1 & -1 \\
1 & 2 & 1 & 3 & 1 \\
0 & 0 & 2 & 2 & -2 \\
-2 & -4 & 2 & -2 & 1
\end{pmatrix}
$$

Basic Vector	$A_{.1}$	$A_{.2}$	$A_{.3}$	$A_{.4}$	$A_{.5}$	
	1	2	0	2	$\boxed{1}$	PR
	2	4	0	4	-1	
	-1	-2	3	1	-1	
	1	2	1	3	1	
	0	0	2	2	-2	
	-2	-4	2	-2	1	
					PC↑	

Basic Vector	$A_{.1}$	$A_{.2}$	$A_{.3}$	$A_{.4}$	$A_{.5}$	
$A_{.5}$	1	2	0	2	1	
	③	6	0	6	0	PR
	0	0	3	3	0	
	0	0	3	3	0	
	0	0	1	1	0	
	2	4	2	6	0	
	−3	−6	2	−4	0	
	PC↑					
$A_{.5}$	0	0	0	0	1	
$A_{.1}$	1	2	0	2	0	
	0	0	③	3	0	PR
	0	0	1	1	0	
	0	0	2	2	0	
	0	0	2	2	0	
			PC↑			

Basic vector	Final Tableau				
	$\bar{A}_{.1}$	$\bar{A}_{.2}$	$\bar{A}_{.3}$	$\bar{A}_{.4}$	$\bar{A}_{.5}$
$A_{.5}$	0	0	0	0	1
$A_{.1}$	1	2	0	2	0
$A_{.3}$	0	0	1	1	0
	0	0	0	0	0
	0	0	0	0	0
	0	0	0	0	0

We have labeled the updated columns in the final tableau by the symbols $\bar{A}_{.j}$. Since a total of three pivot steps were carried out, the rank of matrix A is three. The basis obtained for the column space of A is $(A_{.5}, A_{.1}, A_{.3})$. In the final tableau each basic vector is transformed into the appropriate unit vector. For each nonbasic column vector $A_{.j}$, when the zero entries in zero rows are deleted from its updated column $\bar{A}_{.j}$, we get the representation of $A_{.j}$ in terms of the basis. Thus, the representation of

$A_{.2}$ is $(0, 2, 0)$, i.e., $A_{.2} = 2A_{.1}$
$A_{.4}$ is $(0, 2, 1)$, i.e., $A_{.4} = 2A_{.1} + A_{.3}$.

In linear programming literature, the final tableau with the basic column vectors transformed into the various unit vectors is called **the canonical tableau** WRT the selected basis.

To get an alternate basis for the column space of this matrix, we can select any nonbasic column as an entering column in the final tableau, and select any nonzero entry in its updated column as the pivot element and the row of the pivot element as the pivot row. The entering column replaces the basic column recorded in the pivot row leading to a new basis for the column space of the matrix, Performing the pivot step leads to the canonical tableau WRT the new basis.

Relationship of Nonsingularity to Rank

Let A be a square matrix of order n. According to the definition given in Section 2.8, A is nonsingular iff its determinant is $\neq 0$; singular otherwise. The following result relates this property to the rank of A.

Result 4.5.2: Rank and nonsingularity: *The square matrix A of order n is nonsingular iff its rank is n, singular iff its rank is $\leq n - 1$.*

If rank of A is $\leq n-1$, then one of the row vectors of A can be expressed as a linear combinations of the other row vectors of A. By results 2, 3 of Section 2.8 this implies that the determinant of A is 0, i.e., A is singular.

If rank of A is n, then in the algorithm discussed above for computing the rank, pivot steps can be carried out in every row. This implies that in the algorithm for computing the determinant of A discussed in Section 2.8, the number of G pivot steps that can be carried out is exactly n. Therefore the determinant of A is $= \pm$ (the product of the boxed pivot elements), which is nonzero because every pivot element is nonzero, i.e., A is nonsingular.

Note: In some linear algebra books a square matrix of order n is alternately defined as being nonsingular if its rank is n, singular if its rank is $\leq n - 1$.

Other Meanings for the Word "Basis"

Let A be a matrix of order $m \times n$ and rank m. Then A is said to be a **matrix of full row rank**. Speaking in terms of the column space of such a matrix A, we defined a basis to be a linearly independent subset of m column vectors of A.

In linear programming literature, the word *basis* is often used to denote a submatrix of A of order $m \times m$ consisting of a linearly independent subset of m column vectors of A. Thus, if the set of column vectors, $\{A_{.1}, \ldots, A_{.m}\}$ of A form a linearly independent set, then in our notation this set of column vectors constitutes a basis for the column space of A. In linear programming literature they would use the word *basis* to denote the submatrix of A

$$B = (A_{.1} \vdots \ldots \vdots A_{.m})$$

consisting of these column vectors.

Exercises

4.5.1: If A, B are square matrices of order n satisfying $AB = BA$ and rank$(AB) = n$, show that rank$\begin{pmatrix} A \\ B \end{pmatrix} = n$ also.

4.5.2: Find the ranks of the following matrices, a basis for its column space, and an expression for each nonbasic column as a linear combination of basic vectors.

$$\begin{pmatrix} 1 & 1 & -1 & 2 & 0 \\ -1 & -2 & 3 & -3 & 2 \\ -1 & -4 & 7 & -5 & 6 \\ 0 & 1 & -1 & 2 & 2 \\ 0 & 1 & 0 & 3 & 6 \end{pmatrix}, \quad \begin{pmatrix} 2 & -1 & 3 & 2 \\ -1 & 1 & -4 & -1 \\ 4 & -1 & 1 & 4 \\ 1 & 0 & 2 & 7 \end{pmatrix},$$

$$\begin{pmatrix} 2 & -4 & 6 & 8 & 10 \\ -4 & 9 & -13 & 0 & -21 \\ 6 & -13 & 19 & 8 & 31 \\ 0 & -1 & 2 & 2 & 1 \end{pmatrix}.$$

4.6   Computing the Inverse of a Nonsingular Square Matrix Using GJ Pivot Steps

As defined in Section 2.8, among matrices, the concept of an inverse applies only to square matrices, and it only exists for nonsingular square matrices. The inverse is not defined for rectangular matrices which are not square. When the inverse of a matrix exists, it is always unique. If square matrix A has an inverse, it is denoted by the symbol A^{-1} and not as $1/A$.

The definition of the inverse of a square matrix when it exists, is analogous to the definition of the inverse of a nonzero real number γ as the number δ satisfying $\gamma\delta = 1$. Among square matrices of order n, the unit or identity matrix I of order n, plays the role of the real number 1. And so, for a square matrix A, its inverse is defined to be the square matrix D of order n satisfying

$$DA = AD = I$$

when it exists, which happens only if A is nonsingular, it is denoted by the symbol A^{-1}.

Inverses of Some Special Matrices

Consider the unit matrix I of order n. Its inverse is itself, i.e., $I^{-1} = I$.

Consider a permutation matrix P, which is a matrix obtained from the unit matrix by permuting its row vectors. For example

$$P = \begin{pmatrix} 0 & 0 & 1 & 0 \\ 1 & 0 & 0 & 0 \\ 0 & 0 & 0 & 1 \\ 0 & 1 & 0 & 0 \end{pmatrix}$$

is a permutation matrix of order 4. It can be verified that $P^T P = P P^T = I$. The same thing holds for all permutation matrices. Hence for every permutation matrix P, we have $P^{-1} = P^T$.

Consider a diagonal matrix D with diagonal entries $d_1, \ldots, d_n$. By definition $d_j \neq 0$ for all j. It can be verified that D^{-1} is given by the following.

$$D = \begin{pmatrix} d_1 & 0 & \cdots & 0 \\ 0 & d_2 & \cdots & 0 \\ \vdots & \vdots & & \vdots \\ 0 & 0 & \cdots & d_n \end{pmatrix}, \quad D^{-1} = \begin{pmatrix} 1/d_1 & 0 & \cdots & 0 \\ 0 & 1/d_2 & \cdots & 0 \\ \vdots & \vdots & & \vdots \\ 0 & 0 & \cdots & 1/d_n \end{pmatrix}.$$

Hence the inverse of $\operatorname{diag}(d_1, \ldots, d_n)$ is $\operatorname{diag}(1/d_1, \ldots, 1/d_n)$.

How To Compute the Inverse of a General Nonsingular Matrix Efficiently?

In Section 2.8 we derived a formula for the inverse of a general nonsingular square matrix in terms of its cofactors. That formula points out the theoretical relation between the inverse of a matrix, and the determinants of its submatrices. It points out the many different places in matrix algebra where determinants play an important theoretical role. However, that formula is not practically efficient for computing the inverse. We will now discuss a much more efficient method for computing the inverse using GJ pivot steps.

Let $A = (a_{ij})$ be a square nonsingular matrix of order n. $A_1., \ldots, A_n.$ denote the various row vectors of A. Let I be the unit matrix of order n, and $I_{1.}, \ldots, I_{n.}$ are its row vectors (these are the various unit row vectors in R^n). Suppose

$$A^{-1} = \begin{pmatrix} \beta_{11} & \cdots & \beta_{1n} \\ \vdots & & \vdots \\ \beta_{n1} & \cdots & \beta_{nn} \end{pmatrix}.$$

Then from the definition of A^{-1} we have

$$A^{-1}A = \begin{pmatrix} \beta_{11} & \cdots & \beta_{1n} \\ \vdots & & \vdots \\ \beta_{n1} & \cdots & \beta_{nn} \end{pmatrix} A = I.$$

From the definition of matrix products, this implies that for each $i = 1$ to n

$$(\beta_{i1}, \ldots, \beta_{in})A = \beta_{i1}A_{1.} + \cdots + \beta_{in} = I_{i.}$$

i.e., the ith row vector of A^{-1} is the vector of coefficients in an expression of $I_{i.}$ as a linear combination of the row vectors of A. This observation is the key behind the algorithm for computing A^{-1} using row operations.

The algorithm begins with a tableau with n rows which are the rows of A; $A_{1.}, \ldots, A_{n.}$. As discussed in Section 1.12 we set up a memory matrix on the left of this tableau to maintain the expression of each row in the current tableau as a linear combination of rows in the original tableau. When we carry out row operations to transform the rows of the tableau into the unit rows $I_{.1}, \ldots, I_{.n}$, the memory matrix gets transformed into A^{-1}.

We now provide a statement of various steps in the algorithm. When applied on a general square matrix A of order n, the algorithm terminates by either producing A^{-1}, or a linear dependence relation for the set of row vectors of A showing that A is singular, and hence A^{-1} does not exist.

Algorithm to Compute the Inverse of a Square Matrix A Using GJ Pivot Steps

BEGIN

Step 1: Start with the following initial tableau, where I is the unit matrix of the same order as A.

Memory Matrix	Original Tableau
I	A

We will perform GJ pivot steps in each row of the tableau in some order (say top to bottom). Select the first row for pivoting, as the pivot row in the initial tableau and go to Step 2. Remember that even though row operations are carried out on all the $2n$ columns (both columns on the memory matrix side, and on the original tableau side), pivot elements are always selected only from the original tableau part of the current tableau. Also the word "present tableau" in the description below, refers to the "original tableau part of the current tableau".

Step 2: General Step: Suppose row r is the pivot row in the present tableau.

If all the entries in row r of the present tableau are zero, A is singular and A^{-1} does not exist, terminate. If the entries in the current memory matrix in this row are $(\beta_{r1}, \ldots, \beta_{rn})$, then

$$\beta_{r1}A_{1.} + \cdots + \beta_{rn}A_{n.} = 0$$

is a linear dependence relation for the set of row vectors of A.

If there are nonzero elements in row r of the present tableau, select one of them as the pivot element, and its column as the pivot column, and perform the GJ pivot step. With the new tableau obtained after this pivot step, move to Step 3 if GJ pivot steps have been performed in all the rows; or repeat Step 2 to perform a GJ pivot step in the next row for pivoting in it.

Step 3: Final Step: The present tableau at this stage contains a permutation matrix P. Rearrange the rows of the present tableau so that P becomes the unit matrix (this is equivalent to multiplying the present tableau on the left by P^T) leading to the final tableau. The square matrix contained in the memory matrix portion of the final tableau is A^{-1}, terminate.

END.

Example 1:

As an example, we will compute the inverse of the following matrix A.

$$A = \begin{pmatrix} 0 & -2 & -1 \\ 1 & 0 & -1 \\ 1 & 1 & 0 \end{pmatrix}.$$

We apply the algorithm discussed above. The pivot element in each step is enclosed in a box, and PR, PC indicate pivot row, pivot column respectively.

Memory Matrix			Tableau			
1	0	0	0	$\boxed{-2}$	−1	PR
0	1	0	1	0	−1	
0	0	1	1	1	0	
				PC↑		
−1/2	0	0	0	1	1/2	
0	1	0	$\boxed{1}$	0	−1	PR
1/2	0	1	1	0	−1/2	
			PC↑			
−1/2	0	0	0	1	1/2	
0	1	0	1	0	−1	
1/2	−1	1	0	0	$\boxed{1/2}$	PR
					PC↑	
−1	1	−1	0	1	0	
1	−1	2	1	0	0	
1	−2	2	0	0	1	

Final Tableau

Memory Matrix			Tableau		
1	−1	2	1	0	0
−1	1	−1	0	1	0
1	−2	2	0	0	1

Hence

$$A^{-1} = \begin{pmatrix} 1 & -1 & 2 \\ -1 & 1 & -1 \\ 1 & -2 & 2 \end{pmatrix}.$$

It can be verified that $AA^{-1} = A^{-1}A = I$, the unit matrix of order 3.

Example 2:

As another example, we will find the inverse of the following matrix $\hat{A}$. The algorithm is applied below, boxing the pivot element in each step (PR, PC indicate pivot row, pivot column respectively).

$$\hat{A} = \begin{pmatrix} 1 & 0 & -2 \\ -1 & 1 & 1 \\ 0 & 1 & -1 \end{pmatrix}$$

Memory Matrix			Tableau			
1	0	0	$\boxed{1}$	0	-2	PR
0	1	0	-1	1	1	
0	0	1	0	1	-1	
			PC↑			
1	0	0	1	0	-2	
1	1	0	0	$\boxed{1}$	-1	PR
0	0	1	0	1	-1	
			PC↑			
1	0	0	1	0	-2	
1	1	0	0	1	-1	
-1	-1	1	0	0	0	

Since the last row has all zero entries in the tableau part, we conclude that $\hat{A}$ is singular and hence has no inverse. From the memory matrix part of the last row we obtain the linear dependence relation for the set of row vectors of $\hat{A}$ as

$$-\hat{A}_{1.} - \hat{A}_{2.} + \hat{A}_{3.} = 0.$$

Computational Effort Needed To Find the Inverse

A matrix is said to be a:

> **sparse matrix** if a large proportion of the entries in it are zero
> **dense matrix** otherwise.

If the various tableaus obtained in the inverse finding algorithm are dense, the method described above for finding the inverse needs about $2n^3$ arithmetic operations.

Exercises

4.6.1: Find the inverses of the following nonsingular matrices.

$$\begin{pmatrix} 2 & 8 \\ 4 & -18 \end{pmatrix}, \quad \begin{pmatrix} 13 & -1 \\ -90 & 7 \end{pmatrix}, \quad \begin{pmatrix} 0 & 2 & -4 \\ 1 & 0 & 2 \\ -2 & -1 & 0 \end{pmatrix},$$

$$\begin{pmatrix} 2 & -1 & -3 \\ -5 & 2 & 8 \\ -3 & 1 & 6 \end{pmatrix}, \quad \begin{pmatrix} -2 & -4 & 1 \\ -5 & -8 & 2 \\ 3 & 5 & -1 \end{pmatrix},$$

$$\begin{pmatrix} 0 & 2 & 2 & 2 \\ 2 & 0 & 2 & 2 \\ 2 & 2 & 0 & 2 \\ 2 & 2 & 2 & 0 \end{pmatrix}, \quad \begin{pmatrix} 1 & 0 & 1 & -1 \\ -1 & 1 & -2 & 2 \\ 0 & -1 & 2 & -2 \\ 0 & 0 & -1 & 2 \end{pmatrix}, \quad \begin{pmatrix} 1 & -1 & -1 & 1 \\ -2 & 3 & 3 & -1 \\ 1 & 1 & 1 & 0 \\ -1 & 0 & 0 & 1 \end{pmatrix}.$$

4.6.2: Apply the inverse finding algorithm on the following matrices and show that they are singular. For each matrix, obtain a linear dependence relation for its row vectors.

$$\begin{pmatrix} 2 & 4 \\ 8 & 16 \end{pmatrix}, \quad \begin{pmatrix} -12 & -3 \\ -60 & -15 \end{pmatrix}, \quad \begin{pmatrix} 0 & 2 & -4 \\ 2 & 0 & 2 \\ 4 & 2 & 0 \end{pmatrix}, \quad \begin{pmatrix} 3 & 6 & -9 \\ -3 & -8 & 7 \\ -3 & -12 & 3 \end{pmatrix},$$

$$\begin{pmatrix} 0 & 5 & -15 & 20 \\ 2 & 0 & 4 & -6 \\ 0 & -5 & 0 & -15 \\ 4 & 20 & -7 & 53 \end{pmatrix}, \quad \begin{pmatrix} 0 & 2 & 6 & 8 \\ 3 & 0 & 9 & 6 \\ 6 & 2 & 24 & 20 \\ 3 & -2 & 3 & -2 \end{pmatrix},$$

4.7 Pivot Matrices, Elementary Matrices, Matrix Factorizations, and the Product Form of the Inverse

Consider a pivot step on a matrix $A = (a_{ij})$ of order $m \times n$, with row r as the PR (pivot row), and column s as the PC (pivot column). So, the pivot element a_{rs}, boxed in the following, is $\neq 0$.

Matrix A						$\bar{A}$ = Matrix after pivot step		
a_{11}	$\cdots$	a_{1s}	$\cdots$	a_{1n}		$\bar{a}_{11}$ $\cdots$	0 $\cdots$	$\bar{a}_{1n}$
$\vdots$		$\vdots$		$\vdots$		$\vdots$	$\vdots$	$\vdots$
$a_{r-1,1}$	$\cdots$	$a_{r-1,s}$	$\cdots$	$a_{r-1,n}$		$\bar{a}_{r-1,1}$ $\cdots$	0 $\cdots$	$\bar{a}_{r-1,n}$
a_{r1}	$\cdots$	$\boxed{a_{rs}}$	$\cdots$	a_{rn}	PR	$\bar{a}_{r1}$ $\cdots$	1 $\cdots$	$\bar{a}_{rn}$
$a_{r+1,1}$	$\cdots$	$a_{r+1,s}$	$\cdots$	$a_{r+1,n}$		$\bar{a}_{r1}$ $\cdots$	0 $\cdots$	$\bar{a}_{r+1,n}$
$\vdots$		$\vdots$		$\vdots$		$\vdots$	$\vdots$	$\vdots$
a_{m1}	$\cdots$	a_{ms}	$\cdots$	a_{mn}		$\bar{a}_{m1}$ $\cdots$	0 $\cdots$	$\bar{a}_{mn}$
	PC							

where

$$\bar{a}_{rj} = a_{rj}/a_{rs} \quad \text{for } j = 1 \text{ to } n$$

$$\bar{a}_{ij} = a_{ij} - (a_{rj}/a_{rs})a_{is} \quad \text{for } i \neq r, \ j = 1 \text{ to } n.$$

This pivot step has transformed the pivot column into the rth unit column (r is the number of the pivot row). Now consider the square matrix P of order $m \times m$ which is the same as the unit matrix except for its rth column, known as its **eta column**; obtained by carrying out on the unit matrix I of order m, exactly the same row operations in the pivot step carried out on the matrix A.

$$P = \begin{pmatrix} 1 & 0 & \cdots & 0 & -a_{1s}/a_{rs} & 0 & \cdots & 0 \\ 0 & 1 & \cdots & 0 & -a_{2s}/a_{rs} & 0 & \cdots & 0 \\ \vdots & \vdots & & \vdots & & \vdots & \vdots & & \vdots \\ 0 & 0 & \cdots & 1 & -a_{r-1,s}/a_{rs} & 0 & \cdots & 0 \\ 0 & 0 & \cdots & 0 & 1/a_{rs} & 0 & \cdots & 0 \\ 0 & 0 & \cdots & 0 & -a_{r+1,s}/a_{rs} & 1 & \cdots & 0 \\ \vdots & \vdots & & \vdots & & \vdots & \vdots & & \vdots \\ 0 & 0 & \cdots & 0 & -a_{ms}/a_{rs} & 0 & \cdots & 1 \end{pmatrix}$$

$$\begin{array}{c} r\text{th col.} \\ \text{eta col.} \end{array}$$

It can be verified that $\bar{A} = PA$. Thus performing the pivot step on the matrix A of order $m \times n$ with row r as the pivot row, column s as the pivot column; is equivalent to multiplying A on its left (i.e., premultiplying A) by the square matrix P of order $m \times m$ that differs from the unit matrix in just its rth column (called its **eta column**). That's why this matrix P is called the **pivot matrix** corresponding to this pivot step. The eta column in P is derived from the data in the pivot column, and its position in P is the same as the position of the pivot row.

Example 1:

Consider the following matrix A of order 5×6 and the pivot step on it with the PR (pivot row) and PC (pivot column) as the row and column of the boxed pivot element.

$$A = \begin{pmatrix} 9 & 1 & 0 & -2 & 8 & 2 \\ -7 & -1 & -2 & 0 & -6 & 0 \\ -2 & 0 & 0 & 4 & 2 & 3 \\ 0 & 2 & 4 & 6 & \boxed{2} & -2 \\ -1 & 0 & 2 & -2 & -4 & 1 \end{pmatrix}$$

We put I, the unit matrix of order 5 on the left hand side of A and carry out all the row operations in this pivot step on it too. After this pivot step, the unit matrix becomes the pivot matrix P and the matrix A becomes $\bar{A}$.

Initial tableau

1	0	0	0	0	9	1	0	-2	8	2
0	1	0	0	0	-7	-1	-2	0	-6	0
0	0	1	0	0	-2	0	0	4	2	3
0	0	0	1	0	0	2	4	6	$\boxed{2}$	-2
0	0	0	0	1	-1	0	2	-2	-4	1

After the pivot step

P					$\bar{A}$					
1	0	0	-4	0	9	-7	-16	-26	0	10
0	1	0	3	0	-7	5	10	18	0	-6
0	0	1	-1	0	-2	-2	-4	-2	0	5
0	0	0	1/2	0	0	1	2	3	1	-1
0	0	0	2	1	-1	4	10	10	0	-3

The pivot matrix corresponding to this pivot step is the 5×5 matrix P in the above tableau, and it can be verified that $\bar{A} = PA$. Notice that the eta column in P is its 4th column, its number is the same as that of the pivot row for this pivot step.

Thus carrying out a pivot step on a matrix of order $m \times n$ is equivalent to premultiplying it by the corresponding pivot matrix (which can be derived from the formula given above, using the pivot column) of order m.

Since the pivot element for every pivot step will be nonzero, a pivot matrix is always square and nonsingular. Also, since a pivot matrix always differs from the unit matrix in just one column (its eta column), to store the pivot matrix, it is sufficient to store its eta column, and the position of this eta column in the pivot matrix. Thus even though a pivot matrix is a square matrix, it can be stored very compactly and requires almost the same storage space as a column vector.

Pivot matrices belong to a special class of square matrices known as **elementary matrices**. The main elementary matrices are defined as square matrices obtained by performing a single elementary row operation (or a similarly defined column operation) on a unit matrix. From Sections 1.11, 1.20, we recall that there are three elementary row operations (and corresponding column operations) which are:

1. Scalar multiplication: Multiply each entry in a row (or column) by the same nonzero scalar
2. Add a scalar multiple of a row to another row (or of a column to another column)
3. Row (or column) interchange: Interchange two rows (or columns) in the matrix

and there is an elementary matrix associated with each of these operations. We show these matrices for elementary row operations first, and for column operations next. We use the following notation. A denotes the original matrix of order $m \times n$ on which the elementary operation is carried out, A_i the matrix obtained after the ith operation on A, and E_i the matrix obtained by carrying out the same ith operation on the unit matrix of order m for row operations (or order n for column operations). Then E_i is the elementary matrix associated with the ith operation, and

$A_i = E_i A$ for row operations
$A_i = A E_i$ for column operations.

Thus carrying out a row (column) operation on a matrix is equivalent to premultiplying (postmultiplying) it by the associated elementary matrix. For illustrative examples, we, will use

$$A = \begin{pmatrix} 1 & -3 & -5 & 6 \\ -1 & 4 & -2 & 3 \\ 2 & 5 & 7 & 8 \end{pmatrix}.$$

Example 2:

Scalar multiplication of a row (row 2) of A by scalar α ($\alpha = 10$), leads to

$$A_1 = \begin{pmatrix} 1 & -3 & -5 & 6 \\ -10 & 40 & -20 & 30 \\ 2 & 5 & 7 & 8 \end{pmatrix}, \quad E_1 = \begin{pmatrix} 1 & 0 & 0 \\ 0 & 10 & 0 \\ 0 & 0 & 1 \end{pmatrix}$$

and it can be verified that $A_1 = E_1 A$, E_1 is the elementary matrix associated with this operation.

Example 3:

Adding a scalar, multiple of a row of A to another row: We will consider adding (-2) times row 3 of A to row 1. This leads to

$$A_2 = \begin{pmatrix} -3 & -13 & -19 & -10 \\ -1 & 4 & -2 & 3 \\ 2 & 5 & 7 & 8 \end{pmatrix}, \quad E_2 = \begin{pmatrix} 1 & 0 & -2 \\ 0 & 1 & 0 \\ 0 & 0 & 1 \end{pmatrix}$$

and it can be verified that $A_2 = E_2 A$, E_2 is the elementary matrix associated with this operation.

Example 4:

Interchanging two rows of A: We will consider interchanging rows 1 and 3. This leads to

$$A_3 = \begin{pmatrix} 2 & 5 & 7 & 8 \\ -1 & 4 & -2 & 3 \\ 1 & -3 & -5 & 6 \end{pmatrix}, \quad E_3 = \begin{pmatrix} 0 & 0 & 1 \\ 0 & 1 & 0 \\ 1 & 0 & 0 \end{pmatrix}$$

and it can be verified that $A_3 = E_3 A$, E_3 is the elementary matrix associated with this operation.

Example 5:

Scalar multiplication of a column (column 4) of A by scalar α ($\alpha = -5$), leads to

$$A_4 = \begin{pmatrix} 1 & -3 & -5 & -30 \\ -1 & 4 & -2 & -15 \\ 2 & 5 & 7 & -40 \end{pmatrix}, \quad E_4 = \begin{pmatrix} 1 & 0 & 0 & 0 \\ 0 & 1 & 0 & 0 \\ 0 & 0 & 1 & 0 \\ 0 & 0 & 0 & -5 \end{pmatrix}$$

and it can be verified that $A_4 = E_4 A$, E_4 is the elementary matrix associated with this operation.

Example 6:

Adding a scalar multiple of a column of A to another column: We will consider adding 3 times column 4 of A to column 2. This leads to

$$A_5 = \begin{pmatrix} 1 & 15 & -5 & 6 \\ -1 & 13 & -2 & 3 \\ 2 & 29 & 7 & 8 \end{pmatrix}, \quad E_5 = \begin{pmatrix} 1 & 0 & 0 & 0 \\ 0 & 1 & 0 & 0 \\ 0 & 0 & 1 & 0 \\ 0 & 3 & 0 & 1 \end{pmatrix}$$

and it can be verified that $A_5 = E_5 A$, E_5 is the elementary matrix associated with this operation.

Example 7:

Interchanging two columns of A : We will consider interchanging columns 1 and 4. This leads to

$$A_6 = \begin{pmatrix} 6 & -3 & -5 & 1 \\ 3 & 4 & -2 & -1 \\ 8 & 5 & 7 & 2 \end{pmatrix}, \quad E_6 = \begin{pmatrix} 0 & 0 & 0 & 1 \\ 0 & 1 & 0 & 0 \\ 0 & 0 & 1 & 0 \\ 1 & 0 & 0 & 0 \end{pmatrix}$$

and it can be verified that $A_6 = E_6 A$, E_6 is the elementary matrix associated with this operation. ⋈

These are the elementary matrices normally discussed in linear algebra textbooks, but numerical analysts include a much larger class of square matrices under the name elementary matrices. In this general definition, an **elementary matrix** of order n is any square matrix of the form

$$E = I - \alpha uv^T$$

where I is the unit matrix of order n, α is a scalar, and $u = (u_1, \ldots, u_n)^T$, $v = (v_1, \ldots, v_n)^T$ are column vectors in R^n.

Recall from Sections 1.5, 2.2, 2.5; when u, v are column vectors in R^n, $u^T v$ is a scalar known as the dot (or inner) product of u, v; while uv^T is the square matrix of order n given below, known as the outer (also cross) product of u, v.

$$uv^T = \begin{pmatrix} u_1v_1 & u_1v_2 & \cdots & u_1v_n \\ u_2v_1 & u_2v_2 & \cdots & u_2v_n \\ \vdots & \vdots & & \vdots \\ u_nv_1 & u_nv_2 & \cdots & u_nv_n \end{pmatrix}$$

In uv^T each column vector is a scalar multiple of u, and each row vector is a scalar multiple of v^T; and hence when u, v are both nonzero, uv^T is a matrix of rank one (or a **rank-one matrix**). $I - \alpha uv^T$ is therefore a rank-one modification of the unit(identity) matrix. In this general definition, an elementry matrix is a rank one modification of the unit matrix, and it includes all pivot matrices.

The special property of the elementary matrix $E = I - \alpha uv^T$ is that even though it is a square matrix of order n, it can be stored compactly by storing u, v, and the scalar α; and that computing the product Ex for any column vector x in R^n can be carried out using u, v, α without explicitly computing all the entries in E because

$$Ex = (I - \alpha uv^T)x = x - \alpha u(v^T x) = x - \gamma u$$

where $\gamma = \alpha(v^T x)$, which is α times the dot product of v and x. In the same way, for any row vector $\pi \in R^n$, the product πE can be computed using u, v, α without explicitly computing all the entries in E.

Product Form of the Inverse

The method for computing the inverse described in Section 4.6 is fine for computing the inverses of matrices of small orders like 3, 4 by hand, or less than about 100 on the computer; but implementing it to compute the inverses of matrices of large orders (greater than several hundreds) on a digital computer is beset with many serious difficulties. Some of these are:

The Problem of Fill-in: Matrices encountered in real world applications are very very sparse (i.e., most of the entries in them are 0) so they can be stored very compactly by storing the nonzero entries only with their positions. It has been observed that the explicit inverse of such a matrix is almost always fully dense. This is called the **fill-in** problem. Storing the dense inverse matrix of a large order needs a lot of computer memory space.

Increased Computer Time: Performing any operations with a dense explicit inverse matrix takes a lot of computer time.

Round-off Error Accumulation: Digital computation introduces rounding errors which accumulate during the algorithm, and may result in serious errors in the computed explicit inverse.

To alleviate these problems, numerical analysts have developed substitute forms for the explicit inverse using what are called **matrix factorizations**. Several of these used in computational procedures (such as LU **factorization**, LDL^T **factorization**, etc.) are beyond the scope of this book. But, we will discuss the simplest among these known as the **product form of the inverse** to give the reader a flavor of matrix factorizations. The interested reader should consult the excellent book Gill, Murray and Wright [4.1].

The product form of the inverse is based on the algorithm for computing the inverse using GJ pivot steps discussed in Section 4.6, but it does not obtain the explicit inverse. Instead, it saves the pivot matrices corresponding to the various pivot steps in that algorithm, as a string in the order in which they are generated.

To find the inverse of a square matrix A of order n, n GJ pivot steps are needed in that algorithm. Let

$P_r =$ the pivot matrix corresponding to the rth pivot step in that algorithm, for $r = 1$ to n,

and $P =$ the permutation matrix into which A is reduced by these pivot steps.

Then from the discussion above, we know that

$$A^{-1} = P^T P_n P_{n-1} \ldots P_1.$$

The product on the right has to be carried out exactly in the order specified, as matrix multiplication is not commutative. A^{-1} is not explicitly computed, but $P_1, \ldots, P_n, P^T$ are stored, and the formula on the right of the above equation can be used in place of A^{-1} wherever A^{-1} is needed.

For instance consider solving the square nonsingular system of equations

$$Ax = b$$

for some RHS vector b. If the explicit inverse A^{-1} is known, then multiplying both sides of the above equation by A^{-1} on the left we get

$$A^{-1}Ax = A^{-1}b, \quad \text{i.e.,} \quad Ix = x = A^{-1}b$$

yielding the solution of the system. However, if we are given the inverse in product form as above, then the solution of the system is

$$P^T P_n P_{n-1} \dots P_1 b.$$

This product can be computed very efficiently by a scheme known as **right to left string multiplication**. In this scheme, we first compute $q_1 = P_1 b$, which is a column vector. Then multiply it on the left by P_2 to get $q_2 = P_2 q_1$, and in general for $r = 2$ to n we compute

$$q_r = P_r q_{r-1}$$

in this specific order $r = 2$ to n. When q_n is obtained, one more premultiplication $P^T q_n = x$ yields the solution to the system.

Because the eta vectors in the various pivot matrices tend to preserve sparcity, using the product form of the inverse alleviates the problem of fill-in, at least for matrices of orders upto several hundreds; and some of the round-off error accumulation problem. Because of this, computing $P^T P_n P_{n-1} \dots P_1 b$ through right to left string multiplication usually takes much less computer time than computing $A^{-1}b$ with the explicit inverse A^{-1}.

Algorithm to Compute the Inverse in Product Form

Let A be the matrix of order $n \times n$ whose inverse we need to compute. In this algorithm we do not use any memory matrix, in fact we do not even use the updated tableaus.

The algorithm goes through n pivot steps (one in each column, one in each row). At any stage we denote by C, R the sets of columns, rows in which pivot steps have been carried out already. In each step the next pivot column is computed and a pivot element is selected in this column by the partial pivoting strategy. Then the pivot matrix corresponding to this pivot step is generated and added to the string, and the method moves to the next step.

BEGIN

Initial Step: Select any column, say $A_{.v_1} = (a_{1,v_1}, \ldots, a_{n,v_1})^T$ of the original matrix A as the pivot column for this step.

If $A_{.v_1} = 0$, A^{-1} does not exist, terminate.

If $A_{.v_1} \neq 0$, select one of the elements of maximum absolute value in $A_{.v_1}$ as the pivot element, suppose it is a_{u_1,v_1}. Then row u_1 is the pivot row for this step. Generate the pivot matrix corresponding to the pivot step with $A_{.v_1}$ as the pivot column, row u_1 as the pivot row, and store this matrix as P_1. $C = \{v_1\}, R = \{u_1\}$. Also store the pivot position (u_1, v_1). Go to the next step.

General Step: Suppose r pivot steps have been carried out so far. Let $C = \{v_1, \ldots, v_r\}, R = \{u_1, \ldots, u_r\}$ be the sets of columns, rows, in which pivot steps have been carried out so far, and let $\{(u_1, v_1), \ldots, (u_r, v_r)\}$ be the set of positions of pivot elements in these pivot steps. Let $P_1, \ldots, P_r$ be the string of pivot matrices generated so far, in the order they are generated.

Select a column vector of A in which a pivot step has not been carried out so far (i.e., one from $\{1, \ldots, n\} \backslash C$) for the next pivot step, suppose it is $A_{.v_{r+1}}$. $A_{.v_{r+1}}$ is the column in the original matrix A, the pivot column, $\bar{A}_{.v_{r+1}}$ will be the updated column corresponding to it (i.e., the v_{r+1}th column in the present tableau obtained by performing all the previous pivot steps on A). From previous discussion we see that this is given by

$$\bar{A}_{.v_{r+1}} = P_r P_{r-1} \ldots P_1 A_{.v_{r+1}}.$$

This can be computed quite efficiently by computing the product on the right using right to left string multiplication. Having obtained $\bar{A}_{.v_{r+1}} = (\bar{a}_{1,v_{r+1}}, \ldots, \bar{a}_{n,v_{r+1}})^T$, if

$\bar{a}_{iv_{r+1}} = 0$ for all $i \in \{1, \ldots, n\} \backslash R$, then

$$A_{.v_{r+1}} - \left(\sum_{t=1}^{r} \bar{a}_{u_t, v_{r+1}} A_{.v_t} \right) = 0$$

is a linear dependence relation satisfied by the columns of A, so A^{-1} does not exist, terminate.

If $\bar{a}_{i,v_{r+1}} \neq 0$ for at least one $i \in \{1, \ldots, n\} \backslash R$, then select one of the elements of maximum absolute value among $\{\bar{a}_{i,v_{r+1}} : i \in \{1, \ldots, n\} \backslash R\}$ as the pivot element in $\bar{A}_{.v_{r+1}}$, suppose it is $\bar{a}_{u_{r+1},v_{r+1}}$, then the pivot row for this pivot step is row u_{r+1}. Generate the pivot matrix P_{r+1} corresponding to this pivot step and store it in the string. Include u_{r+1}

in R, v_{r+1} in C, and the position (u_{r+1}, v_{r+1}) in the list of pivot element positions.

> If $r + 1 = n$, let P be the permutation matrix of order n with entries of "1" in the positions (u_t, v_t), $t = 1$ to n; and "0" entries everywhere else. Then the product form of the inverse of A is $P^T P_n \ldots P_1$; where $P_1, \ldots, P_n$ are the stored pivot matrices in the order in which they are generated. Terminate.

> If $r + 1 < n$, go to the next step.

END

Example 8:

Let

$$A = \begin{pmatrix} 1 & 0 & 1 \\ 1 & 1 & 0 \\ -1 & 1 & 2 \end{pmatrix}.$$

We will now compute the product form of the inverse of A. Selecting $A_{.2}$ as the first pivot column, and row 3 as the pivot row for the first pivot step (hence $(3, 2)$ is the pivot element position for this step), leads to the pivot matrix

$$P_1 = \begin{pmatrix} 1 & 0 & 0 \\ 0 & 1 & -1 \\ 0 & 0 & 1 \end{pmatrix}.$$

Selecting column 1 for the next pivot step, the pivot column is

$$\bar{A}_{.1} = P_1 A_{.1} = \begin{pmatrix} 1 & 0 & 0 \\ 0 & 1 & -1 \\ 0 & 0 & 1 \end{pmatrix} \begin{pmatrix} 1 \\ 1 \\ -1 \end{pmatrix} = \begin{pmatrix} 1 \\ 2 \\ -1 \end{pmatrix}.$$

Row 2 is the pivot row for this pivot step by the partial pivoting strategy (since "2" in row 2 is the maximum absolute value entry in the pivot column among rows 1 and 2 in which pivot steps have not been carried out so far). So, the 2nd pivot element position is $(2, 1)$; and the pivot matrix corresponding to this pivot step is

$$P_2 = \begin{pmatrix} 1 & -1/2 & 0 \\ 0 & 1/2 & 0 \\ 0 & 1/2 & 1 \end{pmatrix}.$$

Column 3 is the column for the final pivot step, the pivot column is

$$\bar{A}_{.3} = P_2 P_1 A_{.3} = \begin{pmatrix} 1 & -1/2 & 0 \\ 0 & 1/2 & 0 \\ 0 & 1/2 & 1 \end{pmatrix} \begin{pmatrix} 1 & 0 & 0 \\ 0 & 1 & -1 \\ 0 & 0 & 1 \end{pmatrix} \begin{pmatrix} 1 \\ 0 \\ 2 \end{pmatrix}$$

$$= \begin{pmatrix} 1 & -1/2 & 0 \\ 0 & 1/2 & 0 \\ 0 & 1/2 & 1 \end{pmatrix} \begin{pmatrix} 1 \\ -2 \\ 2 \end{pmatrix} = \begin{pmatrix} 2 \\ -1 \\ 1 \end{pmatrix}.$$

Row 1 is the pivot row, and hence $(1,3)$ is the pivot position for this pivot step. The corresponding pivot matrix is

$$P_3 = \begin{pmatrix} 1/2 & 0 & 0 \\ 1/2 & 1 & 0 \\ -1/2 & 0 & 1 \end{pmatrix}.$$

Finally the permutation matrix defined by the pivot positions is

$$P = \begin{pmatrix} 0 & 0 & 1 \\ 1 & 0 & 0 \\ 0 & 1 & 0 \end{pmatrix}.$$

So, the product form of A^{-1} is $P^T P_3 P_2 P_1$ with these matrices given above.

Example 9:

Let

$$A = \begin{pmatrix} 1 & 1 & -5 \\ 0 & 1 & -2 \\ 1 & 2 & -7 \end{pmatrix}.$$

We will now compute the product form of the inverse of A. Selecting $A_{.2}$ as the first pivot column, and row 3 as the pivot row for the first pivot step (hence $(3,2)$ is the pivot element position for this step), leads to the pivot matrix

$$P_1 = \begin{pmatrix} 1 & 0 & -1/2 \\ 0 & 1 & -1/2 \\ 0 & 0 & 1/2 \end{pmatrix}.$$

Selecting column 1 for the next pivot step, the pivot column is

$$\bar{A}_{.1} = P_1 A_{.1} = \begin{pmatrix} 1 & 0 & -1/2 \\ 0 & 1 & -1/2 \\ 0 & 0 & 1/2 \end{pmatrix} \begin{pmatrix} 1 \\ 0 \\ 1 \end{pmatrix} = \begin{pmatrix} 1/2 \\ -1/2 \\ 1/2 \end{pmatrix}.$$

Selecting Row 2 as the pivot row for this pivot step by the partial pivoting strategy (so, the 2nd pivot element position is $(2,1)$) leads to the pivot matrix corresponding to this pivot step

$$P_2 = \begin{pmatrix} 1 & 1 & 0 \\ 0 & -2 & 0 \\ 0 & 1 & 1 \end{pmatrix}.$$

Column 3 is the column for the final pivot step. The pivot column is

$$\bar{A}_{.3} = P_2 P_1 A_{.3} = \begin{pmatrix} 1 & 1 & 0 \\ 0 & -2 & 0 \\ 0 & 1 & 1 \end{pmatrix} \begin{pmatrix} 1 & 0 & -1/2 \\ 0 & 1 & -1/2 \\ 0 & 0 & 1/2 \end{pmatrix} \begin{pmatrix} -5 \\ -2 \\ -7 \end{pmatrix}$$

$$= \begin{pmatrix} 1 & 1 & 0 \\ 0 & -2 & 0 \\ 0 & 1 & 1 \end{pmatrix} \begin{pmatrix} -3/2 \\ 3/2 \\ -7/2 \end{pmatrix} = \begin{pmatrix} 0 \\ -3 \\ -2 \end{pmatrix}.$$

Row 1, the only remaining row in which a pivot step has not been carried out so far, is the pivot row, but the entry in $\bar{A}_{.3}$ in row 1 is 0. This implies that there is a linear dependence relation among the columns of A, and therefore A^{-1} does not exist. From $\bar{A}_{.3}$ we see that the linear dependence relation is

$$A_{.3} - (-3A_{.1} - 2A_{.2}) = 0.$$

The memory matrix at this stage is $P_2 P_1$ and the row vector in it. Corresponding to row 1 (the row in which we need to carry out the next pivot step) is $I_1 P_2 P_1$, where I is the unit matrix of order 3, and this row is $(1, 1-1)$. It can be verified that $(1, 1, -1)A = A_{1.} + A_{2.} - A_{3.} = 0$, giving the linear dependence relation among the row vectors of A here.

Exercises

4.7.1: Find the inverses of the following nonsingular matrices in product form.

$$\begin{pmatrix} 0 & 2 & -4 \\ 1 & 0 & 2 \\ -2 & -1 & 0 \end{pmatrix}, \quad \begin{pmatrix} 0 & 2 & 2 & 2 \\ 2 & 0 & 2 & 2 \\ 2 & 2 & 0 & 2 \\ 2 & 2 & 2 & 0 \end{pmatrix}.$$

4.7.2: The following matrix is singular. Apply the algorithm to find the inverse in product form on this matrix and show how the singularity of this matrix is detected in that algorithm.

$$\begin{pmatrix} 0 & 2 & -4 \\ 2 & 0 & 2 \\ 4 & 2 & 0 \end{pmatrix}.$$

4.8 How to Update the Inverse When a Column of the Matrix Changes

In some algorithms (for example, the Simplex algorithm for linear programming) we need to compute the inverses of a sequence of square matrices. Consecutive elements of this sequence differ by just one column. To make these algorithms efficient, we need an efficient scheme to update the inverse of a matrix when exactly one of its column vectors changes. We discuss such a scheme here.

Let B be a nonsingular square matrix of order n with its column vectors $B_{.1}, \ldots, B_{.n}$. Let $d = (d_1, \ldots, d_n)^T \neq 0$ be a given column vector, and suppose $\bar{B}$ is the matrix obtained by replacing the rth column vector $B_{.r}$ of B by d. So, the

original matrix $B = (B_{.1} \vdots \ldots \vdots B_{.r-1} B_{.r} B_{.r+1} \vdots \ldots \vdots B_{.n})$

modifed matrix $\bar{B} = (B_{.1} \vdots \ldots \vdots B_{.r-1} d B_{.r+1} \vdots \ldots \vdots B_{.n}).$

Here the symbol r is the number of the column of the original matrix B that is being replaced by the new column vector d. Given B^{-1}, we want an efficient scheme to obtain $(\bar{B})^{-1}$ from B^{-1}. We show that $(\bar{B})^{-1}$ can be obtained from B^{-1} with just one GJ pivot step.

Let us review how B^{-1} is computed in the first place. We start with the initial tableau $(I \vdots B)$, and convert it by GJ pivot steps and a row permutation at the end to the final tableau which is $(B^{-1} \vdots I)$. Now suppose we also put the column vector $d = (d_1, \ldots, d_n)^T$ in the initial tableau, i.e., take the initial tableau to be $(I \vdots B \vdots d)$, and perform exactly the same pivot operations and row permutation needed to get B^{-1} starting with this as the initial tableau. What will the column vector d get transformed into in the final tableau in this process? Let

$$B^{-1} = (\beta_{ij}) = \begin{pmatrix} \beta_{11} & \cdots & \beta_{1n} \\ \vdots & & \vdots \\ \beta_{n1} & \cdots & \beta_{nn} \end{pmatrix}$$

and suppose the final tableau is $(B^{-1} \vdots I \vdots \bar{d})$, where $\bar{d} = (\bar{d}_1, \ldots, \bar{d}_n)^T$. Since $d \neq 0$, by Result 1.11.1(b) of Section 1.11 we know that $\bar{d} \neq 0$. Also, from the memory matrix interpretation of the matrix B^{-1} under columns 1 to n in the final tableau, we know that for $i = 1$ to n

$$\bar{d}_i = \beta_{i1} d_1 + \cdots + \beta_{in} d_n$$

i.e., $\bar{d} = B^{-1}d$ or $d = B\bar{d}$; i.e.,

$$d = \bar{d}_1 B_{.1} + \cdots + \bar{d}_n B_{.n}.$$

So, $\bar{d}$ is the representation of d in terms of the columns of the matrix B. If $\bar{d}_r = 0$, then by rearranging the terms in the above equation, we have

$$\bar{d}_1 B_{.1} + \cdots + \bar{d}_{r-1} B_{.r-1} - d + \bar{d}_{r+1} B_{.r+1} + \cdots + \bar{d}_n B_{.n} = 0$$

a linear dependence relation for the set of column vectors of the matrix $\bar{B}$. So, $(\bar{B})^{-1}$ does not exist if $\bar{d}_r = 0$.

So, we assume that $\bar{d}_r \neq 0$. In the final tableau, the rth column of B, $B_{.r}$ got transformed into $I_{.r}$. Suppose we delete $I_{.r}$ from the final tableau. Then what we have left on the right hand part of the final tableau are the updated columns of the matrix $\bar{B}$, and all we need to do to make it the unit matrix again is to convert the column $\bar{d}$ into $I_{.r}$, i.e., perform a GJ pivot step on the final tableau with $\bar{d}$ as the pivot column and row r as the pivot row, this should convert B^{-1} into $\bar{B}^{-1}$. Since $\bar{B}$ is obtained by replacing $B_{.r}$ by d, we say that in this operation, d is the **entering column** replacing the **dropping column** $B_{.r}$ in B. We now summarize the procedure.

Algorithm to Compute Inverse of $\bar{B}$ Obtained by Replacing $B_{.r}$ in B by d, from B^{-1}

BEGIN

Step 1: Compute $\bar{d} = B^{-1}d$. $\bar{d}$ is called the **updated entering column**, and the operation of computing it is called **updating the entering column**.

If $\bar{d} = (\bar{d}_1, \ldots, \bar{d}_n)^T$ where $\bar{d}_r = 0$, then $\bar{B}$ is singular and its inverse does not exist, terminate.

Suppose $\bar{d}_r \neq 0$. If B^{-1} is given explicitly, to get $(\bar{B})^{-1}$ explicitly, go to Step 2. If B^{-1} is given in product form (discussed in Section 4.7) as $PP_nP_{n-1}\ldots P_1$, where P is a permutation matrix, and $P_1,\ldots,P_n$ are pivot matrices; to get $(\bar{B})^{-1}$ in product form go to Step 3.

Step 2: Here $\bar{d}_r \neq 0$, B^{-1} is given explicitly, and we want to compute $(\bar{B})^{-1}$ explicitly. Put the updated entering column vector $\bar{d}$ by the side of B^{-1} as in the following tableau (PC = pivot column, PR = pivot row)

	PC	
	$\bar{d}_1$	
	$\vdots$	
B^{-1}	$\boxed{\bar{d}_r}$	PR
	$\vdots$	
	$\bar{d}_n$	

and perform the GJ pivot step on this tableau. This pivot step converts B^{-1} into $(\bar{B})^{-1}$.

Step 3: Here B^{-1} is given in product form as $PP_n\ldots P_1$ and we want to compute $(\bar{B})^{-1}$ in product form.

Let P_{n+1} denote the pivot matrix corresponding to the pivot step in Step 2. It is the unit matrix of order n with its rth column replaced by

$$\left(\frac{\bar{d}_1}{\bar{d}_r},\ldots,-\frac{\bar{d}_{r-1}}{\bar{d}_r},\frac{1}{\bar{d}_r},-\frac{\bar{d}_{r+1}}{\bar{d}_r},\ldots,-\frac{\bar{d}_n}{\bar{d}_r}\right)^T.$$

Then the product form for the inverse of $\bar{B}$ is $P_{n+1}PP_nP_{n-1}\ldots P_1$.

END

Example 1:

Let

$$B = \begin{pmatrix} 0 & -2 & -1 \\ 1 & 0 & -1 \\ 1 & 1 & 0 \end{pmatrix}, \quad B^{-1} = \begin{pmatrix} 1 & -1 & 2 \\ -1 & 1 & -1 \\ 1 & -2 & 2 \end{pmatrix}, \quad d = \begin{pmatrix} 1 \\ 2 \\ 1 \end{pmatrix}.$$

Let $\bar{B}$ be the matrix obtained by replacing the third column of B by d; and $\hat{B}$ the matrix obtained by replacing the 2nd column of B by d. So,

$$\hat{B} = \begin{pmatrix} 0 & 1 & -1 \\ 1 & 2 & -1 \\ 1 & 1 & 0 \end{pmatrix}, \quad \bar{B} = \begin{pmatrix} 0 & -2 & 1 \\ 1 & 0 & 2 \\ 1 & 1 & 1 \end{pmatrix}.$$

From B^{-1} given above we want to find the inverses of $\bar{B}, \hat{B}$ by the updating procedure given above. The updated entering column is

$$\bar{d} = B^{-1}d = \begin{pmatrix} 1 & -1 & 2 \\ -1 & 1 & -1 \\ 1 & -2 & 2 \end{pmatrix} \begin{pmatrix} 1 \\ 2 \\ 1 \end{pmatrix} = \begin{pmatrix} 1 \\ 0 \\ -1 \end{pmatrix}.$$

Since the 2nd entry in the updated column $\bar{d}$ is 0, it shows that $\hat{B}$ is singular and does not have an inverse.

To find the inverse of $\bar{B}$, we have to carry out the pivot step in the following tableau (PC = pivot column, PR = pivot row, and the pivot element is boxed).

			PC	
1	−1	2	1	
−1	1	−1	0	
1	−2	2	$\boxed{-1}$	PR
2	−3	4	0	
−1	1	−1	0	
−1	2	−2	1	

So, we have

$$(\bar{B})^{-1} = \begin{pmatrix} 2 & -3 & 4 \\ -1 & 1 & -1 \\ -1 & 2 & -2 \end{pmatrix}.$$

Exercises

4.8.1: For $r = 1, 2, 3$, and 4 in (ii), let B^r refer to the matrix obtained by replacing the rth column vector of the following matrix B by d. Find the inverses of B^1, B^2, B^3 and B^4 in (ii) respectively by updating B^{-1}

given below.

(i) $B = \begin{pmatrix} 0 & -2 & -1 \\ 1 & 0 & -1 \\ 1 & 1 & 0 \end{pmatrix}$, $B^{-1} = \begin{pmatrix} 1 & -1 & 2 \\ -1 & 1 & -1 \\ 1 & -2 & 2 \end{pmatrix}$, $d = \begin{pmatrix} 4 \\ 1 \\ -1 \end{pmatrix}$

(ii) $B = \begin{pmatrix} 1 & -1 & 0 & 1 \\ 0 & 1 & 0 & 1 \\ 0 & 2 & 1 & 1 \\ 0 & 0 & 0 & 1 \end{pmatrix}$, $B^{-1} = \begin{pmatrix} 1 & 1 & 0 & -2 \\ 0 & 1 & 0 & -1 \\ 0 & -2 & 1 & 1 \\ 0 & 0 & 0 & 1 \end{pmatrix}$, $d = \begin{pmatrix} 0 \\ -1 \\ -1 \\ -1 \end{pmatrix}$.

4.9 Results on the General System of Linear Equations $Ax = b$

Here we summarize various results on systems of linear equations, with a brief explanation of each result. We consider the general system of linear equations

$$Ax = b$$

where the coefficient matrix $A = (a_{ij})$ is of order $m \times n$, and the RHS constants vector $b = (b_i)$ is a column vector in R^m. $A_1., \ldots, A_m.$ are the row vectors of A; and $A._1, \ldots, A._n$ are its column vectors. The "system" in the following refers to this system of equations. The $m \times (n+1)$ matrix $(A \vdots b)$ obtained by including the RHS constants vector as an additional column in the coefficient matrix A, is known as the **augmented matrix** of the system. The augmented matrix contains all the *data* in the system.

1. **Constraint Representation:** Each row vector of the coefficient matrix A corresponds to a constraint in the system. These constraints are:

$$A_i. x = b_i \qquad \text{for } i = 1 \text{ to } m.$$

2. **Interpretation of a Solution of the System:** Another way of looking at the system through the columns of the coefficient matrix A yields the following:

$$x_1 A._1 + x_2 A._2 + \cdots + x_n A._n = b$$

i.e., each solution $x = (x_j)$ of the system is the coefficient vector in an expression of b as a linear combination of the columns of A.

, Hence the system has a solution iff b can be expressed as a linear combination of the columns of A.

3. **Conditions for the Existence of a Solution:** System has a solution iff $\text{rank}(A) = \text{rank}$ of the $m \times (n+1)$ augmented matrix $(A\dot{:}b)$.

Rank $(A) = \text{rank}(A\dot{:}b)$ iff a basis for the set of column vectors of A is also a basis for the set of column vectors of $(A\dot{:}b)$, i.e., iff b can be expressed as a linear combination of the columns of A, by Result 2 above, this is the condition for the existence of a solution to the system.

If $\text{rank}(A) \neq \text{rank}(A\dot{:}b)$, then b cannot be expressed as a linear combination of the columns of A, and the system has no solution. In this case, including b as an additional column vector in A increases its rank, i.e., $\text{rank}(A\dot{:}b) = 1 + \text{rank}(A)$.

4. **Condition for No Solution:** System has no solution iff $\text{rank}(A\dot{:}b) = 1 + \text{rank}(A)$.
   Follows from Result 3 above.

5. **Condition for Many Solutions:** If the system has at least one solution, and the set of column vectors of the coefficient matrix A is linearly dependent, then the system has an infinite number of solutions. Conversely, if the system has more than one solution, the set of column vectors of A is linearly dependent, and in this case the system has an infinite number of solutions.

   To see this, suppose $\bar{x} = (\bar{x}_j)$ is a solution for the system. Then, from Result 2 above we have

$$b = \bar{x}_1 A_{.1} + \cdots + \bar{x}_j A_{.j} + \cdots + \bar{x}_n A_{.n}.$$

If $\{A_{.1}, \ldots, A_{.n}\}$ is linearly dependent, let the linear dependence relation for this set be

$$0 = \alpha_1 A_{.1} + \cdots + \alpha_j A_{.j} + \cdots + \alpha_n A_{.n}$$

where $(\alpha_1, \ldots, \alpha_n) \neq 0$. Adding λ times the 2nd equation to the first (here λ is a scalar, an arbitrary real number) yields

$$b = (\bar{x}_1 + \lambda\alpha_1)A_{.1} + \cdots + (\bar{x}_j A_{.j} + \lambda\alpha_j)A_{.j} + \cdots + (\bar{x}_n + \lambda\alpha_n)A_{.n}.$$

So, if we define $x(\lambda) = (x_1(\lambda), \ldots, x_n(\lambda))^T$ where for $j = 1$ to n

$$x_j(\lambda) = \bar{x}_j + \lambda\alpha_j$$

then $x(\lambda)$ is a solution of the system for all λ real. Since $(\alpha_1, \ldots, \alpha_n) \neq 0$, for each λ, $x(\lambda)$ is a distinct vector; hence the system has an infinite number of solutions. Also, as λ takes all real values, $x(\lambda)$ traces a straight line in R^n. This entire straight line is contained in the set of solutions of the system.

Conversely, suppose the system has two distinct solutions $\bar{x}, \hat{x} = (\hat{x}_j)$. Then we have

$$b = \bar{x}_1 A_{.1} + \cdots + \bar{x}_j A_{.j} + \cdots + \bar{x}_n A_{.n}$$
$$b = \hat{x}_1 A_{.1} + \cdots + \hat{x}_j A_{.j} + \cdots + \hat{x}_n A_{.n}.$$

Subtracting the bottom one from the top one, we have

$$0 = (\bar{x}_1 - \hat{x}_1) A_{.1} + \cdots + (\bar{x}_j - \hat{x}_j) A_{.j} + \cdots + (\bar{x}_n - \hat{x}_n) A_{.n}.$$

Since $\bar{x} \neq \hat{x}$, at least one of the coefficients in the above equation is nonzero, hence this equation is a linear dependence relation for $\{A_{.1}, \ldots, A_{.n}\}$, showing that the set of column vectors of A is linearly dependent.

6. **Condition For Unique Solution:** If the system has a solution $\bar{x}$, then $\bar{x}$ is the unique solution of the system iff the set of column vectors of A is linearly independent.

This follows from Result 5 above.

7. **At Least One Solution If A Is of Full Row Rank:** The matrix A is said to be **of full row rank** if its set of row vectors is linearly independent, i.e., if its rank = the number of rows in it, m. In this case the system has at least one solution.

   In this case when GJ method is applied to solve the system, a basic variable will be selected in every row. So, this method will terminate with a solution of the system after performing a pivot step in each row.

8. **In a Consistent System, Number of Constraints Can Be Reduced to $\leq$ Number Of Variables:** If the system has at least one solution, either $m \leq n$, or after eliminating redundant constraints from it, an equivalent system in which the number of constraints is $\leq n$ will be obtained.

   If rank$(A) = r$, then $r \leq \min\{m, n\}$. The result follows because when the system has at least one solution, the GJ method, after eliminating redundant constraints, will reduce it to an equivalent system in which the number of constraints is $r \leq n$. Also, if the system has a unique solution, it becomes a square nonsingular system of equations after the elimination of redundant constraints from it.

Therefore, without any loss of generality, we can assume that in a consistent system of linear equations, the number of constraints is $\leq$ the number of variables.

Exercises

4.9.1: Newton's problem of the fields and cows (from his *Arithmetica universalis*, 1707): There are several fields of growing grass, and cows. Assume that initially all the fields provide the same amount of grass (say x_1 per field), and that the daily growth of grass in all the fields remains constant (say x_2 in each field), and that all of the cows eat exactly the same amount of grass each day (say exactly x_3 per cow per day). The following information is given:

> a cows graze b fields bare in c days
> a' cows graze b' fields bare in c' days
> a'' cows graze b'' fields bare in c'' days.

Then what relationship exists between the 9 data elements a, a', a''; b, b', b''; c, c', c''?

4.10   Optimization, Another Important Distinction Between LA, LP, and IP; and the Duality Theorem for Linear Optimization Subject to Linear Equations

We already discussed in Chapter 1 that Linear Algebra (**LA**) is the subject dealing with the analysis of systems of linear equations, Linear Programming (**LP**) is the subject dealing with the analysis of systems of linear constraints including inequalities, and that Integer Programming (**IP**) is the subject dealing with the analysis of systems of linear constraints with some variables restricted to take integer values only. In this section we will present some differences in the mathematical properties of these various systems when dealing with optimization. We will also present the important duality theorem for linear optimization subject to linear equality constraints, and its relationship to the alternate system for a system of linear equations discussed in Section 1.17.

Linear Inequalities Viewed as Nonlinear Equations

Each linear inequality actually can be viewed as a nonlinear equation. For example, consider the linear inequality

$$x_1 + x_2 + x_3 \leq 100.$$

This inequality is the same as $100 - x_1 - x_2 - x_3 \geq 0$. Every nonnegative number has a nonnegative square root. So, let s_1 denote the nonnegative square root $\sqrt{100 - x_1 - x_2 - x_3}$. Then $100 - x_1 - x_2 - x_3 = s_1^2$, or

$$x_1 + x_2 + x_3 + s_1^2 = 100.$$

This equation is a nonlinear equation because of the square term s_1^2 term in it, and it is clearly equivalent to the above inequality. In the same way the inequality constraint $x_1 + 3x_2 - 2x_3 \geq -10$ is equivalent to the nonlinear equation $x_1 + 3x_2 - 2x_3 - s_2^2 = -10$.

These additional variables s_1^2, s_2^2 that were used to convert the inequalities into equations are called **slack variables associated with those inequalities** in LP literature. So linear inequalities can be converted into equations by introducing slack variables, but the resulting equations are nonlinear. As another example, consider the system of linear constraints

$$2x_1 + 3x_2 - 7x_3 = 60$$
$$x_1 + x_2 + x_3 \leq 100$$
$$x_1 + 3x_2 - 2x_3 \geq -10.$$

This system is the same as the system of equations

$$2x_1 + 3x_2 - 7x_3 = 60$$
$$x_1 + x_2 + x_3 + s_1^2 = 100$$
$$x_1 + 3x_2 - 2x_3 - s_2^2 = -10$$

which is nonlinear.

Linear Optimization Subject to Linear Equality Constraints

In most real world applications one is always interested in finding not just any solution, but the *best solution* among all solutions to a system of constraints, *best* in the sense of optimizing the value of an objective function that is to be optimized (i.e., either minimized or maximized as desired). Since LA considers systems of linear constraints only, it is reasonable to assume that the user may want to optimize an objective function that is itself a linear function of the decision variables. This would lead to a problem of the following form which we state using matrix notation.

$$\text{Minimize} \quad z = z(x) = cx$$
$$\text{subject to} \quad Ax = b \tag{1}$$

instead of finding an arbitrary solution to the system $Ax = b$. Such a problem is called **a linear optimization problem subject to linear equations.** Here let A be an $m \times n$ coefficient matrix, b an $m \times 1$ RHS constants vector, and c is a $1 \times n$ row vector of objective function coefficients.

Instead of minimizing, if it is desired to maximize an objective function $c'x$, it is the same as minimizing its negative $-c'x$ subject to the same constraints. Hence it is sufficient to consider minimization problems only.

If system (1) is either infeasible, or has a unique solution, there is nothing to choose, so we will assume that (1) has many solutions. Even in this case, it turns out that there are only two possibilities for problem (1). They are:

Possibility 1: Under this possibility all solutions of system (1) give the same value for the objective function cx, i.e., cx is a constant on the set of solutions of (1). If this possibility occurs, it is immaterial which solution of (1) is selected, all have the same objective value.

This possibility occurs iff the row vector c can be expressed as a linear combination of the row vectors of the coefficient matrix A, i.e., iff c is in the row space of the matrix A as defined in Section 3.10.

Possibility 2: The minimum value of cx on the set of solutions of (1) is $-\infty$, i.e., (1) has no finite optimum solution.

This possibility occurs iff the row vector c cannot be expressed as a linear combination of the row vectors of the coefficient matrix A.

Algorithm to Determine Which Possibility Occurs for (1)

The GJ or G elimination methods for solving systems of linear equations discussed in Sections 1.16, 1.20 can directly determine which of Possibilities 1, 2 occurs in (1). Here is the simple modification of the GJ or G elimination methods to solve (1).

BEGIN

Step 1: Detached coefficient tableau: Initiate the method with the following detached coefficient tableau

x	$-z$	RHS
A	0	b
c	1	0

where the bottom row corresponds to the additional equation $cx - z = 0$ that just defines the objective function. When the algorithm determines that Possibility 1 occurs, if it is required to obtain the coefficients in an expression of c as a linear combination of rows of A, we need to introduce a memory matrix (a unit matrix of order $m + 1$ with its column vectors labeled as $A_1., \ldots, A_m., c$) on the left side of the tableau as explained in Section 1.12, and carry out all the row operations on the columns of this memory matrix also.

Step 2: Carry out the GJ or G elimination methods of Sections 1.16 or 1.20 on this tableau and perform pivot steps in all the rows of the tableau except the last row. At the end if the infeasibility conclusion is not reached, go to Step 3.

Step 3: In the final tableau suppose the entries in the bottom row are $\bar{c}_1, \ldots, \bar{c}_n; \bar{b}_{m+1}$.

If all of $\bar{c}_1, \ldots, \bar{c}_n$ are $= 0$, then c is a linear combination of the rows of A, and Possibility 1 occurs for (1). In this case the objective function in (1), $cx = -\bar{b}_{m+1}$ at all the solutions of the system $Ax = b$. Delete the last row from the final tableau, and obtain the general solution of the system from the remaining tableau, every one of these solutions is an optimal solution of (1).

If the memory matrix is introduced, its bottom row in the final tableau gives the coefficients in an expression of c as a linear combination of rows of A.

If at least one of $\bar{c}_1, \ldots, \bar{c}_n$ is $\neq 0$, then the minimum value of cx on the set of solutions of (1) is $-\infty$, i.e., Possibility 2 occurs for (1). In this case we can construct a half-line every point on which is feasible to (1), and along which the objective function $cx \to -\infty$. The construction of this half-line from the final tableau is illustrated in Example 2 given below.

END

Example 1:

Consider the following problem.

$$\text{Minimize } z = 6x_1 + 22x_2 + 34x_3 + 6x_4$$
$$\text{subject to } x_1 + x_2 + x_3 + x_4 = 15$$
$$x_1 - x_2 - 2x_3 - x_4 = -5$$
$$-x_1 + 2x_4 = 7.$$

The original tableau for solving this problem by the GJ elimination method with memory matrix (to find the coefficients in an expression of c as a linear combination of the rows of the coefficient matrix A, if it turns out to be so) is the one given at the top. The columns of the memory matrix are labeled A_1, A_2, A_3, and c since these are the various rows in the original tableau. The abbreviations used are: BV = basic variable, and all the pivot elements are boxed. PR, the pivot row; and PC, the pivot column, are always the row, column containing the boxed pivot element.

Memory matrix				BV						RHS
A_1	A_2	A_3	c		x_1	x_2	x_3	x_4	$-z$	
1	0	0	0		1	$\boxed{1}$	1	1	0	15
0	1	0	0		1	-1	-2	-1	0	-5
0	0	1	0		-1	0	0	2	0	7
0	0	0	1		6	22	34	6	1	0
1	0	0	0	x_2	1	1	1	1	0	15
1	1	0	0		2	0	$\boxed{-1}$	0	0	10
0	0	1	0		-1	0	0	2	0	7
-22	0	0	1		-16	0	12	-16	1	-330

Memory matrix				BV						RHS
A_1	A_2	A_3	c		x_1	x_2	x_3	x_4	$-z$	
2	1	0	0	x_2	3	1	0	1	0	25
-1	-1	0	0	x_3	-2	0	1	0	0	-10
0	0	1	0		$\boxed{-1}$	0	0	2	0	7
-10	12	0	1		8	0	0	-16	1	-220
				Final Tableau						
2	1	3	0	x_2	0	1	0	7	0	46
-1	-1	-2	0	x_3	0	0	1	-4	0	-26
0	0	-1	0	x_1	1	0	0	-2	0	-7
-10	12	8	1		0	0	0	0	1	-164

GJ pivot steps have been carried out in all the constraint rows. Since the coefficients of all the original variables x_j are 0 in the last row of the final tableau, this is an indication that c is a linear combination of rows of the coefficient matrix A in this example. From the final row of the memory matrix in the final tableau we find

$$-10A_1 + 12A_2 + 8A_3 + c = 0$$

$$\text{or} \quad c = 10A_1 - 12A_2 - 8A_3.$$

Also the equation corresponding to the last row in the final tableau is $-z = -164$ or $z = 164$. Thus in this example, the objective function z has the same value 164 at all solutions of the system of constraints. From the final tableau we see that the general solution of the system is

$$\begin{pmatrix} x_1 \\ x_2 \\ x_3 \\ x_4 \\ x_5 \end{pmatrix} = \begin{pmatrix} -7 - 2x_3 + 2x_5 \\ 46 + 4x_3 - 7x_5 \\ x_3 \\ -26 - 3x_3 + 4x_5 \\ x_5 \end{pmatrix}$$

where the nonbasic (free) variables x_3, x_5 can be given any real values at the user's choice to get a solution of the system; and all these solutions yield the same value 164 to the objective function z.

Example 2:

Consider the following problem.

$$\text{Minimize } z = x_1 + x_2 - 2x_4 - 4x_5$$
$$\text{subject to } x_1 + x_2 - x_3 + x_5 = -6$$
$$x_1 + 2x_2 + x_4 - 2x_5 = 20$$
$$-x_1 + x_3 - x_4 + 3x_5 = 12.$$

We solve this problem by the G elimination method, but use the same notation as in the previous example. In this example we do not intend to find the expession of c as a linear combination of the row vectors of A even if it turns out to be in the linear hull of the rows of A; so we do not introduce the memory matrix.

BV	x_1	x_2	x_3	x_4	x_5	$-z$	RHS	
	$\boxed{1}$	1	−1	0	1	0	−6	PR
	1	2	0	1	−2	0	20	
	−1	0	1	−1	3	0	12	
	1	1	0	−2	−4	1	0	
	PC↑							
x_1	1	1	−1	0	1	0	−6	
	0	$\boxed{1}$	1	1	−3	0	26	PR
	0	1	0	−1	4	0	6	
	0	0	1	−2	−5	1	6	
		PC↑						

BV	x_1	x_2	x_3	x_4	x_5	$-z$	RHS	
x_1	1	1	-1	0	1	0	-6	
x_2	0	1	1	1	-3	0	26	
	0	0	$\boxed{-1}$	-2	7	0	-20	PR
	0	0	1	-2	-5	1	6	
			PC↑					

Final Tableau

BV	x_1	x_2	x_3	x_4	x_5	$-z$	RHS
x_1	1	1	-1	0	1	0	-6
x_2	0	1	1	1	-3	0	26
x_3	0	0	1	2	-7	0	20
	0	0	0	-4	2	1	-14

The nonbasic variables x_4, x_5 have nonzero coefficients in the objective row in the final tableau. This indicates that the objective function z diverges to $-\infty$ on the set of solutions of the system of constraints in this example, so there is no finite optimum solution in this problem.

To construct a half-line in the set of feasible solutions along which the objective function z diverges to $-\infty$, select any nonbasic variable with a nonzero coefficient in the objective row in the final tableau, call it the **entering variable**, x_s say with $\bar{c}_s \neq 0$ as its coefficient in the objective row in the final tableau. Fix all the nonbasic variables other than the entering variable at 0 in the solution, and make the value of the entering variable $= \lambda$, a real valued parameter. Multiply the column of the entering variable in the final tableau by λ and move it to the right. Now compute the corresponding values of the basic variables in the solution by the back substitution method. The solution is a function of the single parameter λ. As $\lambda \to +\infty$ if $\bar{c}_s < 0$, or $-\infty$ if $\bar{c}_s > 0$, this solution traces a half-line in the solution set along which the objective function $z \to -\infty$.

In this example we select x_4 with $\bar{c}_4 = -4$ as the entering variable. We fix the other nonbasic variable $x_5 = 0$, and entering variable $x_4 = \lambda$. The triangular system to find the corresponding values of the basic variables in the solution is

x_1	x_2	x_3	
0	1	-1	-6
0	1	1	$26 - \lambda$
0	0	1	$20 - 2\lambda$
	$-z$		$-14 + 4\lambda$

The expression for $-z$ is obtained from the equation corresponding to the last row in the final tableau after fixing the entering variable $= \lambda$, and all the other nonbasic variables $= 0$. Solving by back substitution, we find that the solution is

$$x(\lambda) = \begin{pmatrix} x_1(\lambda) \\ x_2(\lambda) \\ x_3(\lambda) \\ x_4(\lambda) \\ x_5(\lambda) \end{pmatrix} = \begin{pmatrix} 8 - 3\lambda \\ 6 + \lambda \\ 20 - 2\lambda \\ \lambda \\ 0 \end{pmatrix}$$

$$z = 14 - 4\lambda.$$

So, as $\lambda \to +\infty$, $x(\lambda)$ traces a half-line in the set of feasible solutions of the system along which the objective function $z \to -\infty$.

Thus, in linear optimization problems subject to linear equality constraints only (of the form (1) above), when feasible solutions exist, either there is no finite optimum solution, or all the feasible solutions are optimal. That's why these problems are considered to be trivial mathematically, and none of the other books on LA discuss optimizing a linear function subject to linear equality constraints.

The Duality Theorem for Linear optimization Problems Subject to Linear Equality Constraints

Consider the general linear optimization problem subject to linear equality constraints (1) above. There is another linear optimization problem subject to linear equality constraints associated with it called its **dual**, sharing the same data A, b, c. In this context the original problem (1) is called the **primal**. It turns out that the dual of the dual problem is the primal; and the special relationship between this pair of problems is called the **duality relationship**, it is a symmetric relationship.*

In this context, the variables in the primal (1), $x = (x_1, \ldots, x_n)^T$, are called the **primal variables**. Since the constraints in the primal problem are a requirement on $Ax = \sum_{j=1}^{n} A_{.j} x_j$, the primal variables x are coefficients in a linear combination of the columns of the data matrix A. Symmetrically, the variables in the dual problem, called the **dual variables**, written as a row vector $y = (y_1, \ldots, y_m)$ usually, are coefficients in a linear combination of the rows of the data matrix A; and the dual constraints will be a requirement on yA. That's why this duality relationship is an instance of **row-column duality** in matrices.

The dual problem of (1) is

$$\text{Minimize} \quad v = v(y) = yb \tag{2}$$
$$\text{subject to} \quad yA = c.$$

Notice that the primal variables do not appear in the statement of the dual problem and vice versa. The primal and the dual problems are distinct problems, but both of them share the same data.

If Possibility 1 occurs for (1), then the algorithm discussed above finds a dual feasible solution $\bar{y}$ satisfying $c = \bar{y}A$. In this case every primal feasible solution x has its primal objective value $= z(x) = cx = \bar{y}Ax = \bar{y}b$. Also, in this case every dual feasible solution y has its dual objective value $= v(y) = yb = yAx = cx$ where x is any primal feasible solution.

Thus if Possibilty 1 occurs for (1), every primal feasible solution has the same primal objective value, every dual feasible solution has the same dual objective value, and the optimum objective values in the two problems are equal.

If Possibility 2 occurs for (1), it is not possible to express c as a linear combination of the rows of A, i.e., the dual has no feasible solution; and we have seen that in this case $z(x)$ is unbounded below in (1).

Now suppose that the dual has a feasible solution $\bar{y}$ say, but the primal has no feasible solution. In this case by the results discussed in Section 1.17, we know that the alternate system for the constraints in (1)

$$\pi A = 0$$

$$\pi b = 1$$

has a feasible solution, $\bar{\pi}$ say. Since $\bar{y}A = c$ and $\bar{\pi}A = 0$, we have $(\bar{y} + \lambda\bar{\pi})A = c$ for any λ real, i.e., $\bar{y} + \lambda\bar{\pi}$ is feasible to the dual problem (2) for all λ. Also $v((\bar{y} + \lambda\bar{\pi}) = (\bar{y} + \lambda\bar{\pi})b = \bar{y}b + \lambda\bar{\pi}b = \bar{y}b + \lambda$, and this tends to $+\infty$ as $\lambda \to +\infty$. Thus the dual objective function is unbounded above in the dual problem in this case. The feasibility of the alternate system for the constraints in (1), is a necessary condition for the dual objective function to be unbounded above on the set of dual feasible solutions in (2).

These results can be summarized in the following statement known as the **duality theorem for linear optimization subject to linear equality constraints**. It is a specialization of the *duality theorem of LP* proved by OR people (John von Neumann, David Gale, Harold Kuhn, and Albert Tucker) soon after George Dantzig anounced his simplex method for LP in 1947, to (1).

Duality Theorem: *A primal, dual pair of linear optimization problems subject to linear equality constraints always satisfy the following properties:*

1. *If both primal and dual are feasible, the optimum objective values in the two problems are equal; and every feasible solution is optimal in both problems.*
2. *If primal [dual] is feasible but the dual [primal] is infeasible, then the primal objective value [dual objective value] is unbounded below [above] in the primal [dual] problem.*
3. *It is possible for both the primal and dual problems to be infeasible.*

Optimizing a Linear Objective Function Subject to Linear Inequality Constraints

This is a problem of the following form stated in matrix notation.

$$\text{Minimize} \quad z = z(x) = cx$$
$$\text{subject to} \quad Ax = b \qquad\qquad (3)$$
$$Dx \geq d.$$

The presence of linear inequality constraints in (3) makes a significant difference in the possible outcomes for (3) as opposed to those of (1) where all the constraints are linear equations only. (3) can have a finite optimum solution, without all the feasible solutions of the problem being optimal. Thus (3) is a nontrivial optimization problem. This is an important distinction between linear equality systems and linear systems including inequalities.

Traditionally LP is defined to be a problem of the form (3) involving optimization, not just that of finding any feasible solution.

But there is a beautiful result in LP theory which states that corresponding to (3), there is an equivalent system of linear constraints involving inequalities such that every feasible solution of it is optimal to (3) and vice versa. This result guarantees that LPs involving optimization are mathematically equivalent to finding a feasible solution to a system of linear constraints containing linear inequalities. In other words any algorithm for finding a feasible solution to a system of linear inequalities can be used directly to solve optimization problems like (3) without doing any optimization, and vice versa.

Optimizing a Linear Objective Function Subject to Linear Constraints and Integer Restrictions on Variables

In matrix notation, this IP is a problem of the following form. Here $x = (x_1, \ldots, x_n)^T$ is the vector of decision variables. $J \subset \{1, \ldots, n\}$ is the set of subscripts of decision variables whose values in the solution are required to be integer.

$$
\begin{aligned}
\text{Minimize} \quad & z = z(x) = cx \\
\text{subject to} \quad & Ax = b \\
& Dx \geq d \\
& x_j \text{ integer for all } j \in J.
\end{aligned}
\tag{4}
$$

In LP all the variables are continuous variables. But in IP some or all the variables are restricted to a discrete set, and hence IPs are also called **discrete optimization problems**.

Traditionally an IP is stated in this problem involving optimization. As in the case of LP, in IP also, an efficient algorithm that can find a feasible solution to a system of constraints like that in (4) can be modified into an efficient algorithm for solving (4) involving optimization too. Also like in LPs, (4) can have a finite optimum without all the feasible solutions being optimal to it. So, IPs also are nontrivial optimization problems.

4.11   Memory Matrix Versions of GJ, G Elimination Methods

The versions of the GJ, G elimination methods discussed in Sections 1.16, 1.20 are the direct simple versions of those methods to solve a system of linear equations dating back to ancient times. Around mid-20th century Operations Researchers have found that the results in Sections 1.12, 4.7 can be used to organize the computation in these methods more efficiently by introducing a memory matrix on the original tableau and restricting the pivot operations to the columns of the memory matrix only, leaving the original system unchanged throughout the computation. The pioneering work for developing this efficient version is due to George Dantzig in his 1953 paper on the revised simplex method for linear programming (G. B. Dantzig, "Computational Algorithm of the Revised Simplex Method", RM-1266, The RAND Corporation, Santa Monica, CA, 1953).

Suppose the original system consists of m equations in n variables. In this version pivot operations are carried out only on $(m+1)$ column vectors (m column vectors of the memory matrix, and the RHS constants vector) instead of the $(n+1)$ column vectors of the original system. Since typically $n \geq m$, the new version is computationally more efficient than the versions discussed in Sections 1.16, 1.20.

We will now provide a complete statement of the new version along with detailed descriptions of all the formulas used in it, and then work out several numerical examples.

Memory Matrix Version of GJ, G Elimination Methods

Let the original system being solved be $Ax = b$ where $A = (a_{ij})$ is an $m \times n$ matrix.

BEGIN

Step 1: Setting up the initial tableau: It is convenient to introduce the columns of the memory matrix (initially, a unit matrix of order m) on the left hand side of the detached coefficient tableau of the original system. Let us denote the constraints in the original system by $R_1, \ldots, R_m$ and enter these labels in a column on the right hand side of the original system. We label the columns of the memory matrix with R_1 to R_m, because as mentioned in Section 1.12, any row vector in the memory matrix is the row vector of coefficients in an expression of the corresponding row in the updated system as a linear combination of rows in the original system. We also set up another column on the right to record the basic variables. We will use the abbreviations: BV = basic variable selected in the row, PC = pivot column, PR = pivot row, **OCL = original constraint label**.

Initial tableau: Tableau 1

BV	OCL	Original system					Memory matrix		
		x_1	$\cdots$	x_j	$\cdots$	x_n	RHS	R_1	$\cdots$	R_m
	R_1	a_{11}	$\cdots$	a_{1j}	$\cdots$	a_{1n}	b_1	1	$\cdots$	0
	$\vdots$	$\vdots$		$\vdots$		$\vdots$	$\vdots$	$\vdots$		$\vdots$
	R_i	a_{i1}	$\cdots$	a_{ij}	$\cdots$	a_{in}	b_i	0	$\ddots$	0
	$\vdots$	$\vdots$		$\vdots$		$\vdots$	$\vdots$	$\vdots$		$\vdots$
	R_m	a_{m1}	$\cdots$	a_{mj}	$\cdots$	a_{mn}	b_m	0	$\cdots$	1

OCL = Original constraint label for the row

Some notes: In this version, the columns of the variables $x_1, \ldots, x_n$ in the original system are left as they are throughout the computation; pivot operations are carried out only on the other columns of the initial tableau. So these other columns constitute the working area in this version, and we will record these only separately in subsequent steps, calling it the **working tableau**. The working tableau in the initial stage is Tableau 2.

Initial working tableau: Tableau 2

BV	Updated RHS	Memory matrix R_1 ... R_m	PC	OCL
b_1		1 ... 0		R_1
$\vdots$		$\vdots$ $\vdots$		$\vdots$
b_i		0 $\ddots$ 0		R_i
$\vdots$		$\vdots$ $\vdots$		$\vdots$
b_m		0 ... 1		R_m

In the GJ elimination method, rows 1 to m can be selected as pivot rows in any order, and the method does not need row interchanges. In the G elimination method however, we perform a row interchange in every step to make sure that all rows which have already served as pivot rows are at the top of the tableau, and those in which pivot steps have yet to be carried out are below them at the bottom. In this version, like pivot steps, row interchanges are only carried out on the working tableau, and never on the original tableau. If some row interchanges have been carried out so far, in the current working tableau rows 1 to m correspond to the original constraints $R_{w_1}, \ldots, R_{w_m}$ respectively, where $(w_1, \ldots, w_m)$ is a permutation of $(1, \ldots, m)$.

Current working tableau: Tableau 3

BV	Updated RHS	Updated mem. matrix R_1 ... R_m	PC	OCL
$\bar{b}_1$		β_{11} ... β_{1m}		R_{w_1}
$\vdots$		$\vdots$ $\vdots$		$\vdots$
$\bar{b}_i$		β_{i1} $\ddots$ β_{im}		R_{w_i}
$\vdots$		$\vdots$ $\vdots$		$\vdots$
$\bar{b}_m$		β_{m1} ... β_{mm}		R_{w_m}

At some general stage suppose the current working tableau is as given in Tableau 3. The matrix $\beta = (\beta_{ij} : i, j = 1, \ldots, m)$ in the current working tableau is the updated memory matrix at this stage. In this version we do not have the current tableau consisting of the updated columns of all the variables, which would have been obtained if all the pivot steps have been carried out on the original tableau as in the versions discussed in Sections 1.16, 1.20. Let us use the following notation for the updated columns of the variables in the current tableau:

OCL	Current tableau*			
	x_1	x_2	x_j	x_n
R_1	$\bar{a}_{11}$	$\bar{a}_{12}$ ...	$\bar{a}_{1j}$ ...	$\bar{a}_{1n}$
R_2	$\bar{a}_{21}$	$\bar{a}_{22}$ ...	$\bar{a}_{2j}$ ...	$\bar{a}_{2n}$
$\vdots$	$\vdots$	$\vdots$	$\vdots$	$\vdots$
R_i	$\bar{a}_{i1}$	$\bar{a}_{i2}$ ...	$\bar{a}_{ij}$ ...	$\bar{a}_{in}$
$\vdots$	$\vdots$	$\vdots$	$\vdots$	$\vdots$
R_m	$\bar{a}_{m1}$	$\bar{a}_{m2}$ ...	$\bar{a}_{mj}$ ...	$\bar{a}_{mn}$

*Not computed in this version

But all the entries in this current tableau can be computed using the updated memory matrix β and the original tableau. We show the formulas for computing them now.

Let $\bar{A} = (\bar{a}_{ij} : i = 1 \text{ to } m, j = 1 \text{ to } n)$ denote the current tableau which is not computed in this version. $\bar{A}_{i.}, \bar{A}_{.j}$ denote the ith row, jth column of $\bar{A}$. $\bar{A}_{.j}$ is the updated column of the variable x_j in the current tableau.

From Section 1.12 we know that the ith row in the current tableau is the linear combination of rows in the original tableau with coefficients from the row of the updated memory matrix corresponding to the OCL R_i (this may not be the ith row in the updated memory matrix if row interchanges have been performed during the algorithm). So, let

p be such that $w_p = i$ in the current working tableau. Then

$$i\text{th row in current tableau} = \bar{A}_{i.} = (\bar{a}_{i1}, \ldots, \bar{a}_{in})$$

$$= \beta_{w_p.} A = (\beta_{w_p 1}, \ldots, \beta_{w_p n}) \begin{pmatrix} a_{11} & \cdots & a_{1n} \\ \vdots & & \vdots \\ a_{m1} & \cdots & a_{mn} \end{pmatrix}.$$

Using this formula any row vector in the current tableau can be computed from β and A as needed.

From the results in Section 4.7, we know that the updated column of any variable x_j in the current tableau with rows rearrranged as in the current working tableau $= (\bar{a}_{w_1 j}, \ldots, \bar{a}_{w_m j})$ is obtained by multiplying the original column of x_j on the left by β, i.e., it is $= \beta A_{.j}$.

Using these formulas, any row vector in the current tableau, or the updated column of any variable can be computed using $\beta, \dot{A}$. We will use these formulae in the following steps.

Step 2: The general pivot step: Select one of the rows in which a pivot step has not been performed, as the row for performing the next pivot step. Suppose it is row w_t in original tableau corresponding to row r in the current working tableau. This updated row is $(\bar{a}_{w_t1}, \ldots, \bar{a}_{w_t n}) = \beta_r.A$.

Case 1: If $\bar{a}_{w_t1} = \cdots = \bar{a}_{w_t n} = 0$, and the updated RHS constant in this row $= \bar{b}_{w_t} = 0$ also, this row represents a redundant constraint. The row $\beta_r.$, the rth row of the updated memory matrix is the evidence for this redundancy conclusion.

So, no pivot step will be carried out in this row in this case. Select one of the remaining rows for performing the next pivot step and repeat this Step 2 with it.

Case 2: If $\bar{a}_{w_t1} = \cdots = \bar{a}_{w_t n} = 0$, and the updated RHS constant in this row $= \bar{b}_{w_t} \neq 0$, this row represents an inconsistent equation, so the original system has no solution. The row $\beta_r.$, the rth row of the updated memory matrix is the evidence for this infeasibility conclusion (i.e., $\bar{b}_r$ times a feasible solution for the alternate system corresponding to the original system discussed in Section 1.17).

If the infeasibility conclusion is all that is needed, terminate the method. At any rate no pivot step will be carried out in this row. If it is required to find a modification of the RHS constants vector b in the original system to make it feasible as discussed in Section 1.22, select one of the remaining rows to perform the next pivot step and repeat this Step 2 with it.

Case 3: Suppose at least one of $\bar{a}_{w_t1}, \ldots, \bar{a}_{w_t n}$ is $\neq 0$. In this case we will perform a pivot step.

If GJ elimination method is being used, select this row, row w_t in the original tableau corresponding to row r in current working tableau as the pivot row for this pivot step. Select an s such that $\bar{a}_{w_t s} \neq 0$ as the pivot element, and the associated variable x_s as the basic variable in this pivot row, i.e., the entering variable in this pivot step.

When doing hand computation, we want to avoid getting difficult fractions, so we look for a pivot element of ± 1 or the smallest available integer as far as possible. But while solving the problem on a computer, we try to select the pivot element to minimize roundoff error accumulation, for this it is better to select the pivot element to be an element of maximum absolute value, i.e., satisfying $|\bar{a}_{w_ts}| = \text{maximum} \{|\bar{a}_{w_tj}| : j = 1 \text{ to } n\}$. The pivot column (PC) for this pivot step is the updated column of the entering variable with rows rearranged as in the current working tableau $= \beta A_{.s}$. Enter the PC on the current working tableau, mark the pivot element in it, and perform the GJ pivot step.

If G elimination method is being used to solve the problem on a computer, we select s such that $\bar{a}_{w_ts}$ is the maximum absolute value element among $\bar{a}_{w_t1}, \ldots, \bar{a}_{w_tn}$, and select the associated variable x_s as the entering variable for this pivot step. Then the pivot column (PC) for this pivot step is the updated column of the entering variable x_s with rows rearranged as in the current working tableau $= \beta A_{.s}$. Enter the PC on the current working tableau. Select the pivot element to be an element of maximum absolute value in the PC among all elements in it in rows r to m (these are the rows in which pivot steps have not been carried out so far). In the working tableau interchange the row containing the pivot element with row r, then make it the pivot row for the pivot step. Then perform the G pivot step.

If there are no more rows in which pivot steps have to be carried out, with the new updated working tableau go to Step 3 if the infeasibility conclusion has not been reached so far, or to Step 4 if the infeasibility conclusion has occured during the algorithm and it is desired to carry out infeasibility analysis.

If there are some more rows in which pivot steps have to be carried out, select one of those to perform the next pivot step in the new working tableau, and repeat this Step 2 with it.

Step 3: Reading the Solution: In the final working tableau every row which has no basic variable recorded in it corresponds to a redundant constraint in the original system, and can be deleted without changing the set of feasible solutions of the original system.

By making all the nonbasic variables 0, and equating each basic variable in the final working tableau to the updated RHS constant in its row, we get a basic solution of the original system if the GJ method is used (if the G method is used instead, a back substitution step is needed to get the values of the basic variables).

The parametric representation of the general solution of the system can be obtained by computing the updated column of each of the nonbasic variables using the formulas discussed above, and then using the formula for the general solution discussed in Sections 1.16, 1.20.

Step 4: Infeasibility analysis: Here the original system is infeasible. It can be made feasible by modifying some of the data in it. One possible modification is to change the RHS constants vector b as discussed in Section 1.22.

For this, if any row interchanges have been carried out during the algorithm, rearrange the rows in the final working tableau according to their order in the original tableau using the information in the OCL column. For each i such that row i corresponds to an infeasible constraint (i.e., updated constraint of the form $0 = \bar{b}_i \neq 0$) change the original RHS constant b_i to $b_i' = b_i - \bar{b}_i$ where $\bar{b}_i$ is the updated RHS constant in this row; this modification has the effect of converting $\bar{b}_i$ in the working tableau to 0. With this change each infeasible constraint becomes a redundant constraint, and the system becomes feasible. A basic solution, and the general solution of the modified system can now be obtained from the final working tableau as in Step 3.

END

We will now provide some numerical examples. In all these examples we will use the following notation:

A, b = Coefficient matrix and RHS constants vector in the original system

$A_{i.}, A_{.j}$ = The ith row, jth column of A

$\beta, \beta_{i.}$ = The current working tableau, its ith row

$\bar{A}, \bar{A}_{i.}, \bar{A}_{.j}$ = The updated coefficient matrix, and its ith row, jth column

$\bar{b} = (\bar{b}_i)$ = Updated RHS constants vector

OCL = Original constraint label

PR, PC; BV = Pivot row, pivot column for the current pivot step, selected basic variable in that row.

Example 1:

This problem is from Example 1 in Section 1.16 solved there by the version of the GJ elimination method discussed there. The original system with the OCL is:

x_1	x_2	x_3	x_4	x_5	RHS	OCL
-1	0	1	1	2	10	R_1
1	-1	2	-1	1	5	R_2
1	-2	5	-1	4	20	R_3
2	0	0	-2	1	25	R_4
3	-2	5	-3	5	45	R_5
5	-2	5	-5	6	70	R_6

We will apply the memory matrix version of the GJ elimination method to find a feasible solution of this system. The initial working tableau is:

BV		Memory matrix β						PC	OCL	
	$\bar{b}$	R_1	R_2	R_3	R_4	R_5	R_6	x_3		
	10	1	0	0	0	0	0	$\boxed{1}$	R_1	PR
	5	0	1	0	0	0	0	2	R_2	
	20	0	0	1	0	0	0	5	R_3	
	25	0	0	0	1	0	0	0	R_4	
	45	0	0	0	0	1	0	5	R_5	
	70	0	0	0	0	0	1	5	R_6	

We select x_3 as the first entering variable and perform a pivot step with row 1 as the PR, leading to the next working tableau.

BV		Memory matrix β						PC	OCL	
	$\bar{b}$	R_1	R_2	R_3	R_4	R_5	R_6	x_2		
x_3	10	1	0	0	0	0	0	0	R_1	
	-15	-2	1	0	0	0	0	$\boxed{-1}$	R_2	PR
	-30	-5	0	1	0	0	0	-2	R_3	
	25	0	0	0	1	0	0	0	R_4	
	-5	-5	0	0	0	1	0	-2	R_5	
	20	-5	0	0	0	0	1	-2	R_6	

Selecting row 2 to perform the next pivot step we find $\bar{A}_{2.} = \beta_{2.}A = (-2, 1, 0, 0, 0, 0)A = (3, -1, 0, -3, -3)$. We select the 2nd entry in it, -1, as the pivot element, x_2 as the entering variable. Its updated column is $\beta A_{.2} = \beta(0, -1, -2, 0, -2, -2)^T = \bar{A}_{.2}$ which is the PC entered on the working tableau. Performing this pivot step leads to the next working tableau.

BV		Memory matrix β						PC	OCL	
	$\bar{b}$	R_1	R_2	R_3	R_4	R_5	R_6	x_5		
x_3	10	1	0	0	0	0	0	2	R_1	
x_2	15	2	-1	0	0	0	0	3	R_2	
	0	-1	-2	1	0	0	0	0	R_3	
	25	0	0	0	1	0	0	$\boxed{1}$	R_4	PR
	25	-1	-2	0	0	1	0	1	R_5	
	50	-1	-2	0	0	0	1	2	R_6	

Selecting row 3 to perform the next piovot step, we find that $\bar{A}_3. = \beta_3.A = (0,0,0,0,0)$, a row vector of zeros. Also, $\bar{b}_3 = 0$. Hence the 3rd constraint in the original system can be treated as a redundant constraint, so we cannot perform a pivot step in it. Moving to row 4, we find that $\bar{A}_4. = \beta_4.A = (2,0,0,-2,1)$ and we select the element 1 in it as the pivot element and x_5 as the entering variable for the next pivot step. The PC, the updated column of x_5 is $\beta A_{.5} = \beta(2,1,4,1,5,6)^T = (2,3,0,1,1,2)^T$ which is entered on the working tableau. Performing this pivot step leads to the next working tableau.

Final working tableau

BV	$\bar{b}$	Memory matrix β						PC	OCL
		R_1	R_2	R_3	R_4	R_5	R_6	x_2	
x_3	-40	1	0	0	-2	0	0		R_1
x_2	-60	2	-1	0	-3	0	0		R_2
	0	-1	-2	1	0	0	0		R_3
x_5	25	0	0	0	1	0	0		R_4
	0	-1	-2	0	-1	1	0		R_5
	0	-1	-2	0	-2	0	1		R_6

Only rows 5, 6 remain for performing pivot steps, but we now find that both sides of their updated rows are 0, so both these rows represent redundant constraints in the original tableau, so no pivot steps need to be performed in them. Hence the above working tableau is the final working tableau. From it we see that a basic solution for the original system is given by

$$\text{Basic variables} \quad \begin{cases} x_3 = -40 \\ x_2 = -60 \\ x_5 = 25 \end{cases}$$

$$\text{nonbasic variables} \quad \begin{cases} x_1 = 0 \\ x_4 = 0 \end{cases}$$

which is $x = (0,-60,-40,0,25)^T$.

To get the general solution of the system we find the updated column of the nonbasic variables x_1, x_4 which are $\bar{A}_{.1} = \beta A_{.1} = (-5,-9,0,2,0,0)^T$, and $\bar{A}_{.4} = \beta A_{.4} = (5,9,0,-2,0,0)^T$. Hence as explained in Section 1.16, the general solution of the system is given by

$$x_3 = -4 + 5x_1 - 5x_4$$

$$x_2 = -60 + 9x_1 - 9x_4,$$

$$x_5 = 25 - 2x_1 + 2x_4$$

$$x_1 = x_1, \quad \text{at user's choice}$$

$$x_4 = x_4, \quad \text{at user's choice}$$

or rearranging the variables in natural order, it is

$$x = \begin{pmatrix} x_1 \\ -60 + 9x_1 - 9x_4 \\ -40 + 5x_1 - 5x_4 \\ x_4 \\ 25 - 2x_1 + 2x_4 \end{pmatrix}.$$

The user can select any real values for the free variables x_1, x_4 and plug them in the above to get a solution of the original system. Giving them values 0 leads to the basic solution mentioned above.

From the final working tableau we also find

Evidence of redundancy for R_3 is $(-1, -2, 1, 0, 0, 0)$
Evidence of redundancy for R_5 is $(-1, -2, 0, -1, 1, 0)$
Evidence of redundancy for R_6 is $(-1, -2, 0, -2, 0, 1)$.

For the original constraint R_3 this means that the linear combination $-R_1 - R_2 + R_3$ is the $0 = 0$ equation, or equivalently, that constraint R_3 is the linear combination $R_1 + R_2$.

Similarly R_5 is $R_1 + 2R_2 + R_4$ and R_6 is $R_1 + 2R_2 + 2R_4$ all of which can be verified easily from the original tableau.

Example 2:

This problem is from Example 3 in Section 1.16. The original system with the OCLs is

x_1	x_2	x_3	x_4	x_5	RHS	OCL
2	1	-1	-2	0	4	R_1
4	1	5	6	4	9	R_2
6	2	4	4	4	12	R_3
10	3	9	10	8	22	R_4

The initial working tableau is

BV	$\bar{b}$	Memory matrix β				PC	OCL	
		R_1	R_2	R_3	R_4	x_2		
	4	1	0	0	0	$\boxed{1}$	R_1	PR
	9	0	1	0	0	1	R_2	
	12	0	0	1	0	2	R_3	
	22	0	0	0	1	3	R_4	

We select row 1 to perform the 1st pivot step, and select x_2 as the entering variable in it. The PC, original column of the entering variable x_2 is entered on the working tableau. Performing the pivot step leads to the next working tableau.

BV	$\bar{b}$	Memory matrix β				PC	OCL	
		R_1	R_2	R_3	R_4	x_1		
x_2	4	1	0	0	0	2	R_1	
	5	-1	1	0	0	$\boxed{2}$	R_2	PR
	4	-2	0	1	0	2	R_3	
	10	-3	0	0	1	4	R_4	

Selecting row 2 for the next pivot step, we find $\bar{A}_{2.} = \beta_{2.}A = (2, 0, 6, 8, 4)$ and we choose 2 in it as the pivot element. The entering variable is x_1, its updated column, $\beta A_{.1} = (2, 2, 2, 4)^T$, is the PC entered on the working tableau. Performing the pivot step leads to the next working tableau.

Final working tableau

BV	$\bar{b}$	Memory matrix β				PC	OCL
		R_1	R_2	R_3	R_4		
x_2	-1	2	-1	0	0		R_1
x_1	5/2	-1/2	1/2	0	0		R_2
	-1	-1	-1	1	0		R_3
	0	-1	-2	0	1		R_4

We verify that the updated row 3 is $\bar{A}_{3.} = \beta_{3.}A = 0$, and the updated RHS constant in it is $-1 \neq 0$. So row 3 represents an infeasible constraint, and the original system has no feasible solution.

Evidence of infeasibility for R_3 is $(-1, -2, 0, 1)$.

This evidence shows that the linear combination of original constraints $-R_1 - R_2 + R_3$ is the infeasible equation $0 = -1$.

This also implies that the linear combination $(-R_1 - R_2 + R_3)/(-1) = R_1 + R_2 - R_3$ is the fundamental infeasible equation $0 = 1$. Hence $(1, 1, -1, 0)$ is a solution of the alternate system corresponding to the original system, discussed in Section 1.17.

We also verify that the updated row 4 is $\beta_4.A = 0$, and since $\bar{b}_4 = 0$ row 4 represents a redundant constraint. The evidence for this is $(-1, -2, 0, 1)$ which implies that $-R_1 - 2R_2 + R_4$ is the $0 = 0$ equation, i.e., R_4 is the linear combination $R_1 + 2R_2$. So no pivot steps will be performed in rows 3, 4; hence the current working tableau is the final one.

We will now perform infeasibility analysis (Section 1.22) on this infeasible system. From the final working tableau, we see that the original system can be made feasible by subtracting $\bar{b}_3 = -1$ from the original $b_3 = 12$, i.e., changing b_3 to $b_3' = 13$. This changes the original RHS constants vector b from $(4, 9, 12, 22)^T$ to $(4, 9, 13, 22)^T$.

If the same pivot steps as above are carried out on the modified original tableau the final working tableau will be

Final working tableau

BV	$\bar{b}$	Memory matrix β				PC	
		R_1	R_2	R_3	R_4		
x_2	-1	2	-1	0	0		R_1
x_1	$5/2$	$-1/2$	$1/2$	0	0		R_2
	0	-1	-1	1	0		R_3
	0	-1	-2	0	1		R_4

The only change here is that $\bar{b}_3$ has now become 0, so row 3 represents a redundant constraint in the modified problem, hence the modified problem is feasible. A basic solution for it, from the final working tableau, is given by $x_2 = -1$, $x_1 = 5/2$, $x_3 = x_4 = 0$; i.e., it is $x = (5/2, -1, 0, 0)^T$.

Example 3:

Here we will solve the problem in Example 1 from Section 1.20 with the memory matrix version of the G elimination method using the partial pivoting strategy for pivot element selection.

The original tableau

BV	x_1	x_2	x_3	x_4	x_5	x_6	RHS	OCL
1	-2	1	1	1	-1	-4	R_1
2	2	0	1	-1	-1	10	R_2
2	-4	-2	0	-2	-2	-8	R_3
0	-6	-2	-1	-1	-1	-18	R_4
0	0	0	4	1	1	20	R_5
1	4	-1	0	-2	0	14	R_6

The Initial working tableau

BV	$\bar{b}$	Memory matrix β						PC	OCL	
		R_1	R_2	R_3	R_4	R_5	R_6	x_1		
	-4	1	0	0	0	0	0	1	R_1	
	10	0	1	0	0	0	0	2	R_2	
	-8	0	0	1	0	0	0	2	R_3	RI to row 1
	-18	0	0	0	1	0	0	0	R_4	
	20	0	0	0	0	1	0	0	R_5	
	14	0	0	0	0	0	1	1	R_6	

Starting with row 1, we select x_1 with an entry of 1 in it as the entering variable. We enter its column as the PC. An entry of maximum absolute value in it is 2 in R_3. So, we interchange R_3 and R_1 leading to the following, in which we perform the G pivot step leading to the 2nd working tableau.

BV	$\bar{b}$	Memory matrix β						PC	OCL	
		R_1	R_2	R_3	R_4	R_5	R_6	x_1		
	-8	0	0	1	0	0	0	$\boxed{2}$	R_3	PR
	10	0	1	0	0	0	0	2	R_2	
	-4	1	0	0	0	0	0	1	R_1	
	-18	0	0	0	1	0	0	0	R_4	
	20	0	0	0	0	1	0	0	R_5	
	14	0	0	0	0	0	1	1	R_6	

The 2nd working tableau is

BV	$\bar{b}$	Memory matrix β						PC	OCL	
		R_1	R_2	R_3	R_4	R_5	R_6	x_3		
x_1	-8	0	0	1	0	0	0	-2	R_3	
	18	0	1	-1	0	0	0	2	R_2	
	0	1	0	$-1/2$	0	0	0	2	R_1	RI to row 2
	-18	0	0	0	1	0	0	-2	R_4	
	20	0	0	0	0	1	0	0	R_5	
	18	0	0	$-1/2$	0	0	1	0	R_6	

Looking at row 2, we see that $\bar{A}_{2.} = \beta_{2.}A = (0,6,2,1,1,1)$ and we select x_3 with a nonzero entry of 2 in it as the entering variable. The PC $= \bar{A}_{.3} = \beta A_{.3} = (-2,2,2,-2,0,0)^T$. To illustrate row interchanges we select the 3rd entry in it as the pivot element, thus requiring the interchange of rows 2 and 3 in the working tableau leading to

BV	$\bar{b}$	Memory matrix β						PC	OCL	
		R_1	R_2	R_3	R_4	R_5	R_6	x_3		
x_1	-8	0	0	1	0	0	0	-2	R_3	
	0	1	0	$-1/2$	0	0	0	$\boxed{2}$	R_1	PR
	18	0	1	-1	0	0	0	2	R_2	
	-18	0	0	0	1	0	0	-2	R_4	
	20	0	0	0	0	1	0	0	R_5	
	18	0	0	$-1/2$	0	0	1	0	R_6	

Performing the G pivot step leads to the 3rd working tableau.

BV	$\bar{b}$	Memory matrix β						PC	OCL	
		R_1	R_2	R_3	R_4	R_5	R_6	x_2		
x_1	-8	0	0	1	0	0	0	-4	R_3	
x_3	0	1	0	$-1/2$	0	0	0	0	R_1	
	18	-1	1	$-1/2$	0	0	0	$\boxed{6}$	R_2	PR
	-18	1	0	$-1/2$	1	0	0	-6	R_4	
	20	0	0	0	0	1	0	0	R_5	
	18	0	0	$-1/2$	0	0	1	6	R_6	

Looking at row 3, the updated row is $\bar{A}_{3.} = \beta_3.A = (0,6,0,0,-1,1)$ and we select the entering variable to be x_2 with an entry of 6 in it. The PC is $\beta A_{.2} = (-4,0,6,-6,0,6)^T$. We make the 6 in row 3 in the PC as the pivot element and row 3 as the PR, so no row interchange is needed in this step. Performing this G pivot step leads to the 4th working tableau.

BV	$\bar{b}$	R_1	R_2	R_3	R_4	R_5	R_6	PC x_4	OCL	
x_1	-8	0	0	1	0	0	0	0	R_3	
x_3	0	1	0	$-1/2$	0	0	0	1	R_1	
x_2	18	-1	1	$-1/2$	0	0	0	0	R_2	
	0	0	1	-1	1	0	0	0	R_4	
	20	0	0	0	0	1	0	$\boxed{4}$	R_5	PR
	0	1	-1	0	0	0	1	0	R_6	

Looking at row 4 we find $\bar{A}_{4.} = \beta_4.A = 0$ and $\bar{b}_4 = 0$ too. So row 4 represents a redundant constraint and no pivot step will be performed in it. Eveidence for this redundancy is $(0,1,-1,1,0,0)$ which implies that $R_4 = -R_2 + R_3$.

Now looking at row 5, the updated row is $\beta_5.A = (0,0,0,4,1,1)$ and we select x_4 with an entry of 4 in it as the entering variable. The PC is $\beta A_{.4} = (0,1,0,0,4,0)^T$. So we keep the 4 in it as the pivot element and row 5 as the PR. Hence no row interchanges now, the G pivot step leads to the 5th working tableau.

Final working tableau

BV	$\bar{b}$	R_1	R_2	R_3	R_4	R_5	R_6	PC	OCL	
x_1	-8	0	0	1	0	0	0		R_3	
x_3	0	1	0	$-1/2$	0	0	0		R_1	
x_2	18	-1	1	$-1/2$	0	0	0		R_2	
	0	0	1	-1	1	0	0		R_4	
x_4	20	0	0	0	0	1	0		R_5	PR
	0	1	-1	0	0	0	1		R_6	

The only remaining row is row 6. The updated row 6 is $\beta_6.A = 0$. Also, the updated RHS constant in this row is 0, hence this row represents a redundant constraint, the evidence for this is $(1,-1,0,0,0,1)$ which shows that $R_6 = -R_1 + R_2$.

From the final working tableau we compute $\beta A_{.1}, \beta A_{.3}, \beta A_{.2}, \beta A_{.4}$ (the updated columns of x_1, x_3, x_2, x_4) and then obtain the following updated system by setting the nonbasic variables x_5, x_6 both at 0.

x_1	x_3	x_2	x_4	RHS
2	-2	-4	0	-8
0	1	0	1	0
0	0	6	0	18
0	0	0	0	0
0	0	0	4	20
0	0	0	0	0

Eliminating the redundant 4th and 6th constraints here leads to the triangular system

x_1	x_3	x_2	x_4	RHS
2	-2	-4	0	-8
0	1	0	1	0
0	0	6	0	18
0	0	0	4	20

from which we obtain the basic solution $x = (-1/2, 3, -5/2, 5, 0, 6)^T$ for the original system by back substitution.

4.12 Orthogonal and Orthonormal Sets of Vectors, Gram–Schmidt Orthogonalization, QR-Factorization, Orthogonal Matrices, Orthonormal Bases

Orthogonal Sets of Vectors

A set of vectors $\{a^1, \ldots, a^p\}$ in R^n is said to be an **orthogonal set of vectors** if every pair of distinct vectors from the set is orthogonal, i.e., if the dot product $\langle a^i, a^j \rangle = 0$ for all $i \neq j$.

Example 1:

The set of n unit vectors in R^n is an orthogonal set. For example, for $n = 3$, these are

$$I_{.1} = \begin{pmatrix} 1 \\ 0 \\ 0 \end{pmatrix}, \quad I_{.2} = \begin{pmatrix} 0 \\ 1 \\ 0 \end{pmatrix}, \quad I_{.3} = \begin{pmatrix} 0 \\ 0 \\ 1 \end{pmatrix}$$

and clearly, $\langle I_{.i}, I_{.j} \rangle = 0$ for all $i \neq j$. As another example, consider the set of vectors $\{a^1, a^2, a^3, a^4\}$ from R^3 where

$$a^1 = \begin{pmatrix} 0 \\ 0 \\ 0 \end{pmatrix}, \quad a^2 = \begin{pmatrix} 6 \\ 2 \\ 2 \end{pmatrix}, \quad a^3 = \begin{pmatrix} -2 \\ 4 \\ 2 \end{pmatrix}, \quad a^4 = \begin{pmatrix} -1 \\ -4 \\ 7 \end{pmatrix}$$

and verify that the dot product $\langle a^i, a^j \rangle = 0$ for all $i \neq j$, so that the set $\{a^1, a^2, a^3, a^4\}$ is an orthogonal set.

Result 4.12.1: Linear independence of an orthogonal set of vectors: *If $\Gamma = \{a^1, \ldots, a^p\}$ is an orthogonal set of nonzero vectors, then Γ is linearly independent.*

To see this, suppose scalars $\alpha_1, \ldots, \alpha_p$ exist such that

$$\alpha_1 a^1 + \cdots + \alpha_p a^p = 0.$$

For any $1 \leq i \leq p$ take the dot product of both sides of this equation with a^i. This gives

$$0 = \langle \alpha_1 a^1 + \cdots + \alpha_p a^p, a^i \rangle = \alpha_1 \langle a^1, a^i \rangle + \cdots + \alpha_{i-1} \langle a^{i-1}, a^i \rangle$$
$$+ \alpha_i \langle a^i, a^i \rangle + \alpha_{i+1} \langle a^{i+1}, a^i \rangle + \cdots + \alpha_p \langle a^p, a^i \rangle$$
$$= \alpha_i \|a^i\|^2, \quad \text{since } \langle a^j, a^i \rangle = 0 \text{ for all } j \neq i \text{ by orthogonality.}$$

Since $a^i \neq 0$ for all i, we have $\|a^i\| > 0$, hence the above equation only holds if $\alpha_i = 0$ for all i; i.e., the only linear combination of vectors in Γ that is 0 is the 0-linear combination. Hence Γ is linearly independent.

Normalizing an Orthogonal Set of Vectors, Orthonormal Sets

If $a \neq 0$ is any vector in R^n, $\|a\| \neq 0$, and hence we can multiply a by the scalar $1/\|a\|$. The process of multiplying the nonzero vector a by $1/\|a\|$ is called **normalizing** a. It yields the **normalized vector** $b = (1/\|a\|)a$ satisfying $\|b\| = 1$.

The zero vector 0 cannot be normalized because $\|0\| = 0$. Every nonzero vector can be normalized.

As an example, consider $a = (-3, 0, 4, 0) \in R^4$.

$$\|a\| = \sqrt{9 + 0 + 16 + 0} = 5.$$

So normalizing a leads to $b = (1/5)a = (-3/5, 0, 4/5, 0)$ whose Euclidean norm is 1.

A set of vectors $\{a^1, \ldots, a^p\}$ in R^n is said to be an **orthonormal set of vectors** if it is an orthogonal set in which each vector has norm 1. For example the following sets of vectors in R^3 are orthonormal sets.

$$\left\{ I_{.1} = \begin{pmatrix} 1 \\ 0 \\ 0 \end{pmatrix}, \; I_{.2} = \begin{pmatrix} 0 \\ 1 \\ 0 \end{pmatrix}, \; I_{.3} = \begin{pmatrix} 0 \\ 0 \\ 1 \end{pmatrix} \right\}$$

$$\left\{ \frac{1}{\sqrt{44}} \begin{pmatrix} 6 \\ 2 \\ 2 \end{pmatrix}, \; \frac{1}{\sqrt{24}} \begin{pmatrix} -2 \\ 4 \\ 2 \end{pmatrix}, \; \frac{1}{\sqrt{66}} \begin{pmatrix} -1 \\ -4 \\ 7 \end{pmatrix} \right\}.$$

An orthogonal set of vectors can be converted into an orthonormal set if all the vectors in it are nonzero, by normalizing each vector in it.

Example 2:

Convert the following sets of vectors into orthonormal sets.

(i) $$\left\{ \begin{pmatrix} 1 \\ -1 \\ 1 \\ 0 \end{pmatrix}, \begin{pmatrix} -1 \\ 1 \\ 0 \\ 1 \end{pmatrix}, \begin{pmatrix} 1 \\ 1 \\ 0 \\ 1 \end{pmatrix} \right\}.$$

This set cannot be converted into an orthonormal set by normalizing, because it is not an orthogonal set to start with (the dot product of the 2nd and the 3rd vectors is $1 \neq 0$).

(ii) $$\left\{ \begin{pmatrix} 1 \\ -1 \\ 1 \\ 0 \end{pmatrix}, \begin{pmatrix} -1 \\ 1 \\ 0 \\ 1 \end{pmatrix}, \begin{pmatrix} 1 \\ 1 \\ 0 \\ 0 \end{pmatrix} \right\}.$$

This set is an orthogonal set. Dividing each vector in the set by its norm leads to the orthonormal set

$$\left\{ \frac{1}{\sqrt{3}} \begin{pmatrix} 1 \\ -1 \\ 1 \\ 0 \end{pmatrix}, \frac{1}{\sqrt{3}} \begin{pmatrix} -1 \\ 1 \\ 0 \\ 1 \end{pmatrix}, \frac{1}{\sqrt{2}} \begin{pmatrix} 1 \\ 1 \\ 0 \\ 0 \end{pmatrix} \right\}$$

(iii) $$\left\{ \begin{pmatrix} 1 \\ -1 \\ 1 \\ 0 \end{pmatrix}, \begin{pmatrix} -1 \\ 1 \\ 0 \\ 1 \end{pmatrix}, \begin{pmatrix} 1 \\ 1 \\ 0 \\ 0 \end{pmatrix}, \begin{pmatrix} 0 \\ 0 \\ 0 \\ 0 \end{pmatrix} \right\}.$$

This is an orthogonal set, but since it contains the 0-vector, it cannot be converted into an orthonormal set by normalizing each vector.

Orthogonal, Orthonormal Bases

A basis for a set of vectors, or for the linear hull of a set of vectors, is said to be an **orthogonal basis** if that basis is an orthogonal set, and an **orthonormal basis** if it is an orthonormal set.

As an example, the set of n unit vectors in R^n is an orthogonal, and also orthonormal basis for R^n. For $n = 3$, this basis is $\{(1,0,0),(0,1,0),(0,0,1)\}$.

Representation of Vectors in Terms of an Orthonormal Basis

Result 4.12.2: Representation of a vector in terms of an orthonormal basis: *Let $\{a^1,\ldots,a^m\}$ be an orthonormal basis for a given set of vectors Γ. Let a be a vector in the linear hull of Γ. Then as defined in Section 4.2, the representation of a in terms of the basis $\{a^1,\ldots,a^m\}$ is the unique vector $\{\alpha_1,\ldots,\alpha_m\}$ satisfying*

$$a = \alpha_1 a^1 + \cdots + \alpha_m a^m.$$

Then, $(\alpha_1,\ldots,\alpha_m) = (\langle a^1,a\rangle,\ldots,\langle a^m,a\rangle).$

To see this, we have for any $i = 1$ to m, $\langle a^i,a\rangle = \langle a^i, \alpha_1 a^1 + \cdots + \alpha_m a^m\rangle = \alpha_1\langle a^i,a^1\rangle + \cdots + \alpha_m\langle a^i,a^m\rangle = \alpha_i\langle a^i,a^i\rangle = \alpha_i$; because $\langle a^i,a^j\rangle = 0$ for all $j \neq i$ and $\langle a^i,a^i\rangle = 1$ by the orthonormality of the set $\{a^1,\ldots,a^m\}$.

Example 3:

As an example, consider the set $\Gamma = \{a^1,\ldots,a^5\}$ where

$$a^1 = \begin{pmatrix} 1 \\ 0 \\ 0 \\ 0 \\ 0 \\ 0 \end{pmatrix}, \quad a^2 = \begin{pmatrix} 0 \\ -2/3 \\ 2/3 \\ 1/3 \\ 0 \\ 0 \end{pmatrix}, \quad a^3 = \begin{pmatrix} 0 \\ 1/2 \\ 1/2 \\ 0 \\ -1/2 \\ -1/2 \end{pmatrix},$$

$$a^4 = \begin{pmatrix} 1 \\ -1 \\ 3 \\ 1 \\ -1 \\ -1 \end{pmatrix}, \quad a^5 = \begin{pmatrix} 1 \\ -3 \\ 5 \\ 2 \\ -1 \\ -1 \end{pmatrix}.$$

$\{a^1, a^2, a^3\}$ is an orthonormal basis for this set. Consider the vector

$$a = a^1 - 6a^2 - 4a^3 + a^4 - a^5 = (1, 4, -8, -3, 2, 2)^T$$

in the linear hull of Γ. Then from Result 4.12.2, the representation of a in terms of the orthonormal basis $\{a^1, a^2, a^3\}$ is $(\langle a^1, a \rangle, \langle a^2, a \rangle, \langle a^3, a \rangle) = (1, -9, -4)$. In fact we verify that $a = a^1 - 9a^2 - 4a^3$.

The Gram–Schmidt Orthogonalization Procedure

Any linearly independent set of vectors, Δ, can be converted into an orthonormal set of vectors with the same linear hull as Δ using this procedure. The procedure is named after J. P. Gram of Denmark, and E. Schmidt of Germany who developed it in the late 19th and early 20th centuries. However, the same procedure was used by other researchers much earlier.

Let $\Delta = \{a^1, \ldots, a^k\}$ be a linearly independent set of vectors in R^n say. The **Gram–Schmidt procedure** converts this into an orthonormal set of vectors $\{q^1, \ldots, q^k\}$ generating the vectors $q^1, \ldots, q^k$ sequentially in this specific order; that's why often the vectors $q^1, \ldots, q^k$ are referred to as the **Gram–Schmidt sequence**. The procedure is:

$$q^1 = (1/\nu_1)a^1$$

and for $r = 2$ to k

$$q^r = (1/\nu_r)\left(a^r - \sum_{i=1}^{r-1} \langle q^i, a^r \rangle q^i\right)$$

where $\nu_1 = \|a^1\|$, $\nu_r = \|(a^r - \sum_{i=1}^{r-1} \langle q^i, a^r \rangle q^i)\|$ for $r = 2$ to k.

For some r between 2 to k, if $(a^r - \sum_{i=1}^{r-1} \langle q^i, a^r \rangle q^i)$ becomes 0, we have a linear dependence relation (because this equation gives a^r as a linear combination of $\{a^1, \ldots, a^{r-1}\}$) which implies that the set Δ is linearly dependent, and the procedure cannot continue further. So, if Δ is linearly independent as originally assumed, the procedure generates the complete Gram–Schmidt sequence.

From the formula for q^r, it is clear that $\|q^r\| = 1$ for all $r = 1$ to k when the entire sequence is obtained. Also, the fact that $\langle q^j, q^r \rangle = 0$ for all $j \leq r - 1$ can be established inductively. For example,

$$\langle q^2, q^1 \rangle = \langle (1/\nu_2)(a^2 - \langle q^1, a^2 \rangle a^1), q^1 \rangle$$

$$= (1/\nu_2)(\langle a^2, q^1 \rangle - \langle q^1, a^2 \rangle \langle a^1, q^1 \rangle)$$

$$= (1/\nu_2)\langle a^2, q^1 \rangle (1 - \langle a^1, q^1 \rangle) = 0$$

because $\langle a^1, q^1 \rangle = 1$.

It can also be verified that the Gram–Schmidt sequence $q^1, \ldots, q^k$ satisfies the following property: for all $r = 1$ to k

linear hull of $\{q^1, \ldots, q^r\} =$ linear hull of $\{a^1, \ldots, a^r\}$.

The Gram–Schmidt procedure given above is mathematically correct, but if implemented in a straightforward way using floating point arithmetic, the sequence produced may be far from being an orthogonal set because of roundoff errors. Numerical analysts have developed modified versions of the Gram–Schmidt algorithm using product from implementations that are much more stable. These are outside the scope of this book, the reader is referred to C. Meyer [4.2] for details.

We will provide a numerical example of Gram–Schmidt orthogonalization, after discussing the QR-factorization of a matrix with a linearly independent set of column vectors.

The QR-Factorization (or Decomposition) of a Matrix

Let $A = (a_{ij})$ be an $m \times n$ matrix of rank n, i.e., the set of column vectors of A, $\{A_{.1}, \ldots, A_{.n}\}$ is linearly independent. Since $\{A_{.1}, \ldots, A_{.n}\}$ is linearly independent, we can convert it into an orthonormal set of n column vectors $\{Q_{.1}, \ldots, Q_{.n}\}$ using the Gram–Schmidt orthogonalization discussed above. This procedure obtains these vectors sequentially using the formulas

$$Q_{.1} = (1/\nu_1)A_{.1}$$

$$Q_{.r} = (1/\nu_r)\left(A_{.r} - \sum_{i=1}^{r-1}\langle Q_{.i}, A_{.r}\rangle Q_{.i}\right), \quad r = 2, \ldots, k$$

where $\nu_1 = \|A_{.1}\|$, and for $r = 2$ to k, $\nu_r = \|A_{.r} - \sum_{i=1}^{r-1}\langle Q_{.i}, A_{.r}\rangle Q_{.i}\|$. Let

$$Q = (Q_{.1} \vdots \ldots \vdots Q_{.n})$$

be the matrix with $Q_{.1}, \ldots, Q_{.n}$ as its column vectors. The above equations can be written as

$$A_{.1} = \nu_1 Q_{.1}$$

$$A_{.r} = \sum_{i=1}^{r-1}\langle Q_{.i}, A_{.r}\rangle Q_{.1} + \nu_r Q_{.r}, \quad r = 2, \ldots, n.$$

In matrix notation, these equations state that

$$A = QR$$

where

$$R = \begin{pmatrix} \nu_1 & \langle Q_{.1}, A_{.2} \rangle & \langle Q_{.1}, A_{.3} \rangle & \cdots & \langle Q_{.1}, A_{.n} \rangle \\ 0 & \nu_2 & \langle Q_{.2}, A_{.3} \rangle & \cdots & \langle Q_{.2}, A_{.n} \rangle \\ 0 & 0 & \nu_3 & \cdots & \langle Q_{.3}, A_{.n} \rangle \\ \vdots & \vdots & \vdots & & \vdots \\ 0 & 0 & 0 & \cdots & \nu_n \end{pmatrix}.$$

So, R is a square upper triangular matrix of order n. The factorization of A as QR is called the QR-**factorization** (sometimes also called QR-**decomposition**) for A, and it is uniquely determined by A.

A matrix has a QR-factorization iff its set of column vectors is linearly independent. If the set of column vectors of $A_{m \times n}$ is linearly independent, it can be expressed as $Q_{m \times n} R_{n \times n}$ where the columns of Q form an orthonormal set, and R is upper triangular with positive diagonal entries.

The QR factorization for $A_{m \times n}$ is called the **rectangular QR-factorization or decomposition** if $m > n$; and the **square QR-factorization or decomposition** if $m = n$.

Example 4:

Find the QR-factorization of

$$A = \begin{pmatrix} -1 & 0 & 0 \\ 0 & 1 & 0 \\ 1 & 1 & 1 \end{pmatrix}.$$

The column vectors of A are

$$A_{.1} = \begin{pmatrix} -1 \\ 0 \\ 1 \end{pmatrix}, \quad A_{.2} = \begin{pmatrix} 0 \\ 1 \\ 1 \end{pmatrix}, \quad A_{.3} = \begin{pmatrix} 0 \\ 0 \\ 1 \end{pmatrix}.$$

Applying the Gram–Schmidt procedure on the set of column vectors of A, we get the orthonormal set of vectors

$$Q_{.1} = (1/\|A_{.1}\|)A_{.1} = (1/\sqrt{2}) \begin{pmatrix} -1 \\ 0 \\ 1 \end{pmatrix}$$

$$Q_{.2} = (A_{.2} - \langle Q_{.1}, A_{.2} \rangle Q_{.1})/(\text{its norm})$$

$$= (\sqrt{(2/3)}) \begin{pmatrix} 1/2 \\ 1 \\ 1/2 \end{pmatrix}$$

$$Q_{.3} = (A_{.3} - \langle Q_{.1}, A_{.3} \rangle Q_{.1} - \langle Q_{.2}, A_{.3} \rangle Q_{.2})/(\text{its norm})$$

$$= \sqrt{3} \begin{pmatrix} 1/3 \\ -1/3 \\ 1/3 \end{pmatrix}.$$

Calculating R by the formula given above, the QR-factorization of A has

$$Q = \begin{pmatrix} -1/\sqrt{2} & 1/\sqrt{6} & 1/\sqrt{3} \\ 0 & \sqrt{2}/\sqrt{3} & -1/\sqrt{3} \\ 1/\sqrt{2} & 1/\sqrt{6} & 1/\sqrt{3} \end{pmatrix}$$

$$R = \begin{pmatrix} \sqrt{2} & 1/\sqrt{2} & 1/\sqrt{2} \\ 0 & \sqrt{3}/\sqrt{2} & 1/\sqrt{6} \\ 0 & 0 & 1/\sqrt{3} \end{pmatrix}.$$

The QR-factorization is an extremely useful tool in numerical computation for computing the eigen values of matrices (discussed in Chapter 6) and for many other applications.

Orthogonal Matrices

An **orthogonal matrix** is a square matrix whose columns form an orthonormal set. Since the columns of a square matrix form an orthonormal set iff its rows form an orthonormal set, the orthogonality of a square matrix can be defined either in terms of orthonormal columns or orthonormal rows.

Thus a square matrix $A_{n \times n}$ is orthogonal iff

$$\|A_{.j}\| = \|A_{i.}\| = 1 \text{ for all } i, j = 1 \text{ to } n$$
$$\langle A_{.i}, A_{.j} \rangle = \langle A_{i.}, A_{j.} \rangle = 0 \text{ for all } i \neq j.$$

i.e., iff $A^T A = A A^T = $ the unit matrix I.

Examples of orthogonal matrices are all permutation matrices, for instance

$$\begin{pmatrix} 0 & 1 & 0 \\ 0 & 0 & 1 \\ 1 & 0 & 0 \end{pmatrix}.$$

Another example is

$$\begin{pmatrix} 1/\sqrt{2} & 1/\sqrt{3} & -1/\sqrt{6} \\ -1/\sqrt{2} & 1/\sqrt{3} & -1/\sqrt{6} \\ 0 & 1/\sqrt{3} & 2/\sqrt{6} \end{pmatrix}.$$

Therefore a square matrix A is orthogonal iff $A^{-1} = A^T$. Orthogonal matrices satisfy the following properties.

1. The product of orthogonal matrices is orthogonal.
2. The inverse of an orthogonal matrix is orthogonal.
3. The determinant of an orthogonal matrix is ± 1.
4. If A is an orthogonal matrix of order n, and x is a column vector in R^n, then $\|Ax\| = \|x\|$.

Orthogonal matrices play a very important role in the problem of diagonalizing square matrices discussed in Chapters 5, 6; and in many other applications.

4.13   Additional Exercises

4.13.1:   Consider the system of linear equations $Ax = b$ where A is a matrix of order $m \times n$.

If the system has a unique solution, what does it imply about rank(A)?

If the system is known to have a unique solution, and $m = n$, then show that the system has a unique solution for all $b \in R^m$.

If the system is known to have a unique solution, and $m > n$ show that the system has either unique solution, or no solution for all $b \in R^m$.

If the system is known to have a unique solution, and $m > n$, show that the system $Ax = 0$ has no nonzero solution.

4.13.2:   The set of vectors $\{a^1, \ldots, a^m, a^{m+1}\}$ in R^n is linearly dependent. Does this mean that a^{m+1} can be expressed as a linear combination of $\{a^1, \ldots, a^m\}$? Either provide an argument why it must be true, or a counterexample.

How does the answer to the above question change if it is also given that $\{a^1, \ldots, a^m\}$ is linearly independent? Why?

4.13.3:   Let $\bar{x}$ be a solution to the system $Ax = b$.

Also, suppose $\bar{y}$ is a solution for the homogeneous system $Ay = 0$.

Then show that $\bar{x} + \lambda \bar{y}$ is a solution for $Ax = b$ for all real values of λ.

4.13.4:   Let $a^1 = (1, 1, -1)^T$, $a^2 = (-1, 1, 1)^T$, $a^3 = (p, q, r)^T$.

Express the conditions that p, q, r have to satisfy for a^3 to be a linear combination, affine combination, convex combination, of $\{a^1, a^2\}$.

4.13.5:   Show that the set of vectors $\{(a, b), (c, d)\}$ in R^2 is linearly independent iff $ad - bc \neq 0$.

4.13.6: Discuss how the rank of the following matrix varies as a, b take all real values.

$$\begin{pmatrix} -1 & 1 & 1 & 1 \\ 2 & -3 & -1 & 2 \\ a & 11 & 5 & -8 \\ b & -5 & -1 & 3 \end{pmatrix}.$$

4.13.7: (From H. G. Campbell, *Linear Algebra With Applications*, Prentice Hall, 1980). Show that the rank of the following matrix takes only two different values as a varies. For what values of a is the rank 3?

$$\begin{pmatrix} 2 & 4 & 2 \\ 2 & 1 & 2 \\ 1 & 0 & a \end{pmatrix}.$$

4.13.8: A, B are orthogonal matrices of the same order.
   What are their inverses? What are the values of their determinants?
   Show that AB is also an orthogonal matrix.
   If A, B are also symmetric, show that $AB = BA$.

References

[4.1] P. E. Gill, W. Murray, and M. H. Wright, *Numerical Linear Algebra and Optimization*, Volume 1 (Addison Wesley, Redwood City, CA, 1991).

[4.2] C. P. Meyer, *Matrix Analysis and Applied Linear Algebra* (SIAM, Philadelphia, PA, 2000).

Chapter 5

Quadratic Forms, Positive, Negative (Semi) Definiteness

5.1 Expressing a Quadratic Function in n Variables Using Matrix Notation

As defined in Section 1.7, a linear function of $x = (x_1, \ldots, x_n)^T$ is a function of the form $c_1 x_1 + \cdots + c_n x_n$ where $c_1, \ldots, c_n$ are constants called the coefficients of the variables in this function. For example, when $n = 4$, $f_1(x) = -2x_2 + 5x_4$ is a linear function of $(x_1, x_2, x_3, x_4)^T$ with the coefficient vector $(0, -2, 0, 5)$.

An affine function of $x = (x_1, \ldots, x_n)^T$ is a constant plus a linear function, i.e., a function of the form $f_2(x) = c_0 + c_1 x_1 + \cdots + c_n x_n$. So, if $f_2(x)$ is an affine function, $f_2(x) - f_2(0)$ is a linear function. As an example, $-10 - 2x_2 + 5x_4$ is an affine function of $(x_1, x_2, x_3, x_4)^T$.

A **quadratic form** in the variables $x = (x_1, \ldots, x_n)^T$ consists of the second degree terms of a second degree polynomial in these variables, i.e., it is a function of the form

$$Q(x) = \sum_{i=1}^{n} q_{ii} x_i^2 + \sum_{i=1}^{n} \sum_{j=i+1}^{n} q_{ij} x_i x_j$$

where the q_{ij} are the coefficients of the terms in the quadratic form.

Define a square matrix $D = (d_{ij})$ of order n where

$$d_{ii} = q_{ii} \quad \text{for } i = 1 \text{ to } n$$

and d_{ij}, d_{ji} are arbitrary real numbers satisfying

$$d_{ij} + d_{ji} = q_{ij} \quad \text{for } j > i.$$

As an example, if $q_{12} = -10$, we could take $(d_{12} = d_{21} = -5)$, or $(d_{12} = 100, d_{21} = -110)$, or $(d_{12} = -4, d_{21} = -6)$, etc. Then

$$Dx = \begin{pmatrix} d_{11}x_1 + d_{12}x_2 + \cdots + d_{1n}x_n \\ d_{21}x_1 + d_{22}x_2 + \cdots + d_{2n}x_n \\ \vdots \\ d_{n1}x_1 + d_{n2}x_2 + \cdots + d_{nn}x_n \end{pmatrix}.$$

So

$$x^T Dx = x_1(d_{11}x_1 + \cdots + d_{1n}x_n) + \cdots + x_n(d_{n1}x_1 + \ldots + d_{nn}x_n)$$

$$= \sum_{i=1}^{n} d_{ii}x_i^2 + \sum_{i=1}^{n} \sum_{j=i+1}^{n} (d_{ij} + d_{ji})x_i x_j$$

$$= \sum_{i=1}^{n} q_{ii}x_i^2 + \sum_{i=1}^{n} \sum_{j=i+1}^{n} q_{ij}x_i x_j$$

$$= Q(x).$$

So, for any matrix $D = (d_{ij})$ satisfying the conditions stated above, we have $x^T Dx = Q(x)$. Now define $\underline{D} = (\underline{d}_{ij})$ by

$$\underline{d}_{ii} = q_{ii} \quad i = 1, \ldots, n$$
$$\underline{d}_{ij} = \underline{d}_{ji} = (1/2)q_{ij} \quad j > i$$

then $\underline{D}$ is a symmetric matrix and $Q(x) = x^T \underline{D}x$. $\underline{D}$ is known as the **symmetric coefficient matrix defining the quadratic form** $Q(x)$.

Clearly for all $D = (d_{ij})$ satisfying the conditions stated above, we have $Q(x) = x^T Dx = x^T D^T x = x^T(\frac{D+D^T}{2})x$ and $\frac{D+D^T}{2} = \underline{D}$, the symmetric coefficient matrix defining the quadratic form $Q(x)$.

As an example consider

$$n = 3, \quad x = (x_1, x_2, x_3)^T, \quad Q(x) = 81x_1^2 - 7x_2^2 + 5x_1x_2 - 6x_1x_3 + 18x_2x_3.$$

Then the following square matrices satisfy the conditions stated above for $Q(x)$.

$$D_1 = \begin{pmatrix} 81 & -10 & 100 \\ 15 & -7 & 10 \\ -106 & 8 & 0 \end{pmatrix}, \quad D_2 = \begin{pmatrix} 81 & 200 & -1006 \\ -195 & -7 & 218 \\ 1000 & -200 & 0 \end{pmatrix},$$

$$D_3 = \begin{pmatrix} 81 & 2 & -2 \\ 3 & -7 & 15 \\ -4 & 3 & 0 \end{pmatrix}, \quad \underline{D} = \begin{pmatrix} 81 & 5/2 & -3 \\ 5/2 & -7 & 9 \\ -3 & 9 & 0 \end{pmatrix}.$$

It can be verified that $x^T D_1 x = x^T D_2 x = x^T D_3 x = x^T \underline{D} x = Q(x)$, and that

$$\underline{D} = \frac{D_1 + D_1^T}{2} = \frac{D_2 + D_2^T}{2} = \frac{D_3 + D_3^T}{2}.$$

Hence a general quadratic form in n variables $x = (x_1, \ldots, x_n)^T$ can be represented in matrix notation as $x^T M x$ where $M = (m_{ij})$ is a square matrix of order n. If M is not symmetric, it can be replaced in the above formula by $(M + M^T)/2$ without changing the quadratic form, and this $(M+M^T)/2$ is known as the symmetric matrix defining this quadratic form.

A **quadratic function** in variables $x = (x_1, \ldots, x_n)^T$ is a function which is the sum of a quadratic form in x and an affine function in x; i.e., it is of the form $x^T D x + cx + c_0$ for some square matrix D of order n, row vector $c \in R^n$ and constant term c_0.

Exercises

5.1.1: Express the following functions in matrix notation with a symmetric coefficient matrix: (i) $-6x_1^2 + 7x_2^2 - 14x_4^2 - 12x_1x_2 + 20x_1x_3 + 6x_1x_4 - 7x_2x_3 + 8x_3x_4 - 9x_1 + 6x_2 - 13x_3 + 100$ (ii) $x_1^2 - x_2^2 + x_3^2 - 18x_1x_3 + 12x_2x_3 - 7x_1 + 18x_2$ (iii) $4x_1^2 + 3x_2^2 - 8x_1x_2$ (iv) $6x_1 + 8x_2 - 11x_3 + 6$.

5.1.2: Express the quadratic form $x^T D x + cx$ as a sum of individual terms in it for the following data:

(i) $D = \begin{pmatrix} 3 & -6 & 9 & 8 \\ 10 & -2 & 0 & 12 \\ 13 & -11 & 0 & 3 \\ -4 & -9 & 9 & 1 \end{pmatrix}, \quad c = \begin{pmatrix} 1 \\ -9 \\ 0 \\ 7 \end{pmatrix}.$

(ii) $D = \begin{pmatrix} 3 & -4 & 8 \\ -4 & 9 & 9 \\ -8 & -9 & 2 \end{pmatrix}, \quad c = \begin{pmatrix} 0 \\ 0 \\ 6 \end{pmatrix}.$

(iii) $D = \begin{pmatrix} 1 & 1 & 1 & 1 & 1 \\ 1 & 1 & 1 & 1 & 1 \\ 1 & 1 & 1 & 1 & 1 \\ 1 & 1 & 1 & 1 & 1 \\ 1 & 1 & 1 & 1 & 1 \end{pmatrix}, \quad c = 0.$

5.2   Convex, Concave Functions; Positive (Negative) (Semi) Definiteness; Indefiniteness

Consider the real valued function $f(x)$ defined over $x \in R^n$, or over some convex subset of R^n. It is said to be a **convex function** if for every x, y for which it is defined, and $0 \le \alpha \le 1$, we have

$$f(\alpha x + (1 - \alpha)y) \le \alpha f(x) + (1 - \alpha)f(y).$$

This inequality defining the convexity of the function $f(x)$ is called **Jensen's inequality** after the Danish mathematician who first defined it. This inequality is easy to visualize when $n = 1$. It says that if you join two points on the graph of the function by a chord, then the function itself lies underneath the chord on the interval joining these points. See Figure 5.1.

The convex function $f(x)$ is said to be a **strictly convex function** if the above inequality holds as a strict inequality for every $x \ne y$ and $0 < \alpha < 1$.

A real valued function $g(x)$ defined over R^n or a convex subset of R^n is said to be a **concave function** if $-g(x)$ is a convex function as defined above, i.e., if for every x, y for which it is defined, and $0 \le \alpha \le 1$, we have

$$g(\alpha x + (1 - \alpha)y) \ge \alpha g(x) + (1 - \alpha)g(y).$$

See Figure 5.2. The concave function $g(x)$ is said to be a **strictly concave function** if $-g(x)$ is a strictly convex function, i.e., if the above inequality holds as a strict inequality for every $x \ne y$ and $0 < \alpha < 1$.

The properties of convex and concave functions are of great importance in optimization theory. From the definition it can be verified that a function $f(x_1)$ of a single variable x_1 is

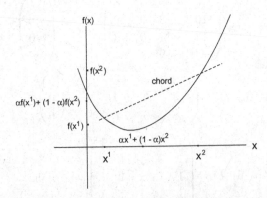

Figure 5.1   A convex function lies beneath any chord.

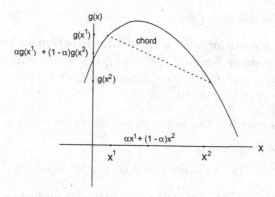

Figure 5.2   A concave function stays above any chord.

Convex iff its slope is nondecreasing, i.e., iff its 2nd derivative is ≥ 0 when it is twice continuously differentiable

Concave iff its slope is nonincreasing, i.e., iff its 2nd derivative is ≤ 0 when it is twice continuously differentiable.

As examples, some convex functions of a single variable x_1 are:

$$x_1^2, x_1^4, e^{-x_1}, -\log(x_1)(\text{in the region } x_1 > 0).$$

Among functions of several variables, linear and affine functions are both convex and concave, but they are not strictly convex or strictly concave. For classifying quadratic functions into the convex and concave classes, we need the following definitions.

A square matrix M of order n, whether symmetric or not, is said to be

Positive semidefinite (PSD)   iff $y^T M y \geq 0$ for all $y \in R^n$

Positive definite (PD)   iff $y^T M y > 0$ for all $y \in R^n$, $y \neq 0$

Negative semidefinite (NSD)   iff $y^T M y \leq 0$ for all $y \in R^n$

Negative definite (ND)   iff $y^T M y < 0$ for all $y \in R^n$, $y \neq 0$

Indefinite   if it is neither PSD nor NSD, i.e., iff there are points $x, y \in R^n$ satisfying $x^T M x > 0$, and $y^T M y < 0$.

We have the following results.

Result 5.2.1: Conditions for a quadratic function to be (strictly) convex: *The quadratic function $f(x) = x^T Dx + cx + c_0$ defined over R^n is a convex function over R^n iff the matrix D is PSD; and a strictly convex function iff the matrix D is PD.*

To see this, take any two points $x, y \in R^n$, and a $0 < \alpha < 1$. It can be verified that

$$\alpha f(x) + (1 - \alpha)f(y) - f(\alpha x + (1 - \alpha)y) = \alpha(1 - \alpha)(x - y)^T D(x - y)$$

by expanding the terms on both sides. So Jensen's inequality for the convexity of $f(x)$ holds iff

$$(x - y)^T D(x - y) \geq 0 \quad \text{for all } x, y \in R^n$$

i.e., iff $z^T Dz \geq 0$ for all $z \in R^n$, i.e., iff D is PSD. Likewise for strict convexity of $f(x)$ we need

$$(x - y)^T D(x - y) > 0 \quad \text{for all } x \neq y \in R^n$$

i.e., iff $z^T Dz > 0$ for all $z \in R^n$, $z \neq 0$, i.e., iff D is PD.

Result 5.2.2: Conditions for a quadratic function to be (strictly) concave: *The quadratic function $f(x) = x^T Dx + cx + c_0$ defined over R^n is a concave function over R^n iff the matrix D is NSD; and a strictly concave function iff the matrix D is ND.*

This follows by applying Result 5.2.1 to $-f(x)$.

Checking whether a given quadratic function is strictly convex, or convex, is a task of great importance in optimization, statistics, and many other sciences. From Result 5.2.1, this task is equivalent to checking whether a given square matrix is PD, PSD. We discuss efficient algorithms for PD, PSD checking based on Gaussian pivot steps in the next section.

5.3 Algorithm to Check Positive (Semi) Definiteness Using G Pivot Steps

Submatrices of a Matrix

Let A be a matrix of order $m \times n$. Let $S \subset \{1, \ldots, m\}$ and $T \subset \{1, \ldots, n\}$. If we delete all the rows of A not in S, and all the columns not in T, what remains is a smaller matrix denoted by $A_{S \times T}$ known as **the submatrix** of A corresponding to the subset of rows S and columns T. As an example, consider the following matrix

$$A = \begin{pmatrix} 6 & 0 & -7 & 8 & 0 & -9 \\ -25 & -1 & 2 & 14 & -15 & -4 \\ 0 & 3 & -2 & -7 & 0 & 10 \\ 11 & 16 & 17 & -5 & 4 & 0 \end{pmatrix}.$$

Let $S_1 = \{1,4\}, S_2 = \{2,3\}, T_1 = \{1,3,4,6\}, T_2 = \{2,5,6\}$. Then the submatrices $A_{S_1 \times T_1}, A_{S_2 \times T_2}$ are

$$A_{S_1 \times T_1} = \begin{pmatrix} 6 & -7 & 8 & -9 \\ 11 & 17 & -5 & 0 \end{pmatrix}, \quad A_{S_2 \times T_2} = \begin{pmatrix} -1 & -15 & -4 \\ 3 & 0 & 10 \end{pmatrix}.$$

Principal Submatrices of a Square Matrix

Square matrices have special submatrices called principal submatrices. Let M be a square matrix of order n, and $S \subset \{1,\dots,n\}$ with r elements in it. Then the submatrix $M_{S \times S}$ obtained by deleting from M all rows and all columns not in the set S is known as the **principal submatrix of the square matrix M corresponding to the subset S**, it is a square matrix of order r. The determinant of the principal submatrix of M corresponding to the subset S is called the **principal minor of M corresponding to the subset S**. As an example, let

$$M = \begin{pmatrix} 4 & 2 & 3 & 5 \\ 1 & 22 & 12 & -1 \\ -4 & -2 & -5 & 6 \\ 0 & -3 & 16 & -7 \end{pmatrix}.$$

Let $S = \{2,4\}$. Then the principal submatrix of M corresponding to S is

$$M_{S \times S} = \begin{pmatrix} 22 & -1 \\ -3 & -7 \end{pmatrix}$$

and the principal minor of M corresponding to S is determinant$(M_{S \times S})$ which is -157.

In this same example, consider the singleton subset $S_1 = \{3\}$. Then the principal submatrix of M corresponding to S_1 is (-5) of order 1, i.e., the third diagonal element of M. In the same way, all diagonal elements of a square matrix are its principal submatrices of order 1.

Some Results on PSD, PD, NSD, ND Matrices

Result 5.3.1: Properties shared by all principal submatrices: *If M is a square matrix of order n which has any of the properties PSD, PD, NSD, ND; then all principal submatrices of M have the same property.*

To see this, consider the subset $S \subset \{1, 2\}$ as an example, and the principal submatrix $M_{S \times S}$ of order 2. Let $\bar{y} = (y_1, y_2, 0, \ldots, 0)^T \in R^n$ with $y_i = 0$ for all $i \notin S$. Since y_3 to y_n are all 0 in $\bar{y}$, we verify that $\bar{y}^T M \bar{y} = (y_1, y_2) M_{S \times S} (y_1, y_2)^T$. So, for all y_1, y_2, if

$\bar{y}^T M \bar{y} \geq 0$ so is $(y_1, y_2) M_{S \times S} (y_1, y_2)^T$
$\bar{y}^T M \bar{y} > 0$ for all $(y_1, y_2) \neq 0$, so is $(y_1, y_2) M_{S \times S} (y_1, y_2)^T$
$\bar{y}^T M \bar{y} \leq 0$ so is $(y_1, y_2) M_{S \times S} (y_1, y_2)^T$
$\bar{y}^T M \bar{y} < 0$ for all $(y_1, y_2) \neq 0$ so is $(y_1, y_2) M_{S \times S} (y_1, y_2)^T$.

Hence, if M is PSD, PD, NSD, ND, $M_{S \times S}$ has the same property. A similar argumant applies to all principal submatrices of M.

Result 5.3.2: Conditions for a 1×1 matrix: *A 1×1 matrix (a) is*

PSD iff $a \geq 0$
PD iff $a > 0$
NSD iff $a \leq 0$
ND iff $a < 0$.

The quadratic form defined by (a) is $a x_1^2$; and it is ≥ 0 for all x_1 iff $a \geq 0$; it is > 0 for all $x_1 \neq 0$ iff $a > 0$; etc. Hence this result follows from the definitions.

Result 5.3.3: Conditions satisfied by diagonal entries: *Let $M = (m_{ij})$ be a square matrix. If M is a*

PSD matrix, all its diagonal entries must be ≥ 0
PD matrix, all its diagonal entries must be > 0
NSD matrix, all its diagonal entries must be ≤ 0
ND matrix, all its diagonal entries must be < 0.

Since all diagonal entries of a square matrix are its 1×1 principal submatrices, this result follows from Results 5.3.1 and 5.3.2.

Result 5.3.4: Conditions satisfied by the row and column of a 0-diagonal entry in a PSD matrix: *Let $M = (m_{ij})$ be a square PSD matrix. If one of its diagonal entries, say $m_{pp} = 0$, then for all j we must have $m_{pj} + m_{jp} = 0$.*

Suppose $m_{11} = 0$, and $m_{12} + m_{21} = \alpha \neq 0$. Let the principal submatrix of M corresponding to the subset $\{1, 2\}$ be

$$\bar{M} = \begin{pmatrix} m_{11} & m_{12} \\ m_{21} & m_{22} \end{pmatrix} = \begin{pmatrix} 0 & m_{12} \\ m_{21} & m_{22} \end{pmatrix}.$$

The quadratic form defined by $\bar{M}$ is $(y_1, y_2)\bar{M}(y_1, y_2)^T = m_{22}y_2^2 + \alpha y_1 y_2$. By fixing

$$y_1 = \frac{-1 - m_{22}}{\alpha}, \quad y_2 = 1$$

we see that this quadratic form has value -1, so $\bar{M}$ is not PSD, and by Result 5.3.1 M cannot be PSD, a contradiction. Hence α must be 0 in this case. The result follows from the same argument.

Result 5.3.5: Conditions satisfied by the row and column of a 0-diagonal entry in a symmetric PSD matrix: *Let $M = (m_{ij})$ be a square matrix. Symmetrize M, i.e., let $D = (d_{ij}) = (M + M^T)/2$. M is PSD iff D is, because the quadratic forms defined by M and D are the same. Also, since D is symmetric, we have $d_{ij} = d_{ji}$ for all i, j. Hence, if D is PSD, and a diagonal entry in D, say $d_{ii} = 0$, then all the entries in the ith row and the ith column of D must be zero, i.e., $D_{i.} = 0$ and $D_{.i} = 0$.*

This follows from Result 5.3.4.

Examples 1:

Consider the following matrices.

$$M_1 = \begin{pmatrix} 10 & 3 & 1 \\ -2 & 0 & 0 \\ 1 & 0 & 4 \end{pmatrix}, \quad M_2 = \begin{pmatrix} 100 & 0 & 2 \\ 0 & 10 & 3 \\ 2 & 3 & -1 \end{pmatrix}$$

$$M_3 = \begin{pmatrix} 10 & 3 & -3 \\ 3 & 10 & 6 \\ 3 & -6 & 0 \end{pmatrix}, \quad M_4 = \begin{pmatrix} 2 & 0 & -7 \\ 0 & 0 & 0 \\ 10 & 0 & 6 \end{pmatrix}.$$

The matrix M_1 is not symmetric, and in it $m_{22} = 0$ but $m_{12} + m_{21} = 3 - 2 = 1 \neq 0$, hence this matrix violates the condition in Result 5.3.4, and is not PSD.

In the matrix M_2, the third diagonal entry is -1, hence by Result 5.3.3 this matrix is not PSD.

In the matrix M_3, $m_{33} = 0$, and $m_{13} + m_{31} = m_{23} + m_{32} = 0$, hence this matrix satisfies the condition in Result 5.3.4.

The matrix M_4 is symmetric, its second diagonal entry is 0, and its 2nd row and 2nd column are both zero vectors. Hence this matrix satisfies the condition in Result 5.3.5.

Result 5.3.6: Conditions satisfied by a principal submatrix obtained after a G pivot step: *Let $D = (d_{ij})$ be a symmetric matrix of order n with its first diagonal entry $d_{11} \neq 0$. Perform a Gaussian pivot step on D with d_{11} as the pivot element, and let $\bar{D}$ be the resulting matrix; i.e. transform*

$$D = \begin{pmatrix} d_{11} & \cdots & d_{1n} \\ d_{21} & \cdots & d_{2n} \\ \vdots & & \vdots \\ d_{n1} & \cdots & d_{nn} \end{pmatrix} \quad into \quad \bar{D} = \begin{pmatrix} d_{11} & d_{12} & \cdots & d_{1n} \\ 0 & \bar{d}_{22} & \cdots & \bar{d}_{2n} \\ \vdots & \vdots & & \vdots \\ 0 & \bar{d}_{n2} & \cdots & \bar{d}_{nn} \end{pmatrix}.$$

Let D_1 be the matrix of order $(n-1) \times (n-1)$ obtained by deleting row 1 and column 1 from $\bar{D}$. Then D_1 is symmetric too, and D is PD (PSD) iff $d_{11} > 0$ and D_1 is PD (PSD).

Since the original matrix D is symmetric, we have $d_{ij} = d_{ji}$ for all i, j. Using this and the formula coming from the Gaussian pivot step, that for $i, j = 2$ to n, $\bar{d}_{ij} = d_{ij} - d_{i1}(d_{1j}/d_{11})$ it can be verified that $\bar{d}_{ij} = \bar{d}_{ji}$, hence D_1 is symmetric.

The other part of this result can be seen from the following.

$$x^T D x = d_{11} \left[x_1^2 + 2(d_{12}/d_{11})x_1 x_2 + \cdots + 2(d_{1n}/d_{11})x_1 x_n \right.$$

$$\left. + (1/d_{11}) \sum_{i=2}^{n} \sum_{j=2}^{n} d_{ij} x_i x_j \right].$$

Let

$$\theta = (d_{12}/d_{11})x_2 + \cdots + (d_{1n}/d_{11})x_n$$

$$\delta = (1/d_{11}) \sum_{i=2}^{n} \sum_{j=2}^{n} d_{ij} x_i x_j.$$

Then $x^T D x = d_{11}[x_1^2 + 2x_1\theta + \delta] = d_{11}[(x_1 + \theta)^2 + (\delta - \theta^2)]$, because $x_1^2 + 2x_1\theta = (x_1 + \theta)^2 - \theta^2$. In mathematical literature this argument is called **completing the square argument**.

After rearranging the terms it can be verified that

$$\delta - \theta^2 = (x_2, \ldots, x_n)D_1(x_2, \ldots, x_n)^T.$$

So,

$$x^T Dx = d_{11}(x_1 + \theta)^2 + (x_2, \ldots, x_n)D_1(x_2, \ldots, x_n)^T.$$

From this it is clear that D is PD (PSD) iff $d_{11} > 0$ and D_1 is PD (PSD).

Based on these results, we now describe algorithms to check whether a given square matrix is PD, PSD. To check if a square matrix is ND, NSD, apply the following algorithms to check if its negative is PD, PSD.

Algorithm to Check if a Given Square Matrix M of Order n is PD

BEGIN

Step 1: First symmetrize M, i.e., compute $D = (M + M^T)/2$. The algorithm works on D.

If any of the diagonal entries in D are ≤ 0, D and M are not PD by Result 5.3.3, terminate.

Step 2: Start performing Gaussian pivot steps on D using the diagonal elements as the pivot elements in the order 1 to n. At any stage of this process, if the current matrix has an entry ≤ 0 in its main diagonal, terminate with the conclusion that D, M are not PD. If all the pivot steps are completed and all the diagonal entries are > 0, terminate with the conclusion that D, M are PD.

END

Example 2:

Check whether the following matrix M is PD.

$$M = \begin{pmatrix} 3 & 1 & 2 & 2 \\ -1 & 2 & 0 & 2 \\ 0 & 4 & 4 & 5/3 \\ 0 & -2 & -13/3 & 6 \end{pmatrix}.$$

Symmetrizing, we get

$$D = \begin{pmatrix} 3 & 0 & 1 & 1 \\ 0 & 2 & 2 & 0 \\ 1 & 2 & 4 & -4/3 \\ 1 & 0 & -4/3 & 6 \end{pmatrix}.$$

All the entries in the principal diagonal of D are >0. So, we apply the algorithm, and obtain the following results. The pivot elements are boxed in the following. PR, PC indicate pivot row, pivot column for each G pivot step.

		PC↓		
PR	$\boxed{3}$	0	1	1
	0	2	2	0
	1	2	4	$-4/3$
	1	0	$-4/3$	6
		PC↓		
	3	0	1	1
PR	0	$\boxed{2}$	2	0
	0	2	11/3	$-5/3$
	0	0	$-5/3$	17/3
			PC↓	
	3	0	1	1
	0	2	2	0
PR	0	0	$\boxed{5/3}$	$-5/3$
	0	0	$-5/3$	17/3
	3	0	1	1
	0	2	2	0
	0	0	5/3	$-5/3$
	0	0	0	4

The algorithm terminates now. Since all the diagonal entries in all the tableaus are > 0, D and hence M are PD.

Example 3:

Check whether the following matrix M is PD.

$$M = \begin{pmatrix} 1 & 0 & 2 & 0 \\ 0 & 2 & 4 & 0 \\ 2 & 4 & 4 & 5 \\ 0 & 0 & 5 & 3 \end{pmatrix}.$$

The matrix M is already symmetric and its diagonal entries are > 0. So, we apply the algorithm on it, and obtain the following results. The pivot elements are boxed in the following. PR, PC indicate pivot row, pivot column for each G pivot step.

PR	PC↓			
	1	0	2	0
	0	2	4	0
	2	4	4	5
	0	0	5	3
	1	0	2	0
	0	2	4	0
	0	4	0	5
	0	0	5	3

Since the third diagonal entry in this tableau is 0, we terminate with the conclusion that this matrix M is not PD.

A square matrix which is not PD, could be PSD. We discuss the algorithm based on G pivot steps, for checking whether a given square matrix is PSD, next.

Algorithm to Check if a Given Square Matrix M of Order n is PSD

BEGIN

Initial step: First symmetrize M, i.e., compute $D = (M + M^T)/2$. The algorithm works on D. Apply the following General step on the matrix D.

General Step: If any of the diagonal entries in the matrix are < 0, terminate with the conclusion that the original matrix is not PSD (Results 5.3.3, 5.3.6). Otherwise continue.

Check if the matrix has any 0-diagonal entries. For each 0-diagonal entry check whether its row and column in the matrix are both completely 0-vectors, if not the original matrix is not PSD (Results 5.3.5, 5.3.6), terminate. If this condition is satisfied, delete the 0-row vector and the 0-column vector of each 0-diagonal entry in the matrix.

If the remaining matrix is of order 1×1, its entry will be > 0, terminate with the conclusion that the original matrix is PSD.

If the remaining matrix is of order ≥ 2, its first diagonal entry will be positive, perform a G pivot step on the matrix with this first diagonal element as the pivot element. After this pivot step, delete row 1, column 1 of the resulting matrix.

Now apply the same general step on the remining matrix, and repeat the same way.

END

Example 4:

Check whether the following matrix M is PSD.

$$M = \begin{pmatrix} 0 & -2 & -3 & -4 & 5 \\ 2 & 3 & 3 & 0 & 0 \\ 3 & 3 & 3 & 0 & 0 \\ 4 & 0 & 0 & 8 & 4 \\ -5 & 0 & 0 & 4 & 2 \end{pmatrix}.$$

Symmetrizing, we get

$$D = \begin{pmatrix} 0 & 0 & 0 & 0 & 0 \\ 0 & 3 & 3 & 0 & 0 \\ 0 & 3 & 3 & 0 & 0 \\ 0 & 0 & 0 & 8 & 4 \\ 0 & 0 & 0 & 4 & 2 \end{pmatrix}.$$

The first diagonal entry in D is 0, and in fact $D_{1.}, D_{.1}$ are both 0, so we eliminate them and apply the general step on the remaining matrix. Since there is a 0-diagonal entry in D, the original matrix M is not PD, but it may possibly be PSD.

	PC↓			
PR	3	3	0	0
	3	3	0	0
	0	0	8	4
	0	0	4	2
	3	3	0	0
	0	0	0	0
	0	0	8	4
	0	0	4	2

Eliminating row 1 and column 1 in the matrix resulting after the first G step leads to the matrix

$$\begin{pmatrix} 0 & 0 & 0 \\ 0 & 8 & 4 \\ 0 & 4 & 2 \end{pmatrix}.$$

The diagonal entries in this matrix are all ≥ 0, and the first diagonal entry is 0. Correspondingly, row 1 and column 1 are both 0-vectors, so we delete them, and apply the general step on the remaining matrix.

PC↓	
PR 8	4
4	2
8	4
0	0

Eliminating row 1 and column 1 in the matrix resulting after this G pivot step leads to the 1×1 matrix (0). Since the entry in it is ≥ 0, we terminate with the conclusion that the original matrix M is PSD but not PD.

Example 5:

Check whether the following matrix M is PSD.

$$M = \begin{pmatrix} 1 & 0 & 1 & 0 \\ 0 & 2 & 4 & 0 \\ 1 & 4 & 1 & 5 \\ 0 & 0 & 5 & 3 \end{pmatrix}.$$

The matrix M is already symmetric, and its diagonal entries are all > 0. So, we apply the general step on it.

PC↓			
PR 1	0	1	0
0	2	4	0
1	4	1	5
0	0	5	3
1	0	1	0
0	2	4	0
0	4	0	5
0	0	5	3

Eliminating row 1 and column 1 in the matrix resulting after the first G step leads to the matrix

$$\begin{pmatrix} 2 & 4 & 0 \\ 4 & 0 & 5 \\ 0 & 5 & 3 \end{pmatrix}.$$

The 2nd diagonal entry in this matrix is 0, but its 2nd row and 2nd column are not 0-vectors. So, we terminate with the conclusion that the original matrix M is not PSD.

Exercises

5.3.1: Find all the conditions that $p, q > 0$ have to satisfy for the following matrix to be PD.

$$\begin{pmatrix} \dfrac{1-p^2}{2p} & \dfrac{1-pq}{p+q} \\ \dfrac{1-pq}{p+q} & \dfrac{1-q^2}{2q} \end{pmatrix}.$$

5.3.2: Show that the following matrices are PD.

$$\begin{pmatrix} 3 & 1 & 2 & 2 \\ -1 & 2 & 0 & 2 \\ 0 & 4 & 4 & 5/3 \\ 0 & -2 & -13/3 & 6 \end{pmatrix}, \quad \begin{pmatrix} 2 & -11 & 3 & 4 \\ 9 & 2 & -39 & 10 \\ -1 & 41 & 5 & 12 \\ -2 & -10 & -12 & 2 \end{pmatrix},$$

$$\begin{pmatrix} 2 & -1 & 1 \\ -1 & 2 & -1 \\ 1 & -1 & 2 \end{pmatrix}, \quad \begin{pmatrix} 2 & 0 & 0 \\ 0 & 3 & 0 \\ 0 & 0 & 6 \end{pmatrix}.$$

5.3.3: Show that the following matrices are not PD.

$$\begin{pmatrix} 1 & 0 & 2 & 0 \\ 0 & 2 & 4 & 0 \\ 2 & 4 & 4 & 5 \\ 0 & 0 & 5 & 3 \end{pmatrix}, \quad \begin{pmatrix} 1 & -1 & -1 & -1 \\ -1 & 1 & -1 & -1 \\ -1 & -1 & 1 & -1 \\ -1 & -1 & -1 & 1 \end{pmatrix},$$

$$\begin{pmatrix} 10 & -2 & -1 \\ -2 & 0 & 1 \\ -1 & 1 & 5 \end{pmatrix}, \quad \begin{pmatrix} 2 & 0 & 3 \\ 0 & 2 & 1 \\ 3 & 1 & -1 \end{pmatrix}, \quad \begin{pmatrix} 2 & 4 & 5 \\ 0 & 2 & 6 \\ 0 & 0 & 2 \end{pmatrix}.$$

5.3.4: Show that the following matrices are PSD.

$$
\begin{pmatrix}
0 & -2 & -3 & -4 & 5 \\
2 & 3 & 3 & 0 & 0 \\
3 & 3 & 3 & 0 & 0 \\
4 & 0 & 0 & 8 & 4 \\
-5 & 0 & 0 & 4 & 2
\end{pmatrix}, \quad
\begin{pmatrix}
2 & 2 & 2 \\
2 & 2 & 2 \\
2 & 2 & 2
\end{pmatrix}, \quad
\begin{pmatrix}
2 & -2 & 1 \\
-2 & 4 & 2 \\
1 & 2 & 6
\end{pmatrix}.
$$

5.3.3: Show that the following matrices are not PSD.

$$
\begin{pmatrix}
1 & 0 & 2 & 0 \\
0 & 2 & 4 & 0 \\
2 & 4 & 4 & 5 \\
0 & 0 & 5 & 3
\end{pmatrix}, \quad
\begin{pmatrix}
0 & 1 & 1 \\
1 & 2 & 1 \\
1 & 1 & 4
\end{pmatrix}.
$$

5.3.5: Let A be a square matrix of order n. Show that $x^T A x$ is a convex function in x iff its minimum value over R^n is 0.

5.3.6: Consider the square matrix of order n with all its diagonal entries = p, and all its off-diagonal entries = q. Determine all the values of p, q for which this matrix is PD.

5.3.7: A square matrix $A = (a_{ij})$ of order n is said to be *skew-symmetric* if $A + A^T = 0$. Show that the quadratic form $x^T A x$ is 0 for all $x \in R^n$ iff A is skew-symmetric.

Also, if A is a skew symmetric matrix of order n, and B is any matrix of order n, show that $x^T(A + B)x = x^T B x$ for all x.

5.3.8: Let A be a square matrix of order n and $f(x) = x^T A x$. For $x, y \in R^n$ show that $f(x + y) - f(x) - f(y)$ is a bilinear function of x, y as defined in Section 1.7.

5.3.9: Find the range of values of α for which the quadratic form $x_1^2 + 4x_1x_2 + 6x_1x_3 + \alpha x_2^2 + \alpha x_3^2$ is a strictly convex function.

5.3.10: Find the range of values of α for which the following matrix is PD.

$$
\begin{pmatrix}
1 & 4 & 2 \\
0 & 5 - \alpha & 8 - 4\alpha \\
0 & 0 & 8 - \alpha
\end{pmatrix}.
$$

5.4 Diagonalization of Quadratic Forms and Square Matrices

Optimization deals with problems of the form: find a $y = (y_1, \ldots, y_n)^T \in R^n$ that minimizes a given real valued function $g(y)$ of y. This problem becomes easier to solve if $g(y)$ is separable, i.e., if it is the sum of n functions each involving one variable only, as in

$$g(y) = g_1(y_1) + \cdots + g_n(y_n).$$

In this case, minimizing $g(y)$ can be achieved by minimizing each $g_j(y_j)$ separately for $j = 1$ to n; and the problem of minimizing a function is much easier if it involves only one variable.

Consider the problem of minimizing a quadratic function

$$f(x) = cx + (1/2)x^T M x$$

where $M = (m_{ij})$ is an $n \times n$ symmetric PD matrix. If some $m_{ij} \neq 0$ for $i \neq j$, the function involves the product term $x_i x_j$ with a nonzero coefficient, and it is not separable. To make this function separable, we need to convert the matrix M into a diagonal matrix. For this we can try to apply a linear transformation of the variables x with the aim of achieving separability in the space of new variables. Consider the transformation

$$x = Py$$

where P is an $n \times n$ nonsingular matrix, and the new variables are $y = (y_1, \ldots, y_n)^T$. Then, in terms of the new variables y, the function is

$$F(y) = f(x = Py) = cPy + (1/2)y^T P^T M P y.$$

$F(y)$ is separable if $P^T M P$ is a diagonal matrix. If $P^T M P = \text{diag}(d_{11}, \ldots, d_{nn})$, then let $cP = \bar{c} = (\bar{c}_1, \ldots, \bar{c}_n)$. In this case $F(y) = \sum_{j=1}^{n}(\bar{c}_j y_j + d_{jj} y_j^2)$. The point minimizing $F(y)$ is $\bar{y} = (\bar{y}_1, \ldots, \bar{y}_n)^T$ where $\bar{y}_j$ is the minimizer of $\bar{c}_j y_j + d_{jj} y_j^2$. Once $\bar{y}$ is found, the minimizer of the original function $f(x)$ is $\bar{x} = P\bar{y}$.

The condition

$$P^T M P = \begin{pmatrix} d_{11} & 0 & \cdots & 0 \\ 0 & d_{22} & \cdots & 0 \\ \vdots & \vdots & \ddots & \vdots \\ 0 & 0 & \cdots & d_{nn} \end{pmatrix}$$

requires

$$(P_{.i})^T M P_{.j} = 0 \quad \text{for all } i \neq j.$$

When M is a PD symmetric matrix, the set of column vectors $\{P_{.1}, \ldots, P_{.n}\}$ of a matrix P satisfying the above condition is said to be a **conjugate set of vectors WRT** M. Optimization methods called **conjugate direction methods** or **conjugate gradient methods** are based on using this type of transformations of variables.

Given a symmetric matrix M, this problem of finding a nonsingular matrix P satisfying $P^T M P$ is a diagonal matrix is called the **problem of diagonalizing a quadratic form**. Conjugate direction methods in nonlinear optimization have very efficient special routines for finding such a matrix P, but a discussion of these methods is outside the scope of this book.

In mathematics they often discuss other matrix diagonalization problems, which we now explain.

The **orthogonal diagonalization problem** is the same as the above, with the additional requirement that the matrix P must be an orthogonal matrix; i.e., it should also satisfy $P^T = P^{-1}$. Hence the orthogonal diagonalization problem is: given a square symmetric matrix M of order n, find an orthogonal matrix P of order n satisfying:

$P^{-1} M P$ is a diagonal matrix D (this is also written sometimes as $MP = PD$).

The **general matrix diagonalization problem** discussed in mathematics is: given a square matrix M of order n, find a nonsingular square matrix P of order n such that $P^{-1} M P$ is a diagonal matrix.

Solving either the orthogonal diagonalization problem, or the matrix diagonalization problem involves finding the eigenvalues and eigenvectors of a square matrix, which are discussed in the next chapter.

Chapter 6

Eigen Values, Eigen Vectors, and Matrix Diagonalization

6.1 Definitions, Some Applications

Eigen values and **eigen vectors** are only defined for square matrices; hence in this chapter we will consider square matrices only.

Let $A = (a_{ij})$ be a square matrix of order n. Let I denote the unit matrix of order n. A scalar λ is said to be an **eigen value** (or **characteristic value**, or **proper value**, or **latent value**) of square matrix A if the matrix $A - \lambda I$ is singular, i.e., if the determinant $\det(A - \lambda I) = 0$.

Let $P(\lambda) = \det(A - \lambda I)$. Expansion of $\det(A - \lambda I)$ produces the polynomial in λ, $P(\lambda)$ which is known as the **characteristic polynomial** for the square matrix A. The degree of $P(\lambda)$ is $n = $ order of the matrix A, and the leading term in it is $(-1)^n \lambda^n$.

The eigen values of A are the solutions of the **characteristic equation** for A

$$P(\lambda) = 0$$

i.e., they are the **roots of the characteristic polynomial** for A.

The set of all eigen values of A, usually denoted by the symbol $\sigma(A)$, is called **the spectrum** of A.

Since the characteristic polynomial is of degree n, it has exactly n roots by the Fundamental Theorem of Algebra. But some of these roots may be complex numbers (i.e., a number of the form $a + ib$, where a, b are real numbers and $i = \sqrt{-1}$), and some of the roots may be repeated.

Since all the entries a_{ij} in the matrix A are real numbers, all the coefficients in the characteristic polynomial are real numbers. By the fundamental theorem of algebra this implies that if $\lambda = a + ib$ is a root of the characteristic polynomial, then so is its complex conjugate $\bar{\lambda} = a - ib$. Thus the complex eigen values of A occur in complex conjugate pairs.

So, if λ_1 is an eigen value of A, then $\lambda = \lambda_1$ is a solution of $P(\lambda) = 0$. From theory of equations, this implies that $(\lambda_1 - \lambda)$ is a factor of the characteristic polynomial $P(\lambda)$. In the same way we conclude that if $\lambda_1, \lambda_2, \ldots, \lambda_n$ are the various eigen values of A, then the characteristic polynomial

$$P(\lambda) = (\lambda_1 - \lambda)(\lambda_2 - \lambda) \ldots (\lambda_n - \lambda).$$

However, as mentioned earlier, some of the eigen values of A may be complex numbers, and some of the eigen values may be equal to each other. If λ_1 is an eigen value of A and $r \geq 1$ is the positive integer satisfying the property that $(\lambda_1 - \lambda)^r$ is a factor of the characteristic polynomial $P(\lambda)$, but $(\lambda_1 - \lambda)^{r+1}$ is not, then we say that r is the **algebraic multiplicity** of the eigen value λ_1 of A. If $r = 1$, λ_1 is called a **simple eigen value** of A; if $r > 1$, λ_1 is called a multiple eigen value of A with **algebraic multiplicity** r.

Thus if $\lambda_1, \lambda_2, \ldots, \lambda_k$ are all the distinct eigen values of A including any complex eigen values, with algebraic multiplicities $r_1, \ldots, r_k$ respectively, then the characteristic polynomial

$$P(\lambda) = (\lambda_1 - \lambda)^{r_1}(\lambda_2 - \lambda)^{r_2} \ldots (\lambda_k - \lambda)^{r_k}$$

and $r_1 + \cdots + r_k = n$. In this case the spectrum of A, $\sigma(A) = \{\lambda_1, \ldots, \lambda_k\}$.

If λ is an eigen value of A, since $A - \lambda I$ is singular, the homogeneous system of equations

$$(A - \lambda I)x = 0 \quad \text{or equivalently} \quad Ax = \lambda x$$

must have a nonzero solution $x \in R^n$. A column vector $x \neq 0$ satisfying this equation is called an **eigen vector** of A corresponding to its eigen value λ; and the pair (λ, x) is called an **eigen pair** for A. Notice that the above homogeneous system may have many solutions x, the set of all solutions of it is the nullspace of the matrix $(A - \lambda I)$ defined in Section 3.10 and denoted by $N(A - \lambda I)$. This nullspace $N(A - \lambda I)$ is called the **eigen space** of A corresponding to its eigen value λ. Therefore, the set of all eigen vectors of A associated with its eigen value λ is $\{x \in R^n : x \neq 0, \text{ and } x \in N(A - \lambda I)\}$, the set of all nonzero vectors in the corresponding eigen space.

The dimension of $N(A - \lambda I)$ (as defined in Section 1.23, this is equal to the number of nonbasic variables in the final tableau when the above homogeneous system $(A - \lambda I)x = 0$ is solved by the GJ method or the G elimination method) is called the **geometric multiplicity** of the eigen value λ of A. We state the following result on multiplicities without proof (for proofs see C. Meyer [6.1]). called the **geometric multiplicity** of the eigen value λ of A.

Result 6.1.1: Multiplicity Inequality: *For simple eigen values λ of A (i.e., those with algebraic multiplicity of 1) the dimension of the eigen space $N(A - \lambda I)$ is 1; i.e., this eigen space is a straight line through the origin.*

If λ is an eigen value of A with algebraic multiplicity > 1, we will always have

geometric multiplicity of $\lambda \leq$ algebriac multiplicity of λ

i.e., the dimension of $N(A - \lambda I) \leq$ algebriac multiplicity of λ.

Result 6.1.2: Linear Independence of Eigen vectors Associated With Distinct Eigen values: *Let $\lambda_1, \ldots, \lambda_h$ be some distinct eigen values of A; and $\{(\lambda_1, x^1), (\lambda_2, x^2), \ldots, (\lambda_h, x^h)\}$ a set of eigen pairs for A. Then the set of eigen vectors $\{x^1, \ldots, x^h\}$ is linearly independent.*

Suppose all the eigen values of A are real, and $\sigma(A) = \{\lambda_1, \ldots, \lambda_k\}$ is this set of distinct eigen values. If, for all $i = 1$ to k, geometric multiplicity of λ_i is equal to its algebraic multiplicity; then it is possible to get a **complete set of n eigen vectors for A** which is linearly independent.

If geometric multiplicity of λ_i is $<$ its algebriac multiplicity for at least one i, then A does not have a complete set of eigen vectors which is linearly independent. Such matrices are called **deficient** or **defective matrices**.

Example 1:

Let

$$A = \begin{pmatrix} 6 & -5 \\ 4 & -3 \end{pmatrix}. \text{ Then } A - \lambda I = \begin{pmatrix} 6 - \lambda & -5 \\ 4 & -3 - \lambda \end{pmatrix}.$$

Therefore the characteristic polynomial of A, $P(\lambda) =$ determinant $(A - \lambda I) = (6 - \lambda)(-3 - \lambda) + 20 = \lambda^2 - 3\lambda + 2 = (2 - \lambda)(1 - \lambda)$. Hence the characteristic equation for A is

$$(2 - \lambda)(1 - \lambda) = 0$$

for which the roots are 2 and 1. So, the eigen values of A are 2 and 1 respectively, both are simple with algebraic multiplicity of one.

To get the eigen space of an eigen value λ_i we need to solve the homogeneous system $(A - \lambda_i I)x = 0$. For $\lambda = 1$ this system in detached coefficient form is the one in the top tableau given below. We solve it using the GJ method discussed in Section 1.23. Pivot elements are enclosed in a box; PR, PC denote the pivot row, pivot column respectively, and BV denotes the basic variable selected in the row. RC denotes redundant constraint to be eliminated.

BV	x_1	x_2		
	5	-5	0	PR
	4	-4	0	
	PC↑			
x_1	1	-1	0	
	0	0	0	RC
Final Tableau				
x_1	1	-1	0	

Since there is one nonbasic variable in the final tableau, the dimension of the eigen space is one. The general solution of the system is $x = (\alpha, \alpha)^T$ for any α real. So, for any $\alpha \neq 0$ $(\alpha, \alpha)^T$ is an eigen vector corresponding to the eigen value 1. Taking $\alpha = 1$ leads to one eigen vector $x^1 = (1, 1)^T$.

For the eigen value $\lambda = 2$, the eigen space is determined by the following system, which we also solve by the GJ method.

BV	x_1	x_2		
	4	-5	0	PR
	4	-5	0	
	PC↑			
x_1	1	$-5/4$	0	
	0	0	0	RC
Final Tableau				
x_1	1	$-5/4$	0	

So, the general solution of this system is $x = (5\alpha/4, \alpha)^T$ which is an eigen vector corresponding to the eigen value 2 for any $\alpha \neq 0$. In particular, taking $\alpha = 1$, yields $x^2 = (5/4, 1)^T$ as an eigen vector corresponding to 2.

So, $\{x^1, x^2\} = \{(1, 1)^T, (5/4, 1)^T\}$ is a complete set of eigen vectors for A.

Example 2:

Let

$$A = \begin{pmatrix} 0 & -1 \\ 1 & 0 \end{pmatrix}. \quad \text{Then } A - \lambda I = \begin{pmatrix} -\lambda & -1 \\ 1 & -\lambda \end{pmatrix}.$$

Therefore the characteristic polynomial of A, $P(\lambda) = $ determinant $(A - \lambda I) = \lambda^2 + 1$. Hence the characteristic equation for A is

$$\lambda^2 + 1 = 0$$

for which the solutions are the imaginary numbers $\lambda_1 = i = \sqrt{-1}$ and $\lambda_2 = -i$. Thus the eigen values of this matrix A are the complex conjugate pair $+i$ and $-i$.

It can also be verified that the eigen vectors corresponding to

$\lambda_1 = +i$ are $(\alpha, -\alpha i)^T$
$\lambda_2 = -i$ are $(\alpha, \alpha i)^T$

for any nonzero complex number α. Thus the eigen vectors of A are both complex vectors in this example, even though the matrix A is real.

Example 3:

Let

$$A = \begin{pmatrix} 0 & 0 & -1 \\ 10 & 1 & 2 \\ 0 & 0 & 1 \end{pmatrix}. \quad \text{Then } A - \lambda I = \begin{pmatrix} \lambda & 0 & -1 \\ 10 & 1-\lambda & 2 \\ 0 & 0 & 1-\lambda \end{pmatrix}.$$

Therefore the characteristic polynomial of A, $P(\lambda) = $ determinant $(A - \lambda I) = -\lambda(1 - \lambda)^2$. Hence the characteristic equation for A is

$$-\lambda(1 - \lambda)^2 = 0$$

for which the roots are 0 and 1, with 1 being a double root. So, the eigen values of A are 0 and 1, 0 is simple with algebraic multiplicity one, and 1 is multiple with algebraic multiplicity 2.

To get the eigen space of eigen value 0 we need to solve the homogeneous system $Ax = 0$. We solve it using the GJ method discussed in Section 1.23. Pivot elements are enclosed in a box; PR, PC denote the pivot row, pivot column respectively, and BV denotes the basic variable selected in the row. RC denotes redundant constraint to be eliminated.

BV	x_1	x_2	x_3		
	0	0	$\boxed{-1}$	0	PR
	10	1	2	0	
	0	0	1	0	
			PC↑		
x_3	0	0	1	0	
	10	$\boxed{1}$	0	0	PR
	0	0	0	0	RC
	PC↑				
Final Tableau					
x_3	0	0	1	0	
x_2	10	1	0	0	

The general solution of this system is $x = (\alpha, -10\alpha, 0)^T$ for any α real. Taking $\alpha = 1$, an eigen vector corresponding to the eigen value 0 is $x^1 = (1, -10, 0)^T$.

For $\lambda = 1$, the eigen space is determined by the following system, which we also solve using the GJ method of Section 1.23.

BV	x_1	x_2	x_3		
	$\boxed{-1}$	0	-1	0	PR
	10	0	2	0	
	0	0	0	0	RC
	PC↑				
x_1	1	0	1	0	
	0	0	$\boxed{-8}$	0	PR
			PC↑		
Final Tableau					
x_1	1	0	0	0	
x_3	0	0	1	0	

The general solution of this system is $x = (0, \alpha, 0)^T$ for any α real; so an eigen vector corresponding to the eigen value 1 is $(0, 1, 0)^T$. The eigen space corresponding to this eigenvalue is a straight line, a one dimensional object.

The algebraic multiplicity of the eigen value 1 is 2, but its geometric multiplicity, the dimension of the eigen space is only 1.

This example illustrates the fact that the geometric multiplicity of an eigen value, the dimension of its eigen space, can be strictly less than its algebraic multiplicity.

Example 4:

Let

$$A = \begin{pmatrix} 0 & 1 & 1 \\ 1 & 0 & 1 \\ 1 & 1 & 0 \end{pmatrix}. \quad \text{Then } A - \lambda I = \begin{pmatrix} -\lambda & 1 & 1 \\ 1 & -\lambda & 1 \\ 1 & 1 & -\lambda \end{pmatrix}.$$

Therefore the characteristic polynomial of A, $P(\lambda) = $ determinant $(A - \lambda I) = (\lambda + 1)^2 (2 - \lambda)$. Hence the characteristic equation for A is

$$(\lambda + 1)^2 (2 - \lambda) = 0$$

for which the roots are -1 and 2, with -1 being a double root. So, the eigen values of A are -1 and 2; 2 is simple with algebraic multiplicity one, and -1 is multiple with algebraic multiplicity 2.

The eigen space corresponding to eigen value -1 is determined by the following homogeneous system which we solve using the GJ method discussed in Section 1.23. Pivot elements are enclosed in a box; PR, PC denote the pivot row, pivot column respectively, and BV denotes the basic variable selected in the row. RC denotes redundant constraint to be eliminated.

BV	x_1	x_2	x_3		
	1	1	$\boxed{1}$	0	PR
	1	1	1	0	
	1	1	1	0	
			PC↑		
x_3	1	1	1	0	
	0	0	0	0	RC
	0	0	0	0	RC
	Final Tableau				
x_3	1	1	1	0	

The general solution of this system is $x = (\alpha, \beta, -(\alpha + \beta))^T$ for any α, β real. So, this eigen space has dimension 2, it is the subspace which is the linear hull of the two vectors $(1, 0, -1)^T$ and $(0, 1, -1)^T$. The set of these two vectors is a basis for this eigen space.

Thus for the matrix A in this example, both the algebraic and geometric multiplicities of its eigen value -1 are two.

The eigen vector corresponding to the eigen value 2 is obtained from the following system.

BV	x_1	x_2	x_3		
	-2	1	$\boxed{1}$	0	PR
	1	-2	1	0	
	1	1	-2	0	
			PC↑		
x_3	-2	1	1	0	
	$\boxed{3}$	-3	0	0	PR
	-3	3	0	0	
	PC↑				
x_3	0	-1	1	0	
x_1	1	-1	0	0	PR
	0	0	0	0	RC
	Final Tableau				
x_3	0	-1	1	0	
x_1	1	-1	0	0	

The general solution of this system is $x = (\alpha, \alpha, \alpha)^T$ for any α real; so an eigen vector corresponding to the eigen value 2 is $(1, 1, 1)^T$.

Example 5: Eigen Values of a Diagonal Matrix:

Let

$$A = \begin{pmatrix} 2 & 0 & 0 \\ 0 & -7 & 0 \\ 0 & 0 & 9 \end{pmatrix}. \quad \text{Then } A - \lambda I = \begin{pmatrix} 2 - \lambda & 0 & 0 \\ 0 & -7 - \lambda & 0 \\ 0 & 0 & 9 - \lambda \end{pmatrix}.$$

Therefore the characteristic polynomial of A, $P(\lambda) =$ determinant $(A - \lambda I) = (2 - \lambda)(-7 - \lambda)(9 - \lambda)$. Hence the characteristic equation for A is

$$(2 - \lambda)(-7 - \lambda)(9 - \lambda) = 0$$

for which the roots are 2, -7, and 9. So, the eigen values of A are 2, -7 and 9; the diagonal entries in the diagonal matrix A.

The eigen space corresponding to eigen value 2 is the solution set of

x_1	x_2	x_3	
0	0	0	0
0	-9	0	0
0	0	7	0

The general solution of this system is $x = (\alpha, 0, 0)^T$ for real α. Hence an eigen vector corresponding to the eigen value 2 is $(1, 0, 0)^T$, the first unit vector.

In the same way it can be verified that for any diagonal matrix, its eigen values are its diagonal entries; and an eigen vector corresponding to its ith diagonal entry is the ith unit vector.

Example 6: Eigen Values of an Upper Triangular Matrix:

Let

$$A = \begin{pmatrix} -1 & -10 & 16 \\ 0 & 2 & -18 \\ 0 & 0 & -3 \end{pmatrix}. \text{ Then } A - \lambda I = \begin{pmatrix} -1-\lambda & -10 & 16 \\ 0 & 2-\lambda & -18 \\ 0 & 0 & -3-\lambda \end{pmatrix}.$$

Therefore the characteristic polynomial of A, $P(\lambda) = $ determinant $(A - \lambda I) = (-1 - \lambda)(2 - \lambda)(-3 - \lambda)$. Hence the characteristic equation for A is

$$(-1 - \lambda)(2 - \lambda)(-3 - \lambda) = 0.$$

Its roots, the eigen values of the upper triangular matrix A, are exactly its diagonal entries.

The eigen space corresponding to eigen value -1 is the set of solutions of

x_1	x_2	x_3	
0	-10	16	0
0	3	-18	0
0	0	-2	0

The general solution of this system is $x = (\alpha, 0, 0)^T$ for α real. Hence an eigen vector corresponding to -1, the first diagonal entry of A, is $(1, 0, 0)^T$, the first unit vector.

In the same way it can be verified that an eigen vector corresponding to the ith diagonal entry of A is the ith unit vector, for all i.

Similarly, for any upper or lower triangular matrix, its eigen values are its diagonal entries, and its eigen vectors are the unit vectors.

Basis for the Eigen Space of a Square Matrix A Corresponding to an Eigen Value λ

Let λ be an eigen value of a square matrix A of order n. Then the eigen space corresponding to λ is $N(A - \lambda I)$, the set of all solutions of the system

$$(A - \lambda I)x = 0.$$

A basis for this eigen space, as defined in Sections 1.23, 4.2, is a maximal linearly independent set of vectors in $N(A - \lambda I)$, it has the property that every point in this eigen space can be expressed as its linear combination in a unique manner.

Example 7:

Consider the matrix

$$A = \begin{pmatrix} 0 & 1 & 1 \\ 1 & 0 & 1 \\ 1 & 1 & 0 \end{pmatrix}$$

from Example 4, and its eigen value -1. The eigen space corresponding to this eigen value is the set of all solutions of

x_1	x_2	x_3	
1	1	1	0
1	1	1	0
1	1	1	0

All the constraints in this system are the same as $x_1 + x_2 + x_3 = 0$. So, a general solution of this system is $(\alpha, \beta, -\alpha - \beta)^T$, where α, β are real valued parameters. A basis for this eigen space is the set $\{(1, 0, -1)^T, (0, 1, -1)^T\}$, the set of two vectors obtained by setting $\alpha = 1$, $\beta = 0$, and $\alpha = 0, \beta = 1$ respectively in the formula for the general solution.

Results on Eigen Values and Eigen Vectors

We now discuss some important results on eigen values and eigen vectors, without proofs.

Result 6.1.3: Linear independence of the set of eigen vectors corresponding to distinct eigen values: Let $(\lambda_1, x^1), \ldots, (\lambda_k, x^k)$ be eigen pairs for a square matrix A, where the eigen values $\lambda_1, \ldots, \lambda_k$ are distinct. Then $\{x^1, \ldots, x^k\}$ is a linearly independent set.

Result 6.1.4: The union of bases of eigen spaces corresponding to distinct eigen values is a basis: Let $\{\lambda_1, \ldots, \lambda_k\}$ be a set of distinct eigen values for a square matrix A. For $i = 1$ to k, let B_i be a basis for the eigen space corresponding to λ_i. Then $B = \cup_{i=1}^{k} B_i$ is a linearly independent set.

A Complete Set of Eigen Vectors

A **complete set of eigen vectors** for a square matrix A of order n is any linearly independent set of n eigen vectors for A. Not all square matrices have complete sets of eigen vectors. Square matrices that do not have a complete set of eigen vectors are called **deficient** or **defective matrices**.

Let A be a square matrix of order n which has distinct real eigen values $\lambda_1, \ldots, \lambda_n$. Let x^i be an eigen vector corresponding to λ_i for $i = 1$ to n. Then by Result 6.1.3, $\{x^1, \ldots, x^n\}$ is a complete set of eigen vectors for A. Thus any square matrix with all eigen values real and distinct has a complete set of eigen vectors.

Let A be a square matrix of order n with distinct real eigen values $\lambda_1, \ldots, \lambda_k$, where for $i = 1$ to k, both the algebraic and geometric multiplicities of λ_i are equal to r_i, satisfying $r_1 + \cdots + r_k = n$. Then the eigen space corresponding to λ_i has a basis B_i consisting of r_i vectors, and $B = \cup_{i=1}^{k} B_i$ is linearly independent by Result 6.1.4, and is a basis for R^n. In this case B is a complete set of eigen vectors for A. Thus if all eigen values of A are real, and for each of them its geometric multiplicity is its algebraic multiplicity, then A has a complete set of eigen vectors.

Thus if a square matrix A has an eigen value for which the geometric multiplicity is $<$ its algebraic multiplicity, then A is deficient in eigen vectors; i.e., it does not have a complete set of eigen vectors.

6.2 Relationship of Eigen Values and Eigen Vectors to the Diagonalizability of a Matrix

Let A be a square matrix of order n. In Section 5.4 we mentioned that the matrix diagonalization problem for matrix A is to find a square non-singular matrix P of order n such that $P^{-1}AP$ is a diagonal matrix

$D = \text{diag}(d_1, \ldots, d_n)$, say. In this case we say that the **matrix P diagonalizes** A, i.e.,

$$P^{-1}AP = \begin{pmatrix} d_1 & 0 & \cdots & 0 \\ 0 & d_2 & \cdots & 0 \\ \vdots & \vdots & & \vdots \\ 0 & 0 & \cdots & d_n \end{pmatrix} = D = \text{diag}(d_1, \ldots, d_n).$$

Then multiplying the above equation on both sides by P, we have

$$AP = PD.$$

$P_{.1}, \ldots, P_{.n}$ are the column vectors of P. Since $D = \text{diag}(d_1, \ldots, d_n)$, we have $PD = (d_1 P_{.1} \vdots \ldots \vdots d_n P_{.n})$, and by the definition of matrix multiplication $AP = (AP_{.1} \vdots \ldots \vdots AP_{.n})$. So, the above equation $AP = PD$ implies

$$(AP_{.1} \vdots \ldots \vdots AP_{.n}) = (d_1 P_{.1} \vdots \ldots \vdots d_n P_{.n})$$

i.e., $\quad AP_{.j} = d_j P_{.j} \quad j = 1, \ldots, n.$

Since P is invertible, $P_{.j} \neq 0$ for all j. So, the above equation $AP_{.j} = d_j P_{.j}$, i.e., $(A - d_j I)P_{.j} = 0$ implies that $(d_j, P_{.j})$ are eigen pairs for A for $j = 1$ to n.

Thus if the matrix P diagonalizes the square matrix A into $\text{diag}(d_1, \ldots, d_n)$ then $d_1, \ldots, d_n$ are the eigen values of A; and $P_{.j}$ is an eigen vector corresponding to d_j for each $j = 1$ to n.

Conversely, suppose a square matrix A of order n has a complete set of eigen vectors $\{P_{.1}, \ldots, P_{.n}\}$ with corresponding eigen values $\lambda_1, \ldots, \lambda_n$. Let $D = \text{diag}(\lambda_1, \ldots, \lambda_n)$. Since $(\lambda_j, P_{.j})$ is an eigen pair for A for $j = 1$ to n; we have $AP_{.j} = \lambda_j P_{.j}$ for $j = 1$ to n, i.e.

$$(AP_{.1} \vdots \ldots \vdots AP_{.n}) = (\lambda_1 P_{.1} \vdots \ldots \vdots AP_{.n}) = PD$$

i.e., $AP = PD$ where P is the square matrix with the eigen vectors $P_{.1}, \ldots, P_{.n}$ as the column vectors. Since $\{P_{.1}, \ldots, P_{.n}\}$ is a complete set of eigen vectors, it is linearly independent, so P is invertible. So, $AP = PD$ implies $P^{-1}AP = D$. Thus P diagonalizes A.

Thus the square matrix A of order n is diagonalizable iff it has a complete set of eigen vectors; and a matrix which diagonalizes A is the matrix whose column vectors are a basis for R^n consisting of eigen vectors of A.

Thus the task of diagonalizing a square matrix A is equivalent to finding its eigen values and a complete set of eigen vectors for it.

We now provide some examples. In this chapter we number all the examples serially. So, continuing the numbering from the previous section, we number the next example as Example 8.

Example 8:

Consider the matrix

$$A = \begin{pmatrix} 0 & 1 & 1 \\ 1 & 0 & 1 \\ 1 & 1 & 0 \end{pmatrix}$$

discussed in Examples 4, 7. From Examples 4, 7, we know that its eigen values are -1 and 2, and that $\{(1,0,-1)^T, (0,1,-1)^T, (1,1,1)^T\}$ is a complete set of eigen vectors for A. So

$$P = \begin{pmatrix} 1 & 0 & 1 \\ 0 & 1 & 1 \\ -1 & -1 & 1 \end{pmatrix}$$

is a matrix that diagonalizes A. In fact $P^{-1}AP = \mathrm{diag}(-1,-1,2)$.

Example 9:

Let

$$A = \begin{pmatrix} 0 & 0 & -1 \\ 10 & 1 & 2 \\ 0 & 0 & 1 \end{pmatrix}.$$

In Example 3 we determined that its eigen values are 0 and 1, with 1 having algebraic multiplicity 2. But the geometric multiplicity of 1 is $1 <$ its algebraic multiplicity. So, this matrix does not have a complete set of eigen vectors, and hence it is not diagonalizable.

Example 10:

Let

$$A = \begin{pmatrix} 6 & -5 \\ 4 & -3 \end{pmatrix}$$

the matrix considered in Example 1. From Example 1 we know that its eigen values are 2, 1 and a complete set of eigen vectors for it is

$\{(1, 4/5)^T, (1, 1)^T\}$. So, the matrix

$$P = \begin{pmatrix} 1 & 1 \\ 4/5 & 1 \end{pmatrix}$$

diagonalizes A, and $P^{-1}AP = \text{diag}(2, 1)$.

We will now present some more important results, most of them without proofs.

Result 6.2.1: Results on Eigen Values and Eigen Vectors for Symmetric Matrices:

(i) **All eigen values are real:** *If A is a square symmetric matrix of order n, all its eigen values are real numbers, i.e., it has no complex eigen values.*

(ii) **Eigen vectors corresponding to distinct eigen values are orthogonal:** *If A is a square symmetric matrix of order n, eigen vectors corresponding to distinct eigen values of A are orthogonal.*

We will provide a proof of this result because it is so simple. Since A is symmetric, $A = A^T$. Let $\lambda_1 \neq \lambda_2$ be eigen values associated with eigen vectors v^1, v^2 respectively. So, $Av^1 = \lambda_1 v^1$ and $Av^2 = \lambda_2 v^2$. So

$$\lambda_1(v^1)^T v^2 = (\lambda_1 v^1)^T v^2 = (Av^1)^T v^2 = (v^1)^T A^T v^2$$
$$= (v^1)^T A v^2 = \lambda_2(v^1)^T v^2.$$

Hence, $\lambda_1(v^1)^T v^2 = \lambda_2(v^1)^T v^2$, i.e., $(\lambda_1 - \lambda_2)(v^1)^T v^2 = 0$. Since $\lambda_1 \neq \lambda_2$, $\lambda_1 - \lambda_2 \neq 0$. Hence $(v^1)^T v^2$ must be zero; i.e., v^1 and v^2 are orthogonal.

(iii) **Special eigen properties of symmetric PD, PSD, ND, NSD, Indefinite matrices:** *A square symmetric matrix is*

Positive definite	*iff all its eigen values are > 0*
Positive semidefinite	*iff all its eigen values are ≥ 0*
Negative definite	*iff all its eigen values are < 0*
Negative semidefinite	*iff all its eigen values are ≤ 0*
Indefinite	*iff it has both a positive and a negative eigen value.*

See Section 5.2 for definitions of these various classes of matrices.

(iv) **Geometric and algebraic multiplicities are equal:** *If A is a square symmetric matrix of order n, for every one of its eigen values, its* *geometric multiplicity is equal to its algebraic multiplicity. Hence symmetric matrices can always be diagonalized.*

(v) **Orthonormal set of eigen vectors:** *If A is a square matrix of order n, it has an orthonormal set of n eigen vectors iff it is symmetric.*

Orthogonal Diagonalization of a Square Matrix

As discussed in Section 5.4, **orthogonal diagonalization** of a square matrix A deals with the problem of finding an orthogonal matrix P such that

$$P^{-1}AP = \text{a diagonal matrix diag}(\lambda_1, \ldots, \lambda_n) \quad \text{say.}$$

When P is orthogonal $P^{-1} = P^T$. The orthogonal matrix P is said to **orthogonally diagonalize** the square matrix A if $P^{-1}AP = P^TAP$ is a diagonal matrix.

Result 6.2.2: Orthogonally diagonalizable iff symmetric: *A square matrix can be orthogonally diagonolized iff it is symmetric.*

To see this, suppose the orthogonal matrix P orthogonally diagonalizes the square matrix A. So, $P^{-1} = P^T$ and $P^{-1}AP = $ a diagonal matrix D. Since D is diagonal, $D^T = D$. Now, $P^{-1}AP = D$ implies $A = PDP^{-1} = PDP^T$. So,

$$A^T = (PDP^T)^T = PD^TP^T = PDP^T = A$$

i.e., A is symmetric.

To orthogonally diagonalize a square symmetric matrix A of order n, one needs to find its eigen values. If they are all distinct, let $(\lambda_i, P_{.i})$ be the various eigen pairs where each $P_{.i}$ is normalized to satisfy $\|P_{.i}\| = 1$. Then by Result 6.2.1 (ii), $\{P_{.1}, \ldots, P_{.n}\}$ is an orthonormal set of eigen vectors for A. Let P be the matrix with $P_{.1}, \ldots, P_{.n}$ as its column vectors, it orthogonally diagonalizes A.

If the eigen values of the symmetric matrix A are not all distinct, find a basis for the eigenspace of each eigen value. Apply the Gram–Schmidt process to each of these bases to obtain an orthonormal basis for each eigen space. Then form the square matrix P whose columns are the vectors in these various orthonormal bases. Then P orthogonally diagonalizes A.

Result 6.2.3: Similar Matrices Have the Same Eigen Spectrum: *Two square matrices A, B of order n are said to be **similar** if there exists a nonsingular matrix P of order n satisfying $B = P^{-1}AP$. In this case we say that B is obtained by a **similarity transformation** on A. Since*

$B = P^{-1}AP$ implies $A = PBP^{-1}$, if B can be obtained by a similarity transformation on A, then A can be obtained by a similarity transformation on B. When $B = P^{-1}AP$, $det(B - \lambda I) = det(P^{-1}AP - \lambda I) = det(P^{-1}(AP - \lambda P)) = det(P^{-1})det(A - \lambda I)det(P) = det(A - \lambda I)$. Thus similar matrices all have the same charateristic polynomial, so they all have the same eigen values with the same multiplicities. However, their eigen vectors may be different.

Result 6.2.4: Eigen values of powers of a matrix: *Suppose the square matrix A is diagonalizable, and let $P^{-1}AP = diag(\lambda_1, \ldots, \lambda_n)$. Then for any positive integer s,*

$$A^s = P^{-1}(\text{diag}(\lambda_1^s, \ldots, \lambda_n^s))P.$$

Since $\lambda_1, \ldots, \lambda_n$ are real numbers, computing $\lambda_1^s, \ldots, \lambda_n^s$ is much easier than computing A^s because of the complexity of matrix multiplication.

So, calculating higher powers A^s of matrix A for large values of s can be carried out efficiently if it can be diagonalized.

Result 6.2.5: Spectral Theorem for Diagonalizable Matrices: *Let A be a square matrix of order n with spectrum $\sigma(A) = \{\lambda_1, \ldots, \lambda_k\}$. Suppose A is diagonalizable, and for $i = 1$ to k, let r_i = algebraic multiplicity of λ_i = geometric multiplicity of λ_i. Then there exist square matrices of order n, $G_1, \ldots, G_k$ satisfying*

(i) $G_i G_j = 0$ for all $i \neq j$
(ii) $G_1 + \cdots + G_k = I$
(iii) $\lambda_1 G_1 + \cdots + \lambda_k G_k = A$
(iv) $G_i^2 = G_i$ for all i
(v) $Rank(G_i) = r_i$ for all i.

Let X_i be the matrix of order $n \times r_i$ whose column vectors form a basis for the eigen space $N(A - \lambda_i I)$. Let $P = (X_1 \vdots X_2 \vdots \ldots \vdots X_k)$. Then P is nonsingular by Result 6.1.2. Partition P^{-1} as follows

$$P^{-1} = \begin{pmatrix} Y_1 \\ \cdots \\ Y_2 \\ \cdots \\ \vdots \\ \cdots \\ Y_k \end{pmatrix}$$

where Y_i is an $r_i \times n$ matrix. Then $G_i = X_i Y_i$ for $i = 1$ to k satisfies all the above properties. All these properties follow from the following facts

$$PP^{-1} = P^{-1}P = I$$

$$A = PDP^{-1} = (X_1 \vdots X_2 \vdots \ldots \vdots X_k) \begin{pmatrix} \lambda_1 I_{r_1} & 0 & \cdots & 0 \\ 0 & \lambda_2 I_{r_2} & \cdots & 0 \\ \vdots & \vdots & & \vdots \\ 0 & 0 & \cdots & \lambda_k I_{r_k} \end{pmatrix} \begin{pmatrix} Y_1 \\ \cdots \\ Y_2 \\ \cdots \\ \vdots \\ \cdots \\ Y_k \end{pmatrix}$$

where I_{r_i} is the unit matrix of order r_i for $i = 1$ to k.

The decomposition of A into $\lambda_1 G_1 + \cdots + \lambda_k G_k$ is known as the **spectral decomposition** of A, and the G_is are called the **spectral projectors** associated with· A.

Example 11:

Consider the matrix

$$A = \begin{pmatrix} 0 & 1 & 1 \\ 1 & 0 & 1 \\ 1 & 1 & 0 \end{pmatrix}$$

discussed in Example 8. Its eigen values are $\lambda_1 = -1$ with an algebraic and geometric multiplicity of two, and $\lambda_2 = 2$ with an algebraic and geometric multiplicity of one. A basis for the eigen space correponding to -1 is $\{(1, 0, -1)^T, (0, 1, -1)^T\}$; and an eigen vector corresponding to 2 is $(1, 1, 1)^T$. So, using the notation in Result 6.2.5, we have

$$X_1 = \begin{pmatrix} 1 & 0 \\ 0 & 1 \\ -1 & -1 \end{pmatrix}, \quad X_2 = \begin{pmatrix} 1 \\ 1 \\ 1 \end{pmatrix}, \quad P = \begin{pmatrix} 1 & 0 & \vdots & 1 \\ 0 & 1 & \vdots & 1 \\ -1 & -1 & \vdots & 1 \end{pmatrix}.$$

We compute P^{-1} and obtain

$$P^{-1} = \begin{pmatrix} 2/3 & -1/3 & -1/3 \\ -1/3 & 2/3 & -1/3 \\ \cdots & \cdots & \cdots \\ 1/3 & 1/3 & 1/3 \end{pmatrix}, \quad Y_1 = \begin{pmatrix} 2/3 & -1/3 & -1/3 \\ -1/3 & 2/3 & -1/3 \end{pmatrix},$$

$$Y_2 = \begin{pmatrix} 1/3 & 1/3 & 1/3 \end{pmatrix}.$$

So

$$G^1 = X_1 Y_1 = \begin{pmatrix} 2/3 & -1/3 & -1/3 \\ -1/3 & 2/3 & -1/3 \\ -1/3 & -1/3 & 2/3 \end{pmatrix}$$

$$G_2 = X_2 Y_2 = \begin{pmatrix} 1/3 & 1/3 & 1/3 \\ 1/3 & 1/3 & 1/3 \\ 1/3 & 1/3 & 1/3 \end{pmatrix}.$$

It can be verified that $\text{rank}(G_1) = 2 =$ algebraic multiplicity of the eigen value -1, and $\text{rank}(G_2) = 1 =$ algebraic multiplicity of the eigen value 2. Also, we verify that

$G_1 + G_2 = I =$ unit matrix of order 3
$2G_1 + G_2 = A$, the spectral decomposition of A.

G_1, G_2 are the spectral projectors associated with A.

Result 6.2.6: Eigen pairs for A and A^{-1} *If (λ, x) is an eigen pair for a nonsingular square matrix A, then $(1/\lambda, x)$ is an eigen pair for A^{-1}.*

By hypothesis we have $Ax = \lambda x$, and since A is nonsingular, we have $\lambda \neq 0$. Multiplying both sides of the equation $Ax = \lambda x$ on the left by $(1/\lambda)A^{-1}$ we get $(A^{-1} - (1/\lambda)I)x = 0$. Therefore $((1/\lambda), x)$ is an eigen pair for A^{-1}.

Result 6.2.7: Sylvester's Law of Inertia: *Let A be a square symmetric matrix of order n. Let C be any nonsingular square matrix of order n, and $B = C^T A C$. Then we say that B is obtained from A by a **congruent transformation**, and that A, B are **congruent**. Since C is nonsingular, $B = C^T A C$ implies $A = (C^T)^{-1} B C^{-1} = (C^{-1})^T B (C^{-1})$, so if B can be obtained by a congruent transformation on A, then A can be obtained by a congruent transformation on B.*

*The **inertia** of a real symmetric matrix is defined to be the triple (ρ, ν, ζ) where*

ρ, ν, ζ are the numbers of positive, negative, and zero eigen values of the matrix respectively.

Sylvester's law states that two symmetric matrices are congruent iff they have the same inertia.

6.3 An Illustrative Application of Eigen Values, Eigen Vectors, and Diagonalization in Differential Equations

Eigen values, eigen vectors, and diagonalization of square matrices find many applications in all branches of science. Here we provide one illustrative application in differential equations, which are equations involving functions and their derivatives. Mathematical models involving differential equations appear very often in studies in chemical engineering and other branches of engineering, physics, chemistry, biology, pharmacy, and other sciences.

Let $u_1(t), \ldots, u_n(t)$ be n real valued differentiable functions of a single real variable t; and let $u_i'(t)$ denote the derivative of $u_i(t)$. Suppose the formulas for these functions are unknown, but we know that these functions satisfy the following equations.

$$u_1'(t) = a_{11}u_1(t) + \cdots + a_{1n}u_n(t)$$
$$\vdots \quad \vdots$$
$$u_n'(t) = a_{n1}u_1(t) + \cdots + a_{nn}u_n(t)$$

$$\text{and} \quad u_i(0) = c_i \quad i = 1, \ldots, n$$

where $A = (a_{ij})$ is a given square matrix of order n, and $c = (c_1, \ldots, c_n)^T$ is a given vector of function values corresponding to $t = 0$.

A system like this is known as a **homogeneous system of first order linear differential equations with constant coefficients**. Solving this system means to find the formulas for the functions $u_1(t), \ldots, u_n(t)$ satisfying these equations. Let

$$A = \begin{pmatrix} a_{11} & \cdots & a_{1n} \\ \vdots & & \vdots \\ a_{n1} & \cdots & a_{nn} \end{pmatrix}, \quad u(t) = \begin{pmatrix} u_1(t) \\ \vdots \\ u_n(t) \end{pmatrix}, \quad u'(t) = \begin{pmatrix} u_1'(t) \\ \vdots \\ u_n'(t) \end{pmatrix}.$$

Then the system can be written in matrix form as

$$u'(t) = Au(t), \quad u(0) = c.$$

For simplicity, we only consider the case where the matrix A is diagonalizable. Let the spectrum of A be $\sigma(A) = \{\lambda_1, \ldots, \lambda_k\}$, and let the spectral projector associated with λ_i be G_i for $i = 1$ to k. Then it can be shown that the unique solution of this system is

$$u(t) = e^{\lambda_1 t} v^1 + \cdots + e^{\lambda_k t} v^k$$

where $v^i = G_i c$ for $i = 1$ to k.

Example 12:

Consider the following linear differential equation involving three functions $u_1(t), u_2(t), u_3(t)$ and their derivatives $u_1'(t), u_2'(t), u_3'(t)$.

$$
\begin{aligned}
u_1'(t) &= & u_2(t) &+u_3(t) \\
u_2'(t) &= u_1(t) & &+u_3(t) \\
u_3'(t) &= u_1(t) &+u_2(t) & \\
(u_1(0), u_2(0), u_3(0))^T &= & (12, -15, 18)^T &
\end{aligned}
$$

Letting $u(t) = (u_1(t), u_2(t), u_3(t))^T$, $u'(t) = (u_1'(t), u_2'(t), u_3'(t))^T$, $c = (12, -15, 18)^T$, the system in matrix notation is

$$u'(t) = Au(t)$$
$$u(0) = c$$

where

$$A = \begin{pmatrix} 0 & 1 & 1 \\ 1 & 0 & 1 \\ 1 & 1 & 0 \end{pmatrix}$$

the matrix discussed in Example 8, 11. Its eigen values are $\lambda_1 = -1$ and $\lambda_2 = 2$, and the associated spectral projectors are

$$G^1 = \begin{pmatrix} 2/3 & -1/3 & -1/3 \\ -1/3 & 2/3 & -1/3 \\ -1/3 & -1/3 & 2/3 \end{pmatrix}, \quad G_2 = \begin{pmatrix} 1/3 & 1/3 & 1/3 \\ 1/3 & 1/3 & 1/3 \\ 1/3 & 1/3 & 1/3 \end{pmatrix}$$

from Example 11. So, in the notation given above, we have

$$v^1 = G^1 c = \begin{pmatrix} 2/3 & -1/3 & -1/3 \\ -1/3 & 2/3 & -1/3 \\ -1/3 & -1/3 & 2/3 \end{pmatrix} \begin{pmatrix} 12 \\ -15 \\ 18 \end{pmatrix} = \begin{pmatrix} 7 \\ -20 \\ 13 \end{pmatrix}$$

$$v^2 = G_2 c = \begin{pmatrix} 1/3 & 1/3 & 1/3 \\ 1/3 & 1/3 & 1/3 \\ 1/3 & 1/3 & 1/3 \end{pmatrix} \begin{pmatrix} 12 \\ -15 \\ 18 \end{pmatrix} = \begin{pmatrix} 5 \\ 5 \\ 5 \end{pmatrix}.$$

So, from the above, we know that the unique solution of this system is

$$u(t) = \begin{pmatrix} u_1(t) \\ u_2(t) \\ u_3(t) \end{pmatrix} = e^{-t} \begin{pmatrix} 7 \\ -20 \\ 13 \end{pmatrix} + e^{2t} \begin{pmatrix} 5 \\ 5 \\ 5 \end{pmatrix} = \begin{pmatrix} 7e^{-t} + 5e^{2t} \\ -20e^{-t} + 5e^{2t} \\ 13e^{-t} + 5e^{2t} \end{pmatrix}.$$

i.e., $u_1(t) = 7e^{-t} + 5e^{2t}$, $u_2(t) = -20e^{-t} + 5e^{2t}$, $u_3(t) = 13e^{-t} + 5e^{2t}$. It can be verified that these functions do satisfy the given differential equations.

6.4 How to Compute Eigen Values?

Computing the eigen values of a matrix A of order n using the main definition boils down to finding the roots of a polynomial equation of degree n. Finding the roots of a polynomial equation of degree ≥ 5 is not a simple task, it has to be carried out using iterative methods. The most practical eigen value computation method is the QR iteration algorithm, an iterative method that computes the QR factors (see Section 4.12) of a matrix in every iteration. Providing a mathematical description of this algorithm, and the details of the work needed to convert it into a practical implementation, are beyond the scope of this book. For details see the references given below.

Exercises

6.4.1: Obtain the characteristic polynomial of the following matrices. Using it obtain their eigen values, and then the corresponding eigen vectors and eigen spaces.

$$\begin{pmatrix} 3 & 2 \\ 8 & 3 \end{pmatrix}, \quad \begin{pmatrix} 3 & 1/3 \\ 31/3 & 5 \end{pmatrix}, \begin{pmatrix} 5 & 1 & 2 \\ 6 & 5 & 3 \\ 6 & 2 & 6 \end{pmatrix}, \quad \begin{pmatrix} 3 & 1 & 2 \\ 2 & 4 & 4 \\ 3 & 3 & 8 \end{pmatrix}.$$

6.4.2: Compute eigen values and eigen vectors of the following matrices.

$$\begin{pmatrix} 1 & 0 & 1 & 0 \\ 0 & 1 & 0 & 1 \\ 1 & 0 & 1 & 0 \\ 0 & 1 & 0 & 1 \end{pmatrix}, \quad \begin{pmatrix} 0 & 0 & 1 & 0 \\ 0 & 0 & 0 & 1 \\ 1 & 0 & 0 & 0 \\ 0 & 1 & 0 & 0 \end{pmatrix}, \quad \begin{pmatrix} 2 & -1 & 1 \\ 0 & 3 & -1 \\ 2 & 1 & 3 \end{pmatrix}.$$

6.4.3: Let A be a square matrix of order n in which all diagonal entries are $= \alpha$, and all off-diagonal entries are $= \beta$. Show that it has an eigen value $\alpha - \beta$ with algebraic multiplicity $n - 1$, and $\alpha + (n-1)\beta$ as the only other eigen value.

6.4.4: If λ is an eigen value of A associated with the eigen vector x, show that λ^k is an eigen value of A^k associated with the same eigen vector x for all positive integers k.

6.4.5: If A, B are square matrices of order n, show that the eigen values of AB, BA are the same.

References

[6.1] C. P. Meyer, *Matrix Analysis and Applied Linear Algebra* (SIAM, Philadelphia, PA, 2000).

[6.2] J. H. Wilkinson, *The Algebraic Eigenvalue Problem* (Oxford University Press, 1965).

Chapter 7

Software Systems for Linear Algebra Problems

Nowadays there are several commercially available and highly popular software systems for solving the problems discussed in previous chapters. Among them the best known are: MATLAB, EXCEL, MATHEMATICA, MAPLE, and MACAULAY. We will provide illustrative numerical examples of how to solve these problems using MATLAB (this name is derived from *Matrix Laboratory*) in this chapter. For using the other software systems, see the references cited on them.

7.1 Difference Between Solving a Small Problem by Hand and Scientific Computing Using a Computer

When we solve small numerical problems by hand, we always use **exact arithmetic**, so all the numbers in the various stages of the work are exactly what they are supposed to be (for example, a fraction like "1/3" is maintained as it is). Hand computation is not feasible for large size problems encountered in applications. Such large problems are usually solved using a software package on a digital computer.

Digital computation uses **finite precision arithmetic**, not exact arithmetic; for example a fraction like "1/3" is **rounded** to 0.3333 or 0.3...3 depending on the number of **significant digits of precision** maintained, introducing a small **rounding error**. These rounding errors

431

abound in scientific computation, and they accumulate during the course of computation. So results obtained in digital computation usually contain some rounding errors, and may only be approximately correct. These errors make it nontrivial to implement many of the algorithms discussed in earlier chapters to get answers of reasonable precision. That's why detailed descriptions of numerical implementations of algorithms are very different and much more complicated than their mathematical descriptions given in earlier chapters.

As an example consider the GJ pivotal method for solving a system of linear equations discussed in Section 1.16. Under exact arithmetic, we know that a redundant constraint in the system corresponds to a "0 = 0" equation in the final tableau and vice versa. However, under finite precision arithmetic some of the "0" entries in it may become nonzero numbers of hopefully small absolute value. So, what is actually a "0 = 0" equation in the final tableau, may in reality become an equation with nonzero entries of small absolute value, and it is very difficult to decide whether it is a redundant equation or not. In computer implementations, one difficult question faced at every stage is whether a nonzero entry of small absolute value is actually a nonzero entry, or a zero entry that has become nonzero due to rounding error accumulation. A practical rule that is often used selects a small positive tolerance, and replaces any entry in the tableau whose absolute value is less than this tolerance by zero. Under such a rule, we can only conclude that a "0 = 0" equation in the final tableau is *redundant to working precision*.

In the same way, when a square matrix has a determinant of very small absolute value, using finite precision arithmetic it is very difficult to decide whether it is actually nonsingular or singular, and software systems will conclude that this *matrix is singular to working precision*. Similarly while the definitions of linear independence or dependence of a set of vectors and its rank defined in Chapter 4 are conceptually precise and mathematically unambiguous under exact arithmetic, using finite precision arithmetic they can only be determined *correct to the working precision*.

Numerical analysts have developed a variety of techniques to reduce the effects of rounding errors, some of these like partial pivoting or complete pivoting for pivot element selection are discussed very briefly in Chapter 1. As a detailed discussion of these techniques is beyond the scope of this book, the interested reader should refer to books in that area (for example, see Gill, Murray, Wright [4.1]).

We will see the effects of rounding errors in some of the illustrative numerical examples given below.

7.2   How to Enter Vectors and Matrices in MATLAB

First you need to begin a MATLAB session. It should be noted that there are many versions of MATLAB in the market, and input prompts, and error and warning messages may vary from version to version. Also, different computer systems may have different commands for initiating a MATLAB session on their system.

After getting the MATLAB prompt, if you want to enter a row vector, $a = (1, 6, 7, -9)$ say, type

$$a = [1 \quad 6 \quad 7 \quad -9] \quad \text{or} \quad a = [1, 6, 7, -9]$$

with a blank space or a comma (,) between entries, and MATLAB responds with

$$a = 1 \quad 6 \quad 7 \quad -9.$$

To enter the same vector as a column vector, type

$$a = [1, 6, 7, -9]' \quad \text{or} \quad a = [1 \quad 6 \quad 7 \quad -9]' \quad \text{or} \quad a = [1; 6; 7; -9]$$

and MATLAB responds with

$$a = \begin{array}{c} 1 \\ 6 \\ 7 \\ -9 \end{array}$$

To suppress the system's response, a semicolon (;) is placed as the last character of the expression. For example, after typing $a = [1, 6, 7, -9];$ MATLAB responds with a new line awaiting a new command.

MATLAB also lets one place several expressions on one line, a line being terminated by pressing the *Enter* button on the computer. In this case, each expression is separated by either a comma (,) or a semicolon (;). The comma results in the system echoing the output; the semicolon suppresses it. For example, by typing

$$a = [1, 2, 6, 9], c = [2, -3, 1, -4]$$

the system responds with

$$a = 1 \quad 2 \quad 6 \quad 9$$
$$c = 2 \quad -3 \quad 1 \quad -4.$$

Matrices are entered into MATLAB row by row, with rows separated by either semicolons, or by pressing *Enter* button on the computer. For example, to enter the matrix

$$A = \begin{pmatrix} 10 & -2 & 4 \\ 0 & 8 & -7 \end{pmatrix}$$

type $A = [10, -2, 4; 0, 8, -7]$

or type $A=[\begin{matrix} 10 & -2 & 4 \\ 0 & 8 & -7 \end{matrix}]$ using *Enter* to indicate the end of 1st row

or type $A=[\begin{matrix} 10 & -2 & 4; \dots \\ 0 & 8 & -7 \end{matrix}]$ the ... called ellipses is a continuation of a MATLAB expression to next line. Used to create more readable code.

MATLAB has very convenient ways to address the entries of a matrix. For example, to display the $(2, 3)$ entry in the above matrix A type $A(2, 3)$, to display 2nd column of A type $A(:, 2)$, and to display 2nd row of A type $A(2, :)$.

Some of the commands used for special matrices are:

$eye(n)$	identity matrix of order n
$diag([a_1 \quad a_2 \quad \dots \quad a_n])$	diagonal matrix with diagonal entries $a_1, \dots, a_n$
$zeros(m, n)$	zero matrix of order $m \times n$
$zero(n)$	zero matrix of order $n \times n$
A'	transpose of matrix A

If B is a square matrix defined in MATLAB, $diag(B)$ returns the diagonal elements of B. To see the contents of a vector or matrix A defined earlier, just type A and MATLAB responds by displaying it.

The arithmetic operators to perform addition, subtraction, multiplication, division, and exponentiation are: $+$, $-$, $*$, $/$, $\hat{}$ respectively. As an example, suppose the vectors $x = (1, 0, 0)$, $y = (0, 1, 0)$, $z = (0, 0, 1)$ have been defined earlier. If you type $2*x+3*y-6*z$ MATLAB responds with

ans $= 2 \quad 3 \quad -6$.

If you type $w = 2*x + 3*y - 6*z$ MATLAB responds with

$w = 2 \quad 3 \quad -6$.

If you type an arithmetical expression that is not defined, for example $x + y'$ with x, y as defined above (i.e., the sum of a row vector and a column vector), MATLAB yields the warning

$???Error\ using \Longrightarrow +$

Matrix dimensions must agree.

Other arithmetical operations on matrices and vectors can be defined in MATLAB in the same way. Here are some MATLAB commands that we will use, and the outputs they produce. For a more complete list of MATLAB commands, see the MATLAB references given at the end of this chapter. Let A, B be two matrices, and x, y be two vectors.

$length(x)$	the number of the elements in the vector x
$size(A)$	$m\ n$ where $m \times n$ is the order of A
$dot(x, y)$	dot product of x and y (whether each is either a row or a column) provided they have same no. of elements. This can also be obtained by typing $x * y$ provided x is a row vector & y is a col. vector
$norm(x)$	the Eucledean norm $\|x\|$ of the vector x
$det(A)$	determinant of A, if it is a square matrix
$rank(A)$	rank of the matrix A
$inv(A)$	inverse of matrix A, if A is an invertible square matrix
$A * B$	matrix product AB if it is defined.
$A \backslash b$	solves the system of linear equations with A as the coefficient matrix, and b as the RHS constants vector
$null(A)$	outputs 0-vector if system $Ax = 0$ has no nonzero solution, or an orthonormal basis for the subspace which is the set of all solutions of $Ax = 0$ obtained by a method called singular value decomposition not discussed in the book if $Ax = 0$ has nonzero solutions.
$rref(A)$	outputs the reduced row echelon form of matrix A.
$rrefmovie(A)$	shows all the operations that MATLAB performs to obtain the RREF of the matrix A. This command used to be there in the past, but not in newer versions of MATLAB. To make this command work, go to http://www.mathworks.com/matlabcentral/fileexchange/23039-matlab-in-physics-matrices/content/Lecture3/rrefmovie.m to download "function rrefmovie (A, tol)" to create this function in your MATLAB program. After downloading this function in your MATLAB, change the stopping criterion in it "while $(i <= m)\&(j <= n)$" to "while $(i <= m)\&(j <= n - 1)$".

$null(A,'r')$	same as $null(A)$ except that in the 2nd case it produces the basis for the set of solutions of $Ax = 0$ obtained by the GJ or the G elimination method as discussed in Section 1.23.
$eig(A)$	produces the eigen values of the square matrix A.
$[V, D] = eig(A)$	produces $D = $ diagonal matrix of eigen values of A, and $V = $ matrix whose column vectors are corresponding eigen vectors.
$all(eig(A + A') > 0)$	for a square matrix A, this checks whether all eigenvalues of $(A + A')$ are strictly positive. The output is 1 if the answer is yes, in this case A is a positive definite matrix; or 0 if the answer is no. In the latter case, using the command described above, one can generate all the eigenvalues and associated eigen vectors of $(A + A')$. The eigen vector x associated with a nonpositive eigen value of $(A + A')$ satisfies $x^T A x \leq 0$.
$all(eig(A + A') >= 0)$	for a square matrix A, this checks whether all eigenvalues of $(A + A')$ are nonnegative. The output is 1 if the answer is yes, in this case A is a positive semidefinite matrix; or 0 if the answer is no and A is not PSD.
$chol(A)$	for a symmetric square matrix A, this command outputs the Cholesky factor of A (Cholesky factorization is not discussed in this book) if A is PD. If A is not PD, it outputs an error message that *Matrix must be positive definite* to use chol. So, this command can also be used to check if a given square symmetric matrix is PD.

When a system of linear equations $Ax = b$ has more than one solution, there are two ways of obtaining an expression of the general solution of the system using MATLAB. One is to get the RREF (a canonical tableau WRT a basic vector) of the system using the *rrefmovie* command (see Examples 2, 3 given below), and then construct an expression of the general solution of the system using it as explained in Section 1.8. The other is to get one solution, $\bar{x}$ say, using the $A\backslash b$ command; and then a basis for the null space of the coefficient matrix A using the $null(A)$ command. If this basis is

$\{x^1, \ldots, x^s\}$, then the general solution of the system is $\bar{x} + \alpha_1 x^1 + \cdots + \alpha_s x^s$ where $\alpha_1, \ldots, \alpha_s$ are parameters that take real values.

7.3 Illustrative Numerical Examples

1. Solving a Square System of Linear Equations: Suppose we want to find a solution for the following system and check whether it is unique. We provide the MATLAB session for doing this. The coefficient matrix is called A, and the RHS constants vector is called b, and the solution vector is called x.

$$2x_1 + x_2 - 3x_3 - x_4 = 1$$
$$-x_1 - 2x_2 + 4x_3 + 5x_4 = 6$$
$$7x_1 - 6x_4 = 7$$
$$-3x_1 - 8x_2 - 9x_3 + 3x_4 = 9$$

Type $A = [2, 1, -3, -1; -1, -2, 4, 5; 7, 0, 0, -6; -3, -8, -9, 3]$

Response $A = \begin{array}{rrrr} 2 & 1 & -3 & -1 \\ -1 & -2 & 4 & 5 \\ 7 & 0 & 0 & -6 \\ -3 & -8 & -9 & 3 \end{array}$

Type $b = [1, 6, 7, 9]'$

Response $b = \begin{array}{r} 1 \\ 6 \\ 7 \\ 9 \end{array}$

Type $x = A \backslash b$

Response $x = \begin{array}{r} 1.7843 \\ -1.5062 \\ 0.0491 \\ 0.9151 \end{array}$

The solution vector x is provided by MATLAB, this indicates that this is the unique solution, and that the coefficient matrix A is nonsingular.

In MATLAB the final canonical tableau is called RREF (reduced row echelon form). The command: rref(A, b), outputs the RREF for the augmented matrix $(A \dot{:} b)$ of the system. Here is the session to obtain that for this example.

Type $d = \text{rref}([A, b])$

Response $d =$

1.0000	0	0	0	1.7843
0	1.0000	0	0	−1.5063
0	0	1.0000	0	0.0491
0	0	0	1.0000	0.9151

2. Another Square System of Linear Equations: The system is given below. We call the coefficient matrix A, the RHS constants vector b, and the solution vector x.

$$2x_1 + x_2 - 3x_3 + x_4 = 4$$
$$-x_1 - 2x_2 + 4x_3 + 5x_4 = -3$$
$$2x_1 - 2x_2 + 2x_3 + 12x_4 = 6$$
$$x_1 - x_2 + x_3 + 6x_4 = 3$$

The MATLAB session is carried out exactly as above, but this time MATLAB gives the following final response.

Warning: Matrix is singular to working precision

$x =$ NaN
      NaN
      NaN
      NaN

This indicates that the coefficient matrix A is singular, and that the system has no solution.

In MATLAB the final canonical tableau is called RREF (reduced row echelon form). The command: $\text{rref}([A, b])$, outputs the RREF for the augmented matrix $(A\dot{:}b)$ of the system. When $\text{rank}(A\dot{:}b) = 1 + \text{rank}(A)$, as in this case, MATLAB performs also a final pivot step in the column vector b, which we do not want. In this case the command:

$\text{rrefmovie}([A, b])$

helps you to observe all the operations that MATLAB performs to obtain the RREF of the augmented matrix ($[A, b]$). The matrix before the final pivot step in the column of b is the RREF to our system. For this example, here is that output (here we did all the computations in fractional form).

$$
\begin{matrix}
1 & 0 & -2/3 & 7/3 & 7/3 \\
0 & 1 & -5/3 & -11/3 & -2/3 \\
0 & 0 & 0 & 0 & -2 \\
0 & 0 & 0 & 0 & 0
\end{matrix}
$$

The rows in this final tableau do not directly correspond to rows (constraints) in the original problem, because of row interchanges performed during the algorithm. To find out which original constraint each row in the final tableau corresponds to, you can use the row interchange information that is displayed in the rrefmovie output.

3. A Rectangular System of Linear Equations: The system is given below. The symbols A, b, x refer to the coefficient matrix, RHS constants vector, solution vector respectively.

x_1	x_2	x_3	x_4	x_5	x_6	x_7	b
2	1	0	0	-1	-3	4	3
0	-2	1	2	0	4	3	-2
4	0	1	2	-2	-2	11	4
-1	2	3	1	-2	0	0	5
5	1	5	5	-5	-1	18	10

The session is carried out exactly as above, and MATLAB produced the following final response.

Warning: Rank deficient, rank = 3   tol = 3.3697e-014.

$$
\begin{aligned}
x = \quad & 0.7647 \\
& 1.4706 \\
& 0.9412 \\
& 0 \\
& 0 \\
& 0 \\
& 0
\end{aligned}
$$

Here since the system has a solution, the final canonical tableau, RREF can be produced with the command: $rref([A,b])$. Here it is:

$$
\begin{array}{ccccccc}
1 & 0 & 0 & 0.2941 & -0.3529 & -0.7059 & 2.4118 & 0.7647 \\
0 & 1 & 0 & -0.5882 & 0.2941 & -1.5882 & -0.8235 & 1.4706 \\
0 & 0 & 1 & 0.8235 & -0.5882 & 0.8235 & 1.3529 & 0.9412 \\
0 & 0 & 0 & 0 & 0 & 0 & 0 & 0 \\
0 & 0 & 0 & 0 & 0 & 0 & 0 & 0
\end{array}
$$

As in the above example, to trace the correspondence of rows in the RREF to the constraints in the original statement of the problem, one has to take into account the row interchanges performed by MATLAB in the process of getting this RREF by using the command: $rrefmovie([A,b])$.

4. Another Rectangular System of Linear Equations: The system is given below. The symbols A, b, x refer to the coefficient matrix, RHS constants vector, solution vector respectively.

x_1	x_2	x_3	x_4	x_5	x_6	x_7	b
1	0	-1	1	-2	1	-3	-4
0	-1	2	3	1	-1	2	2
1	-2	3	7	0	-1	1	-2
-3	1	2	-1	0	0	2	3
-2	0	3	3	-1	0	1	3

The canonical tableau for this system, RREF, produced using the command: $rrefmovie(A,b)$ is given below.

x_1	x_2	x_3	x_4	x_5	x_6	x_7	b
1	0	0	6	-7	3	-8	-12
0	1	0	7	-11	5	-12	-17
0	0	1	5	-5	2	-5	-8
0	0	0	0	0	0	0	1
0	0	0	0	0	0	0	3

The last two rows in the canonical tableau represent inconsistent equations, "$0 = \alpha$" for some $\alpha \neq 0$; so this system has no solution. To trace the correspondence of rows in the RREF to the constraints in the original statement of the problem, one has to take into account the row interchanges performed by MATLAB in the process of getting this RREF.

5. A Homogeneous System of Linear Equations: Calling the coefficient matrix, RHS constants vector, solution vector as A, b, x respectively, here is the session.

Type $A = [1,1,0; 0,1,1; 1, 0, 1; 2, 2, 2]$

Response $A =$
$$
\begin{matrix}
1 & 1 & 0 \\
0 & 1 & 1 \\
1 & 0 & 1 \\
2 & 2 & 2
\end{matrix}
$$

Type $b = zeros(4, 1)$ (Note: this command generates a 0-matrix of order 4×1)

Response $b =$
$$
\begin{matrix}
0 \\
0 \\
0 \\
0
\end{matrix}
$$

Type $x = A \backslash b$

Response $x =$
$$
\begin{matrix}
0 \\
0 \\
0
\end{matrix}
$$

Type $rank(A)$

Response 3

Since rank of the coefficient matrix is 3, from the results in Sections 4.5 and 1.22, we conclude that this system has no nonzero solution.

6. Another Homogeneous System of Linear Equations: Calling the coefficient matrix, RHS constants vector, solution vector as A, b, x respectively, here is the session.

Type $A = [1,-1,0, 2, 3, -2, 1; 0,1, -2, 1, -1, 3, -2; 1, 0, -2, 3, 2, 1, -1; -1, 2, 1, -2, 1, 3, 2]$

Response $A =$
$$
\begin{matrix}
1 & -1 & 0 & 2 & 3 & -2 & 1 \\
0 & 1 & -2 & 1 & -1 & 3 & -2 \\
1 & 0 & -2 & 3 & 2 & 1 & -1 \\
-1 & 2 & 1 & -2 & 1 & 3 & 2
\end{matrix}
$$

Type $b = zeros(4, 1)$ (Note: this command generates a 0-matrix of order 4×1)

Response $b =$ 0
                    0
                    0
                    0

Type $d = \text{rref}([A, b])$

Response $d =$

1	0	0	7/3	16/3	−1/3	7/3	0
0	1	0	1/3	7/3	5/3	4/3	0
0	0	1	−1/3	5/3	−2/3	5/3	0
0	0	0	0	0	0	0	0

From the RREF we can see that this homogeneous system has nonzero solutions. From the RREF a basic set of nonzero solutions for the system can be constructed as discussed in Section 1.23.

7. Checking Linear Independence: The most convenient way to check linear independence of a given set of vectors (either all row vectors or all column vectors) using MATLAB is to write each of these vectors as a column vector of a matrix, A say. Then check whether the homogeneous system of equations $Ax = 0$ has a nonzero solution. If this system has no nonzero solution, the set is linearly independent. If $\bar{x}$ is a nonzero solution of the system, the set is linearly dependent; and $\bar{x}$ is the vector of coefficients in a linear dependence relation for it in the order in which the vectors are entered as column vectors in the matrix A.

As an example, consider the set of 4 column vectors, $\{A_{.1}.A_{.2}, A_{.3}, A_{.4}\}$ of the matrix A in Example 1. Here is the session for checking its linear independence; we suppress the display of the matrix in this session.

Type $A = [2, 1, -3, -1; -1, -2, 4, 5; 7, 0,0, -6; -3, -8, -9, 3];$
null(A)

Response   ans = Empty matrix: 4-by-0.

This indicates that the system $Ax = 0$ has no nonzero solution, i.e., the set of column vectors of A is linearly independent.

8. Checking Linear Independence, Another Example: Consider the set of 4 column vectors, $\{A_{.1}.A_{.2}, A_{.3}, A_{.4}\}$ of the matrix A in Example 2. Here is the session for checking its linear independence; we suppress the display of the matrix in this session.

Type $A = [2, 1, -3, 1; -1, -2, 4, 5; 2, -2, 2, 12; 1, -1, 1, 6]$;
null(A)

$$
\text{Response} \quad \text{ans} = \begin{array}{rr} -0.7818 & -0.3411 \\ -0.2469 & 0.9111 \\ -0.5436 & 0.1382 \\ 0.1797 & 0.1857 \end{array}
$$

Each of the column vectors in the output above is a nonzero solution of $Ax = 0$ and these two vectors together form an orthonormal basis for the subspace which is the set of all solutions of this homogeneous system, obtained by using a method called singular value decomposition, which is not discussed in the book. If you do not require an orthonormal basis, but want the basis obtained by the GJ or G elimination methods as discussed in Section 1.22, change the command null(A) to null($A,' r'$), then the output is (each column vector is a nonzero solution, together they form a basis for the subspace which is the set of all solutions of $Ax = 0$):

$$
\text{Response} \quad \text{ans} = \begin{array}{rr} 0.6667 & -2.3333 \\ 1.6667 & 3.6667 \\ 1.0000 & 0 \\ 0 & 1.0000. \end{array}
$$

Each of the column vectors in these outputs is the vector of coefficients in a linear dependence relation for the set of column vectors of the matrix A in this example. For instance the first column vector in the output under the command null(A) yields the linear dependence relation

$$-0.7818A_{.1} - 0.2469A_{.2} - 0.5436A_{.3} + 0.1797A_{.4} = 0.$$

9. Matrix Inversion: Calling the matrix A, here is the session.

Type $A = [1,2,1, -1; -1, 1, 1, 2; 0, -1, 2, 1; 2, 2, -1, 0]$

$$
\text{Response} \quad A = \begin{array}{rrrr} 1 & 2 & 1 & -1 \\ -1 & 1 & 1 & 2 \\ 0 & -1 & 2 & 1 \\ 2 & 2 & -1 & 0 \end{array}
$$

Type   inv(A)

Response   ans = $\begin{array}{rrrr} -0.1071 & -0.2500 & 0.3929 & 0.4286 \\ 0.2500 & 0.2500 & -0.2500 & 0.0000 \\ 0.2857 & 0.0000 & 0.2857 & -0.1429 \\ -0.3214 & 0.2500 & 0.1786 & 0.2857 \end{array}$

10. Another Matrix Inversion: Calling the matrix A, here is the session.

Type $A = [0, 1, 1, 2; 1, 0, 2, 1; 2, 1, 0, 1; 3, 2, 3, 4]$

Response $A =$ $\begin{array}{rrrr} 0 & 1 & 1 & 2 \\ 1 & 0 & 2 & 1 \\ 2 & 1 & 0 & 1 \\ 3 & 2 & 3 & 4 \end{array}$

Type $inv(A)$
Response   Warning: Matrix is close to singular or badly scaled. Results may be inaccurate. RCOND = $4.336809e - 018$.

ans = $1.0e + 016*$ $\begin{array}{rrrr} 0.1801 & 0.1801 & 0.1801 & -0.1801 \\ -1.2610 & -1.2610 & -1.2610 & 1.2610 \\ -0.5404 & -0.5404 & -0.5404 & 0.5404 \\ 0.9007 & 0.9007 & 0.9007 & -0.9007 \end{array}$

According to the warning the input matrix A in this example appears to be singular, but reaching this exact conclusion is made difficult due to rounding errors. Of course a singular matrix does not have an inverse, but an inverse is obtained because rounding errors have altered the outcome of singularity. The warning message indicates that the outputted inverse is probably not accurate due to the singularity of the original matrix. Most likely, an inverse is still computed due to roundoff error introduced into the computation by computer arithmetic.

11. Nearest Point to $\bar{x}$ On a Given Straight Line: Find the nearest point, x^*, (in terms of the Euclidean distance), to the given point $\bar{x}$, on the straight line L given in parametric form: $L = \{x : x = a + \lambda c, \text{ where } a = (2, 1, 3, 3)^T, c = (0, 1, -1, 2)^T, \text{ and } \lambda \text{ is the real valued parameter}\}$.

We denote the point $\bar{x}$ by $x1$ to avoid confusion, and we call the nearest point xn. Here is the session.

Type $a = [2, 1, 3, 3];$
$c = [0, 1, -1, 2]';$
$x1 = [1, -1, 2, 0]';$
$lambda = c' * (x1 - a')/[c(1, 1)^2 + c(2, 1)^2 + c(3, 1)^2 + c(4, 1)^2]$

Response   *lambda* $= -1.1667$

Type $xn = a + lambda * c$

Response $xn =$

> 2.0000
> −0.1667
> 4.1667
> 0.6667

Type $dist = [[x1(1,1) - xn(1,1)]^2 + [x1(2,1) - xn(2,1)]^2 + [x1(3,1) - xn(3,1)]^2 + [x1(4,1) - xn(4,1)]^2](1/2)$

Response $dist = 2.6141$

"dist" is the Euclidean distance between $x1$ and the nearest point to $x1$ on the given straight line. All these formulae are from Section 3.20.

12. **Nearest Point to $\bar{x}$ On a Given Hyperplane:** Find the nearest point, x^*, (in terms of the Euclidean distance), to the given point $\bar{x}$, on the hyperplane $H = \{x : ax = a_0$ where $a = (1, -2, 3, -4), a_0 = -13\}$.

Again we denote $\bar{x}$ by $x1$, and the nearest point $x*$ by xn to avoid confusion. Here is the session.

Type $a = [1, -2, 3, -4]; a_0 = -13; x1 = [1, -1, 2, 0]';$
$xn = x1 + a' * [-(a * x1 - a_0)/(a(1,1)^2 + a(1,2)^2 + a(1,3)^2 + a(1,4)^2)]$

Response $xn =$

> 0.2667
> 0.4667
> −0.2000
> 2.9333

Type $dist = [[x1(1,1) - xn(1,1)]^2 + [x1(2,1) - xn(2,1)]^2 + [x1(3,1) - xn(3,1)]^2 + [x1(4,1) - xn(4,1)]^2](1/2)$

Response $dist = 8.0067$

"dist" is the Euclidean distance between $x1$ and the nearest point to $x1$ on the given hyperplane. All these formulae are from Section 3.16.

13. **Nearest Point to $\bar{x}$ In a Given Affine Space:** Find the nearest point, x^*, (in terms of the Euclidean distance), to the given point $\bar{x} = (1, 1, 0)^T$, in

the affine space $F = \{x : Ax = b \text{ where } A, b \text{ are given below}\}$.

$$A = \begin{pmatrix} 1 & 1 & -1 \\ 2 & 1 & 1 \end{pmatrix}, \quad b = \begin{pmatrix} 2 \\ 9 \end{pmatrix}.$$

Here is the portion of the session after reading in $A, b, x1$ where we are calling the given point $\bar{x}$ as $x1$. The nearest point to $x1$ in the given affine space will be denoted by xn.

Type $xn = x1 - A' * inv(A * A') * (A * x1 - b)$

Response $xn =$

  2.7143
  1.4286
  2.1429

The Euclidean distance between $x1$ and xn using the same command as in the above examples is 3.8571.

14. Checking PD, PSD: Check whether the following matrices A, B, C are PD, PSD, or neither.

$$A = \begin{pmatrix} 3 & 1 & 2 & 2 \\ -1 & 2 & 0 & 2 \\ 0 & 4 & 4 & 5/3 \\ 0 & 2 & -13/3 & 6 \end{pmatrix}, \quad B = \begin{pmatrix} 8 & 2 & 1 & 0 \\ 4 & 7 & -1 & 1 \\ 2 & 1 & 1 & 2 \\ 2 & -2 & 1 & 1 \end{pmatrix}$$

$$C = \begin{pmatrix} 0 & -2 & -3 & -4 & 5 \\ 2 & 3 & 3 & 0 & 0 \\ 3 & 3 & 3 & 0 & 0 \\ 4 & 0 & 0 & 8 & 4 \\ -5 & 0 & 0 & 4 & 2 \end{pmatrix}.$$

Here are the sessions after reading in the matrices.

Type $all(eig(A + A') > 0)$

Response   ans = 0. This means that A is not PD.

We show how the same thing is done using the *chol* cammand.

Type $chol(A + A')$

Response   ??? Error using $\Longrightarrow$ chol. Matrix must be positive definite.

We check that B is not PD using same procedures. Now to check whether B is PSD, we do the following.

Type   all(eig($B + B'$) >= 0)

Response   1. This means that B is PSD.

Since C has a 0 diagonal element, it is clearly not PD. To check whether it is PSD, we use the command $[V, D] = eig(Cs)$ command where $Cs = C + C'$. We get

$$D = \begin{matrix} 0 & 0 & 0 & 0 & 0 \\ 0 & 12 & 0 & 0 & 0 \\ 0 & 0 & 0 & 0 & 0 \\ 0 & 0 & 0 & 20 & 0 \\ 0 & 0 & 0 & 0 & 0 \end{matrix}$$

Since all the eigenvalues are nonnegative C is PSD.

15. Eigen Values and Eigen Vectors: Calling the matrix A, here is the session.

Type $A = [-3, 1, -3; 20, 3, 10; 2, -2, 4]$

Response $A = \begin{matrix} -3 & 1 & -3 \\ 20 & 3 & 10 \\ 2 & -2 & 4 \end{matrix}$

Type   eig(A)

Response   ans   = $\begin{matrix} -2.0000 \\ 3.0000 \\ 3.0000 \end{matrix}$

$[V, D] = eig(A)$

Response $V = \begin{matrix} 0.4082 & -0.4472 & 0.4472 \\ -0.8165 & 0.0000 & 0.0000 \\ -0.4082 & 0.8944 & -0.8944 \end{matrix}$

Response $D = \begin{matrix} -2.0000 & 0 & 0 \\ 0 & 3.0000 & 0 \\ 0 & 0 & 3.0000 \end{matrix}$

16. Matrix Diagonalization: Diagonalize the following matrices A, B.

$$A = \begin{pmatrix} 1 & 3 & 3 \\ -3 & -5 & -3 \\ 3 & 3 & 1 \end{pmatrix}, \quad B = \begin{pmatrix} 2 & 4 & 3 \\ -4 & -6 & -3 \\ 3 & 3 & 1 \end{pmatrix}.$$

The command $[V, D] = eig(A)$ gives the following output.

$$V = \begin{matrix} 0.5774 & 0 & -0.7459 \\ -0.5774 & -0.7071 & 0.0853 \\ 0.5774 & 0.7071 & 0.6606 \end{matrix}$$

$$D = \begin{matrix} 1 & 0 & 0 \\ 0 & -2 & 0 \\ 0 & 0 & -2 \end{matrix}$$

-2 is an eigenvalue of A with algebraic multiplicity 2. Also, since two eigenvectors (which are not scalar multiples of each other) are associated with this eigenvalue, its geometric multiplicity is also 2. So, A has a complete set of eigenvectors which is linearly independent. So, A can be diagonalized. In fact by the results discussed in Chapter 6, the matrix V whose column vectors are the distinct eigenvectors of A diagonalizes A, i.e., $V^{-1}AV = D$.

The command $[V, D] = eig(B)$ gives the following output.

$$V = \begin{matrix} -0.5774 & 0.7071 & 0.7071 \\ 0.5774 & -0.7071 & -0.7071 \\ -0.5774 & 0.0000 & 0.0000 \end{matrix}$$

$$D = \begin{matrix} 1 & 0 & 0 \\ 0 & -2 & 0 \\ 0 & 0 & -2 \end{matrix}$$

Since the two eigenvectors associated with the eigenvalue -2 are the same, it indicates that the geometric multiplicity of the eigenvalue -2 is 1, while its algebraic multiplicity is 2. So, B does not have a complete set of eigenvectors, and hence it is not diagonalizable.

17. Orthogonal Diagonalization: Orthogonally diagonalize the following matrix A.

$$A = \begin{pmatrix} 5 & 2 & 9 & -6 \\ 2 & 5 & -6 & 9 \\ 9 & -6 & 5 & 2 \\ -6 & 9 & 2 & 5 \end{pmatrix}.$$

Since A is symmetric, it is possible to orthogonally diagonalize A by the results in Chapter 6. The command $[V, D] = eig(A)$ produced the following output:

$$V = \begin{bmatrix} 0.5 & 0.5 & -0.5 & 0.5 \\ 0.5 & 0.5 & 0.5 & -0.5 \\ 0.5 & -0.5 & -0.5 & -0.5 \\ 0.5 & -0.5 & 0.5 & 0.5 \end{bmatrix}$$

$$D = \begin{bmatrix} 10 & 0 & 0 & 0 \\ 0 & 4 & 0 & 0 \\ 0 & 0 & 18 & 0 \\ 0 & 0 & 0 & -12 \end{bmatrix}$$

By the results in Chapter 6, the matrix V whose column vectors form a complete set of eigenvectors for A orthogonally diagonalize A, i.e., $V^T A V = D$.

References

References on MATLAB ·

[7.1] M. Golubitsky and M. Dellnitz, *Linear Algebra and Differential Equations Using MATLAB* (Brooks/Cole Publishing Co., NY, 1999).

[7.2] E. B. Magrab and others, *An Engineer's Guide to MATLAB* (Prentice Hall, Upper Saddle River, NJ, 2000).

[7.3] C. F. Van Loan, *Introduction to Scientific Computing*, MATLAB Curriculum Series (Prentice Hall, Upper Saddle River, NJ, 1997).

[7.4] *The Student Edition of MATLAB, Users Guide*, The MathWorks Inc., Natick, MA.

References on Other Software Systems

[7.5] B. W. Char, et al. *First Leaves: A Tutorial Introduction to Maple V* (Springer Verlag, NY, 1992).

[7.6] J. H. Davenport, *Computer Algebra: Systems and Algorithms for Algebraic Computation* (Academic Press, San diago, 1993).

[7.7] D. Eugene, *Schaum's Outline of Theory and Problems of Mathematica* (McGraw-Hill, NY, 2001).

[7.8] M. B. Monagan, *et al.*, *Maple V Programming Guide* (Springer, NY, 1998).

[7.9] Microsoft Corp., *Microsoft Excel User's Guide, Version 5.0* (Redmond, WA, 1993).

[7.10] S. Wolfram, *The Mathematica Book*, 3rd ed., Wolfram Media (Cambridge University Press, 1996).

Index

Following abbreviations are used below:
algo. (algorithm), bet. (between), eq. (equation), eqs. (equations),
st. line (straight line)

Printed in the United States
By Bookmasters